Remote Sensing Observations of Continental Surfaces Set

coordinated by
André Mariotti

Land Surface Remote Sensing in Agriculture and Forest

Edited by

Nicolas Baghdadi
Mehrez Zribi

ISTE PRESS

ELSEVIER

First published 2016 in Great Britain and the United States by ISTE Press Ltd and Elsevier Ltd

ISTE Press Ltd
27-37 St George's Road
London SW19 4EU
UK

www.iste.co.uk

Elsevier Ltd
The Boulevard, Langford Lane
Kidlington, Oxford, OX5 1GB
UK

www.elsevier.com

British Library Cataloguing-in-Publication Data
A CIP record for this book is available from the British Library
Library of Congress Cataloging in Publication Data
A catalog record for this book is available from the Library of Congress
ISBN 978-1-78548-103-1

Contents

Foreword

I have been entrusted by ISTE Science Publishing with the responsibility for a multidisciplinary editorial line: Earth System – Environment, and within this framework it gives me great pleasure today to present a set of books dedicated to the topic of remote sensing, compiled and edited by Nicolas Baghdadi and Mehrez Zribi.

Both the content and the organization of this collection have largely been inspired by reflections, analyses and prospective works conducted by almost 200 authors and researchers with a high level of international expertise in this discipline.

This community, which is recognized for its scientific merit, has sought to expand its research activities under the direction the two editing authors founded on a solid effort in the area of acquisition and wider dissemination of knowledge within this field.

This represents a community characterized by the firm commitment to adopting a holistic or even an ecosystem approach within the context of an interdisciplinary science of the Earth system. In this scientific context where the complexity of natural systems is compounded with the complexity of societies, the authors have given careful consideration to depicting a finalizable and "public" type of discipline, open to decision makers, managers and all those beyond the scientific community who are interested in the future of our planet.

Two main tools are necessary in order to satisfy the requirements in terms of understanding and characterizing our environment and its evolution: process modeling and observation.

Remote sensing observations in conjunction with measurements and modeling constitute a discipline that makes it possible to understand the functional properties of the observed system and their dependence on its structural properties. This is one of the key disciplines that allow the analysis and provide access to the understanding of the functioning of our environment: in general, this is dedicated to aspects such as the analysis of climate change, the effects of anthropogenic and demographic pressure, natural disasters, the increasing decline in resources (water, etc.), the degradation of biodiversity across all environments, desertification, the need to nourish the planet (for example, mapping of crops and yield prediction), etc.

Analyzing and understanding these different types of problems is rendered possible:

– by analyzing the detected, structural and functional objects (soils, hydrosystems, vegetation, etc.);

– by understanding the main basic processes, which incorporate these main elements: water flow covering all scales and compartments, erosion, meteorology, crop development, soil pollution, etc;

– by developing indicators in order to evaluate the short-, medium- and long-term evolution of all environmental compartments and variables.

The importance of these scientific questions has led to a general mobilization of the international organizations by means of various international conventions and agreements to protect the environment and meet the specific requirements in terms of observation. Various international networks have been developed over the past few years with the purpose of conducting continuous measurements. However, these punctual measurements could not provide sufficient spatiotemporal monitoring, in particular in difficult-to-access regions. Within this context, spatial observation could be implemented to its full potential, both by means of considerable progress in terms of instrumentation and by means of the development of effective data processing and analysis methods, data whose provision becomes increasingly free of rights.

Under the initiative of numerous space agencies (in particular European, North-American, Japanese, etc.), important space missions were launched for the purpose of conducting Earth observations, among which the following may be mentioned:

– Sentinel, within the framework of the Copernius program (formerly referred to as GMES for Global Monitoring for Environment and Security) implemented in numerous areas such as land and marine environment monitoring, emergency management (for example, natural disasters) and climate change monitoring (radar and optical imaging);

– Landsat;

– ALOS, launched by the Japan Aerospace Exploration Agency, in particular for deforestation monitoring;

– SMOS and SMAP, in particular for the global mapping of soil moisture, etc.

Although remote sensing represents a field in which specialist knowledge is required in order to conduct a better analysis and interpretation of data, this programmatic development is undoubtedly associated with a significant progress with respect to the implementation of space-based Earth observations at the level of an increasing number of laboratories across both developed and emerging nations. This development is likewise associated with new disciplines and thematic backgrounds, among which numerous areas of the humanities and social sciences which enrich and extend the primarily physical foundations of remote sensing may be mentioned.

It would be superfluous to list all remote sensing applications along with the disciplines and scientific questions which adopted this concept, as this would also inevitably result in regrettable omissions: nonetheless, the extensive implementation of spatial observation grants the latter a strong interdisciplinary status.

The launch of new large-scale space missions, the higher degree of convenience, including financial convenience, as well as the access to data will facilitate an intensification and generalization of the use of spatial observation data and products: new scientific subjects, new users (managers, decision makers, etc).

The high demand for educational material containing updated information on the various remote sensing concepts and methods and the main applications thereof, in particular at the level of continental surfaces, are derived therefrom.

It is within this framework that this collection of books is proposed, which aims to provide researchers, students in masters, engineer and PhD programs, as well as decision makers, engineers specialized in management services on a territorial, departmental, regional or national scale and players in the decision-making authorities with a tool which incorporates both the foundations of the physical principles underlying various spatial applications and the implementation methods and exemplification at the level of various applications based on spatial observation.

In these six volumes, Nicolas Baghdadi and Mehrez Zribi have mobilized almost 200 internationally recognized researchers to propose a comprehensive "toolkit", describing the latest scientific methods and actions in terms of the implementation of spatial observation.

The first two volumes describe the physical principles underlying various techniques which cover the frequency spectrum ranging from visible to microwaves. The third volume illustrates the agricultural and forestry applications of spatial observation. The fourth volume presents the applications of spatial observation in the field of continental hydrology. The fifth volume is dedicated to the observation of urban and coastal areas, whereas the final volume presents the implementation of spatial observation within the context of risk assessment and understanding.

Thanks are due to Nicolas Baghdadi and Mehrez Zribi for taking the time to draft, harmonize and partially edit these volumes and committing to this effort in terms of putting this modern and high-quality knowledge across and making it accessible to a diverse and vast scientific audience.

I wish to thank them both for their altruism, perseverance and devotion in service of the success of this endeavor.

André MARIOTTI

June 2016

Acronyms

2D	Two dimensions
3D	Three dimensions
4AOP	Automatized Atmospheric Absorption Atlas Operational
6S	Second Simulation of Satellite Signal in the Solar Spectrum
AATSR	Advanced Along-Track Scanning Radiometer
ACORN	Atmospheric Correction Now
ADC	Analog-to-digital converter
ADCP	Acoustic Doppler Current Profiler
ADEME	French Environment and Energy Management Agency
ADEOS	Advanced Earth Observing Satellite
AERONET	Aerosol Robotic Network
AET	Actual Evapotranspiration
AFRITRON	African Tropical Rainforest Observation Network
AGB	Above-Ground Biomass
AGNES	Agglomerative Nesting
AHS	Airborne Hyperspectral Scanner
AHT	Astronomical High Tide
AIEM	Advanced Integral Equation Model
AirSAR	Airborne Synthetic Aperture Radar
ALB	Airborne LiDAR Bathymeter
ALEXI	Atmosphere-Land Exchange Inverse
ALS	Airborne Laser Scanning

AltBOC	Alternate Binary Offset Carrier
AMARTIS	Advanced Modeling of the Atmospheric Radiative Transfer for Inhomogeneous Surfaces
AMMA	African Monsoon Multidisciplinary Analysis
AMSR	Advanced Microwave Scanning Radiometer
ANA	*Agência Nacionalde Aguas* (Brazilian National Water Agency)
AOL	Airborne Oceanographic LiDAR
APD	Avalanche Photodiode
API	Antecedent Precipitation Index
APOM	Aerosol Plume Optical Model
ARVI	Atmospherically Resistant Vegetation Index
ASAR	Advanced Synthetic Aperture Radar
ASCAT	Advanced Scatterometer
ASDF	Averaged Square Difference Function
ASI	*Agenzia Spaziale Italiana* (Italian Space Agency)
ASI	Agriculture Stress Index
ASTER	Advanced Spaceborne Thermal Emission and Reflection Radiometer
ATCOR	Atmospheric and Topographic Correction
ATLAS	Advanced Topographic Laser Altimeter System
ATM	Airborne Topographic Mapper
ATREM	Atmospheric Removal (atmospheric correction method)
ATSR	Along Track Scanning Radiometers
AUC	Area Under Curve
AVHRR	Advanced Very High Resolution Radiometer
AVIRIS	Airborne Visible/Infrared Imaging Spectrometer
AWiFS	Advanced Wide Field Sensor
BEIDOU/ COMPASS	Chinese Navigation Satellite System
BELA	BepiColombo Laser Altimeter
BGB	Below-Ground Biomass
BIL	Band Interleaved by Line
BIOMOD	Biomolecular Design (Computer platform implemented in R)
BIP	Band Interleaved by Pixel

BLUE	Best Linear Unbiased Estimator
BOC	Binary Offset Carrier
BPSK	Binary Phase Shift Keying
BPT	Binary Partition Tree
BRDF	Bidirectional Reflectance Distribution Function
BRGM	*Bureau de Recherches Géologiques et Minières* (French Geological Survey)
BSA	Back Scatter Alignment
BSQ	Band Sequential
BUI	Build-up Index
BWI	Bassist Wetness Index
C/A	Coarse Acquisition
CaCO3	Calcium Carbonate
CAIC	Consistent Akaike's Information Criterion
CALIOP	Cloud-Aerosol LiDAR with Orthogonal Polarization
CALIPSO	Cloud-Aerosol LiDAR Pathfinder Satellite Observation
CanEx-SM	Canadian Experiment for Soil Moisture
CART	Classification and Regression Trees
CASI	Compact Airborne Spectrographic Imager
CBERS	China–Brazil Earth Resources Satellite
CBOC	Composite Binary Offset Carrier
CCD	Charged Coupled Devices
CCDAS	Carbon Cycle Data Assimilation System
CCI	Climate Change Initiative
CCRS	Canada Center for Remote Sensing
CDF	Cumulative Density Function
CDMA	Code Division Multiple Access
CDOM	Colored Dissolved Organic Matter
CEC	Cation-Exchange Capacity
CEM	Constrained Energy Minimization
CEOS	Committee on Earth Observation Satellites
CERES	Crop Environment Resource Synthesis

CESBIO	*Centre d'Études Spatiales de la Biosphère* (Center for the Study of the Biosphere from Space)
CFC	Chlorofluorocarbons
CFD	Constant Fraction Discriminator
CFFDRS	Canadian Forest Fire Danger Rating System
cGNSS	Conventional Global Navigation Satellite Systems
CHAMP	Challenging Minisatellite Payload (German geosciences satellite)
CHL	Chlorophyll Content
CHM	Canopy Height Model
CHRIS	Compact High Resolution Imaging Spectrometer
CIRAD	Agricultural research for development (France)
CLASlite	Carnegie Landsat Analyse System Lite
CLM	Community Land Model
CLMGW	Community Land Model with a Ground Water Parameterization
CLPX	Cold Land Processes Field Experiment
CLS	Collecte Localisation Satellite
CLSM	Catchment Land Surface Model
CM	Code Moderate (for GNSS)
CMC	Canadian Meteorological Centre
CMEM	Community Microwave Emission Model
CNES	*Centre National d'Études Spatiales* (French Space Agency)
CNRS	*Centre National de Recherche Scientifique* (National French Center for Scientific Research)
COD	Controlled Origin Designation
COG	Center of Gravity
CONUS	Contiguous United States
COST	Cosine Estimation of Atmospheric Transmittance
CR	Continuum Removed
CRI	Carotenoid Reflectance Index
Cryosat	Satellite Radar Altimeter
CSA	Canadian Space Agency
CSR	Centre for Space Research
CTA	Classification Tree Analysis

CTFC	Center for Tropical Forest Science (Cameroon)
CTFS	Center for Tropical Forest Science
CTMF	Cluster Tuned Matched Filter
CW	Continuous Waves
CWFIS	Canadian Wildland Fire Information System
CyGNSS	NASA's Cyclone Global Navigation Satellite System
CZCS	Coastal Zone Color Scanner
CZMIL	Coastal Zone Mapping and Imaging LiDAR
DAM	Dry Aerial Mass
DART	Discrete Anisotropic Radiative Transfer
DATAR	Interministerial Delegation for Territorial Planning and Regional Attractiveness
DBH	Diameter at Breast Height
DC	Drought Code
DCA	Dual Channel Approach
DCM	Digital Canopy Model
DDM	Delay Doppler Map
DEGRAD	Forest Degradation program
DEIMOS	Deep Imaging Multi-Object Spectrograph
DEM	Digital Elevation Model
DEOS	Delft Institute of Earth Observations and Space Systems
DERD	Double bounce Eigenvalue Relative Difference
DETER	*Detecção de Desmatamento em Tempo Real* (Near real-time deforestation detection system)
DGPS	Differential Global Positioning System
DHSVM	Distributed Hydrology Soil-Vegetation Model
DIACT	Inter-ministerial Agency for Spatial Planning and Competitiveness
DIANA	Divise Analysis
DIC	Digital Image Correlation
DIMAP	Digital Image MAP
DInSAR	Differential SAR Interferometry
DisALEXI	Disaggregated Atmospheric Land Exchange Inverse
DLIS	Desert Locust Information Service

DLR	*Deutsches Zentrum für Luft und Raumfahrt* (German Space Agency)
DM	Dry Matter
DMC	Disaster Monitoring Constellation or Duff Moisture Code, depending on the application
DMRT	Dense Media Radiative Transfer
DMSP	U.S. Air Force Defense Meteorological Satellite Program
DMU	De Monfort University
DOAS	Differential Optical Absorption Spectroscopy
DOS	Dark Object Substraction
DOY	Day-of-Year
DPSS	Diode-Pumped Solid-State
DS	Diffuse Scatterer
DSM	Digital Soil Mapping
DSM	Digital Surface Model
DTC	Dry Troposphere Correction
DTED	Digital Terrain Elevation Data
DTM	Digital Terrain Model
DVI	Difference Vegetation Index
DW	Dry Weight
DWBA	Distorted Wave Born Approximation
EAARL	Experimental Advanced Airborne Research LiDAR
ECDC	European Centre for Disease Prevention and Control
ECMWFMMT	European Centre for Medium-Range Weather Forecasts Mobile Mapping Technology
ECV	Essential Climate Variables
EDF	*Électricité de France* (French electric company)
EEA	European Environment Agency
EFA	Effective Field Approximation
EFFIS	European Forest Fire Information System
EGNOS	European Geostationary Navigation Overlay Service
EID-Méditerranée	Interdepartmental Agreement for Mosquito Control on the Mediterranean Coast
EKF	Extended Kalman Filter

ELBARA	ETH L-Band Radiometer
ELUE	Effective Light Use Efficiency
EM	Electromagnetic
ENEA	*Energia Nucleare ed Energie Alternative* (Italian National Agency for New Technologies, Energy and Sustainable Economic Development)
EnKF	Ensemble Kalman Filter
ENMAP	Environmental Monitoring and Analysis Program
ENSO	El Niño Southern Oscillation
ENVEO	Environmental Earth Observation Information
ENVISAT	Environmental Satellite
EOS	Earth Observing System
EPICA	European Project for Ice Coring in Antarctica
EPS	European Polar System
EQeau	Model developed by INRS-Ete with the objective to extract the soil water content from SAR images
ERM	Exact Repeat Missions
ERS	European Remote-sensing Satellite
ESA	European Space Agency
ESCAT	ERS Scatterometer
ESSA	Environmental Science Services Administration Satellite
ET	Evapotranspiration
ETM	Enhanced Thematic Mapper
EUFAR	European Facility for Airborne Research
EUMETSAT	European Organization for the Exploitation of Meteorological Satellites
EVASPA	Evapotranspiration Assessment from Space
EVI	Enhanced Vegetation Index
EWT	Equivalent Water Thickness
EZW	Embedded Zerotrees of Wavelet transforms
FAA	Federal Aviation Administration (USA)
FAI	Floating Algae Index
FAO	Food and Agriculture Organization of the United Nations
FAPAR	Fraction of Absorbed Photosynthetically Active Radiation

FBD	Fine-Beam Double Polarization
FBP	Fire Behavior Prediction
Fcover	Fraction of Vegetation Cover
FDTD	Finite Difference Time Domain
FFMC	Fine Fuel Moisture Code
FFT	Fast Fourier Transform
FI	Fine Particles Index
FIPAR	Fraction of Intercepted Photosynthetically Active Radiation
FLAASH	Fast Line-of-sight Atmospheric Analysis of Spectral Hypercubes (Atmospheric correction)
FLOE	Fish LiDAR Oceanographic Experimental
FMA	Fore-Mid-Aft
FMCW	Frequency Modulated Continuous Waves
FOTO	Fourier-based Textural Ordination
FOV	Field of View
FSA	Forward Scatter Alignment
FST	Spectro-Transfert functions
FT	Fourier Transform
FW	Full Waveform or Fresh Weight, depending on the application
FWHM	Full Width at Height Maximum
FWI	Fire Weather Index
FWS	Fraction of Water Surface
GAGAN	GPS Aided GEO Augmented Navigation system
GAI	Green Area Index
GALILEO	European Global navigation satellite system
GCOS	Global Climate Observing System
GCP	Ground Control Points
GDAL	Geospatial Data Abstraction Library
GDR	Geophysical Data Record
GEO	Group on Earth Observations
GEOCAPI	Geostationary Ocean Color Advanced Permanent Imager
GEOGLAM	Group on Earth Observations Global Agricultural Monitoring
GEOSS	Global Earth Observing System of Systems

GEOSUD	Geoinformation for Sustainable Development (France)
GEOTIFF	Geostationary Earth Orbit Tagged Image File Format
GEROS-ISS	GNSS Reflectometry, Radio Occultation and Scatterometry Onboard International Space Station
GF	Green Fraction
GFO	Geosat Follow-On (radar altimeter satellite)
GFZ	*GeoForschungs Zentrum* (German research centre for Geosciences)
GHG	Green House Gase
GIAM	Global Irrigated Area Map
GIEWS	Global Information and Early Warning System on Food and Agriculture
GIM	Global Ionospheric Model
GIMMS	Global Inventory Modeling and Mapping Studies
GIS	Geographic Information System
GLAI	Green Leaf Area Index
GLAS	Geoscience Laser Altimeter System
GLCM	Grey Level Co-occurrence Matrix
GLDAS	Global Land Data Assimilation System
GlobCover	Global Land-Cover Map
GLONASS	Global Navigation Satellite System (Russia)
GLP	Global Land Project
GLRT	Generalized Likelihood Ratio Test
GLS	Global Land Survey (Landsat)
GM	Geodetic Mission
GMES	Global Monitoring for Environment and Security
GMFS	Global Monitoring of Food Security
GMM	Gaussian Mixture Model
GNSS	Global Navigation Satellite System
GNSS-R	Global Navigation Satellite System – Reflectometry
Go	Giga-Octet (1,000,000,000 octets)
GOCE	Gravity field and Steady-state Ocean Circulation Explorer (ESA)
GOES	Geostationary Operational Environmental Satellite
GOFC Fire IT	Global Observation of Forest Cover Fire Implementation Team

GOM	Geometrical Optics Model
GORS	GNSS Occultation Reflectometry Scatterometry
GPCC	Global Precipitation Climatology Centre
GPP	Gross Primary Production
GPR	Gaussian Process Regression
GPS	Global Positioning System
GRACE	Gravity Recovery and Climate Experiment (satellite)
GRDC	Global Runoff Data Center
GRGS	Space Geodesy Research Group
GtC	Gigatonnes of Carbon (10^9 tons)
GUS	Ground Uplink Stations
GVI	Difference Vegetation Index
GWIS	Global Wildfire Information System
Ha	Hectare (= 10,000 m^2)
HCFC	Hydro-chlorofluorocarbons
HEC-RAS	Hydrologic Engineering Center River Analysis System (hydraulic model)
HgCdTe	Mercury Cadmium Telluride
HiRI	High Resolution Optical Imager
HITRAN	High-Resolution Transmission Database
HOG	Histogram of Oriented Gradient
HPC	High Performance Computing
HRG	High Resolution Geometry
HSC	Height-Scaled Crown
HSCOI	Height-Scaled Crown Openness Index
HSR	High Spatial Resolution
HYMAP	Hyperspectral Mapper (airborne hyperspectral sensor)
HYPXIM	Hyperspectral-X Imagery
HYSPEX	Hyperspectral Imaging System
HyspIRI	NASA's Hyperspectral Infrared Imager
IASI	Infrared Atmospheric Sounding Interferometer
ICA	Independent Component Analysis
ICARE	Inversion Code for urban Areas Reflectance Extraction

ICESat	Ice, Cloud and Land Elevation Satellite
ICF	Interferometric Complex Field
ICP	Iterative Closest Point technique
IDAN	Intensity Driven Adaptative Neighborhood
IEM	Integral Equation Model
IFN	French National Forest Inventory
IFOV	Instantaneous Field of View
IGN	French National Geographic Institute and Forest Information
iGNSS	Interferometric GNSS
IHS	Intensity, Hue, Saturation
IMU	Inertial Measurement Unit
INERIS	French National Institute for Environmental Protection and Industrial Risks
InGaAs	Indium Gallium Arsenide
INPE	*Instituto Nacional de Pesquisas Espaciais* (Brazilian Institute of Space Research)
INRA	French National Institute for Agricultural Research
INSAR	Interferometric Synthetic Aperture Radar
InSb	Indium antimonide
INSEE	French National Institute for Statistics and Economic Studies
IOD	Indian Ocean Dipole
IPCC	Intergovernmental Panel on Climate Change
IPT	Interference Pattern Technique
IRD	French Research Institute for Development
IRNSS	Indian Regional Navigational Satellite System
IRSTEA	French National Research Institute of Science and Technology for Environment and Agriculture
ISBA	Interactions Soil-Biosphere-Atmosphere (model)
ISDC	Integrated System Data Center
ISI	Initial Spread Index
ISODATA	Iterative Self-Organizing Data Analysis Technique
ISRO	Indian Space Research Organisation
ITC	Individual Tree Crown
ITCZ	Intertropical Convergence Zone

ITG	Institute of Theoretical Geodesy
IWPB	Institute of Water Problem of Bishkek, Kyrgyzstan
Jason	Radar altimeter
JAXA	Japan Aerospace Exploration Agency
JECAM	Joint Experiment for Crop Assessment and Monitoring
JPEG	Joint Photographic Experts Group (image format)
JPL	Jet Propulsion Laboratory
JRC	European Commission's Joint Research Centre
KBR	K-Band Microwave Ranging
KLT	Karhunen–Loeve transform
LaDAR	Laser Detection And Ranging
LADS	Laser Airborne Depth Sounder
LAGEOS	Laser Geodynamics Satellite
LAI	Leaf Area Index
LANDSAT	LAND + Satellite
LAUVA	Airborne Ultraviolet Aerosol LiDAR
LAX	Maximum LAI
LBAS	Local Based Augmentation System
LCCS	Land Cover Classification System
LCLU	Land Cover / Land Use
LDAS	Land Data Assimilation System
LEGOS	Laboratory for Studies in Geophysics and Spatial Oceanography (France)
LEnKS	Local EnKF Smoother
LEO	Low Earth Orbit
LEP	Leading Edge Position
LEWIS	L-band for Estimating Water In Soils
LFMC	Live Fuel Moisture Content (%)
LHCP	Left Hand Circular Polarization
LiDAR	Light Detection and Ranging
LISAH	Laboratory for Soil, Agrosystems and Water Systems (France)
LISFLOOD-FP	Two-dimensional Hydrodynamic Model
LMA	Leaf Mass per Area

LMM	Linear Mixed Model
LOADDT	Spatial planning and territorial development
LOLA	Lunar Orbiter Laser Altimeter
LOV	Villefranche Oceanography Laboratory (France)
LPCA	Laboratory for Physico-Chemistry of the Atmosphere
LPRM	Land Parameter Retrieval Model
LRM	Low Resolution Mode
LSCE	Climate and Environment Sciences Laboratory (France)
LSM	Land Surface Model
LSSM	Least Squares 3D Surface Matching
LST	Land Surface Temperature
LULCC	Land Use and Land Cover Change
LUT	Look-Up Table
LWIR	Long-Wave Infrared
LZW	*Lempel-Ziv-Welch* (compression algorithm)
MACCS	Multisensor Atmospheric Correction and Cloud Screening processor
MARS	Monitoring of Agriculture with Remote Sensing
MATISSE	Advenced Earth Modeling for Imaging and Scene Simulation
MaxEnt	Maximum Entropy Method
MBOC	Multiplexed Binary Offset Carrier
MCT	Mercury Cadmium Telluride
MEB	Microwave Emission of the Biosphere
MERIS	Medium Resolution Imaging Spectrometer
MESA	Monitoring of Environment and Security in Africa
METEOSAT	METEO + Satellite
METOP	Meteorological Operational Satellite Programme/Advanced Scatterometer
METRIC	Mapping Evapotranspiration at High Resolution with Internalized Calibration
MGVI	MERIS Global Vegetation Index
MHz	Mega-Hertz (= 1,000,000 Hz)
MIMR	Multichannel Microwave Imaging Radiometer

MIPERS	Multistatic Interferometric Polarimetric Electromagnetic model for Remote Sensing
MIR	Middle Infrared
MISDc	*Modello Idrologico Semi Distribuito in Continuo* (Continuous rainfall-runoff model)
MISR	Multi-angle Imaging Spectro Radiometer
MISTIGRI	Microsatellite for Thermal Infrared Ground Surface Imaging (CNES, France)
MLR	Multiple Linear Regression
MLS	Mobile LiDAR Scanner
MMD	Minimum–Maximum Difference
MMS	Mobile Mapping Systems
MMU	Minimum Mapping Units
MMV	Mobile Mapping Vehicle
MNDWI	Modified Normalized Difference Water index
MNF	Maximum Noise Fraction
Mo	Mega-octet (1,000,000 octets)
MODCOU	Hydrogeological model
MODIS	Moderate Resolution Imaging Spectroradiometer
MODTRAN	Moderate Resolution Atmospheric Transmission
MOLA	Mars Orbiter Laser Altimeter
MPE	Maximum Permissible Exposure
MSAS	Multi-functional Satellite-based Augmentation System
MSG	Meteosat Second Generation
MSI	Moisture Stress Index
MSI	Multispectral Instrument (Sentinel-2)
MTSAT	Multi-functional Transport Satellites
MVSA	Minimum Volume Simplex Analysis
MWIR	Mid-Wavelength Infrared
NAOMI	New AstroSat Optical Modular Instrument
NASA	National Aeronautics and Space Administration (USA)
NBR	Normalized Burn Ratio
NCC	Normalized Cross-correlation
NCEP	National Centers for Environmental Prediction

Nd:YAG	Neodymium-doped Yttrium Aluminium Garnet
NDSI	Normalized Difference Snow Index
NDVI	Normalized Difference Vegetation Index
NDVITM	Normalized Difference Vegetation Index Threshold Method
NDWI	Normalized Difference Water Index
NEBN	Noise Equivalent Beta Naught
NEDT	Noise Equivalent Delta Temperature
NEE	Net Ecosystem Exchange
NEF	Noise Equivalent Flux
NEM	Normalized Emissivity Method
NEP	Net Ecosystem Productivity
NEP	Noise Equivalent Power
NIR	Near Infrared
NLES	Navigation Land Earth Station
NLRI	Near Laser Ranging Investigation
NMC	National Meteorological Center (USA)
NMF	Non-negative Matrix Factorization
NOAA	National Oceanic and Atmospheric Administration
NOHD	Nominal Ocular Hazard Distance
NORUT	Norut Northern Research Institute (Norway)
NPV	Non-photosynthetic Vegetation
NPW	Numerical Weather Prediction models
NSC	NarynSyrdarya Cascade
NSCAT	NASA Scatterometer
NSIDC	National Snow and Ice Data Center (USA)
NWP	Numerical Weather Prediction
OA	Overall Accuracy
OBIA	Object Based Image Analysis
OLCI	Ocean and Land Colour Instrument
OLI	Operational Land Imager
OLS	Operational Linescan System
OM	Organic Matter
OMI	Ozone Monitoring Instrument

ONERA	French Aerospace Research Agency
ONF	French National Forest Office
OS	Open Service
OSCAT	OceanSat-2 Scatterometer
OTB	OrfeoToolBox
PA	Producer's Accuracy
PACE	Pre-Aerosol, Clouds and Ocean Ecosystems
PAH	Polycyclic Aromatic Hydrocarbons
PAI	Plant Area Index
PARIS-IoD	Passive Reflectometry and Interferometry System In Orbit Demonstrator
PCA	Principal Components Analysis
PERSIANN	Precipitation Estimation from Remotely Sensed Information using Artificial Neural Networks
PET	Potential Evapotranspiration
PFT	Phytoplankton Functional Types
PGN	Permanent GNSS Network
PHR	Panchromatic High Resolution
PIF	Pseudo-Invariant Features
piGNSS-R	Partial Interferometric GNSS-R
PLF	Polarization Loss Factor
PLOF	Local Land Tenure Plan (France)
PLSR	Partial Least Squares Regression
PMT	Photomultiplier Tubesv
POA	Polarization Orientation Angle
PODAAC	Physical Oceanography Distributive Active Data Center
POLDER	Polarization and Directionality of the Earth's Reflectances
PolInSAR	Polarimetry-interferometry Synthetic Aperture Radar
PolSAR	Polarimetry SAR
POLYMER	Polynomial Based Algorithm Applied to Meris
POM	Physical Optics Model
PPCDAM	Plan for Preventing and Controlling Deforestation in Amazônia Legal
PRESS	Prediction Sum of Squares

PRF	Pulse Repeat Frequency
PRISMA	*Precursore Iperspettrale della Missione Applicativa* (Italian hyperspecral mission)
PRN	Pseudo Random Noise
PROBA	Project for On-board Autonomy
PRODES	*Programa de Cálculo do Desflorestamento da Amazônia* (Brazilian Amazon Deforestation Monitoring Program)
PROSPECT	Leaf Optical Properties Spectra (radiative transfer model)
PRS	Public Regulated Service
PS	Persistent Scatterers
PSI	Persistent Scatterers Interferometry
PSR	Penalized-Spline Regression
PV	Photosynthetic Vegetation
PVI	Perpendicular Vegetation Index
PWC	Plant Water Content
QBO	Quasi-Biennal Oscillation
QCA	Quasi-Crystalline Approximation
Qgis	GIS software (open source)
QPSK	Quadrature Phase Shift Keying
QUIKSCAT	Quick Scatterometer (NASA)
QZSS	Quasi-Zenith Satellite System
RADAR	Radiodetection and Ranging
RAF	French Altimetric System
RAINFOR	Amazon Forest Inventory Network
RAMSES	ONERA Airborne Multi-frequency SAR Imaging System (France)
RAN	Royal Australian Navy
RANSAC	Random Sample Consensus
RCA	Radio Corporation of America
RCM	RADARSAT Constellation Mission
RCS	Radiometric Control Sets
REDD	Reduction of Emissions from Deforestation and Forest Degradation

REDDAF	Reducing Emissions from Deforestation and Degradation in Africa (European Project)
RENAG	French National GNSS Permanent Networks
RESIF	French Seismological and Geological Network
RF	Random Forests (classifier)
RFI	Radio Frequency Interference
RG	Relative Greenness
RGB	Red Green Blue
rGNSS-R	Reconstructed GNSS-R
RHCP	Right Hand Circular Polarization
RMSD	Root Mean Square Difference
RMSE	Root Mean Square Error
RPAS	Remotely Piloted Aircraft System
RPC	Rational Polynomial Coefficients
RPD	Ratio of Performance to Deviation
RPIQ	Ratio of Performance to Inter-Quartile range
RTK	Real Time Kinematic
RTM	Radiative Transfer Model
RUE	Rain Use Efficiency
RVI	Radar Vegetation Index
RVoG	Random Volume over Ground
RXD	Reed-Xiaoli Detector
SAFY	Simple Algorithm For Yield estimate
SAM	Spectral Angle Mapper
SAMIR	Satellite Monitoring of Irrigation (model)
SAR	Synthetic Aperture Radar
SAR in	SAR interferometric
Saral/Altika	Radar Altimeter (French–Indian altimetry mission)
SASS	SEASAT Advanced Scatterometer System
SAVI	Soil-Adjusted Vegetation Index
SBAS	Satellite Local-based Augmentation System
SBAS	Small Baselines
SCA	Snow Cover Area
SCARAB	Scanning Radiometer for Radiation Balance

SCF	Snow Cover Fraction
SDC	Snow Depletion Curve
SDS	Science Data System (USA)
SEAS	Survey of the Environment Assisted by Satellite
SeaWiFS	Sea-viewing Wide Field-of-view Sensor (satellite)
SEBAL	Surface Energy Balance Algorithm for Land
SEBS	Surface Energy Balance System
SEC	Standard Error of Calibration
SEKF	Self-extended Kalman Filter
SEP	Standards Error of Prediction
SER	Section Efficace Radar
SERD	Single bounce Eigenvalue Relative Difference
SEVIRI	Spinning Enhanced Visible and Infrared Imager
SFCW	Stepped Frequency Continuous Waves
SfM	Surface-from-Motion
SFT	Strong Fluctuation Theory
SGBM	Semi-Global Block Matching algorithm
SHALOM	Spaceborne Hyperspectral Applicative Land and Ocean Mission
SHI	State Hydrological Institute (St. Petersburg, Russia)
SHOALS	Scanning Hydrographic Operational Airborne LiDAR Survey
SHOM	French Navy's Hydrographic and Oceanographic Service
SID	Spectral Information Divergence
SIERRA	Spectral Reflectance Image Extraction from Radiance with Relief and Atmospheric Correction
SIFT	Scale Invariant Feature Transform
SIGMA	Simulation Innovation for Global Monitoring of Agriculture
SLA	*Scanner LiDAR aérien* (aerial LiDAR scanner)
SLC	Single Look Complex
SLR	Single-Lens Reflex
SM	Soil Moisture
SMA	Spectral Mixture Analysis
SMAC	Simplified Method for Atmospheric Correction

SMAP	Soil Moisture Active and Passive mission (Radiometer)
SMAPVEX	SMAP Validation Experiment
SMEX02	Soil Moisture Experiment 2002
SMF	Spectral Matched Filter
SMLR	Stepwise Multiple Linear Regression
SMMR	Scanning Multichannel Microwave Radiometer
SMOS	Soil Moisture and Ocean Salinity mission (satellite)
SNAS	Chinese Satellite Navigation Augmentation System
SNR	Signal to Noise Ratio
SNSB	Swedish National Space Board
SNV	Standard Normal Variate
SPAD	Single-Photon Avalanche Diode
SPM	Small Perturbation Model
SPM	Suspended Particulate Matter
SPOT	Satellites for Earth Observation
SRM	Snowmelt-Runoff Model
SRTM	Shuttle Radar Topography Mission
SSA	Surface Specific Area
SSC	Soil Surface Characteristics
S-SEBI	Simplified Surface Energy Balance Index
SSM	Soil Surface Moisture
SSMI	Special Sensor Microwave Imager (satellite)
SST	Sea Surface Temperatures
STF	Spectrotransfer Functions
STICA	Socio-technical Information and Communication Arrangements
STICS	Crop model
SUCROS	Simple and Universal Crop Growth Simulator
SUHI	Surface Urban Heat Islands
Suomi-NPP	Suomi National Polar-Orbiting Partnership
SURFEX	Surface model platform (Météo France)
SVAT	Soil–Vegetation–Atmospheric Transfer
SVM	Support Vector Machine
SVMR	Support Vector Machine Regression

SVR	Support Vector Regression
SWAT	Soil and Water Assessment Tool
SWE	Snow Water Equivalent
SWI	Soil Wetness Index
SWIR	Short-wave Infrared
SWOT	Surface Water Ocean Topography (satellite)
SYSIPHE	Airborne hyperspectral imaging system
T/P	Topex/Poséidon (Franco-American altimeter)
TDR	Time Domaine Reflectometry
TEC	Total Electron Content
TES	Emissivity Separation algorithm
THEIA	French Land Data Centre
THIRSTY	Thermal Infrared Spatial System (satellite project by CNES and NASA)
TIFF	Tag Image File Format
TIN	Triangular Irregular Network
TiO_2	Titanium dioxide
TIR	Thermal Infrared
TISI	Temperature Independant Spectral Indices
TLS	Terrestrial Laser Scanning
TM	Thematic Mapper
TMBOC	Time Multiplexed Binary Offset Carrier
TNT2	Topography based Nitrogen Transfer and Transformation
To	Tera-octet (1,000,000,000,000 octets)
TOA	Top of Atmosphere
TOC	Top of Canopy
TomoSAR	Tomography SAR
Topex/ Poseidon	Radar altimeter
TOPLATS	Topographic Land Atmosphere Transfer Scheme
TRIP	Total Runoff Integrating Pathways
TRMM	Tropical Rainfall Measuring Mission (satellite)
TSAVI	Transformed Soil Adjusted Vegetation Index

TSEB	Two-Source Energy Balance (model)
TTL	Transistor-Transistor Logic
TWAP	Transboundary Water Assessment Program
TWS	Terrestrial Water Storage
UA	User's Accuracy
UAA	Utilized Agricultural Area
UAV	Unmanned Aerial Vehicles
UHI	Urban Heat Islands
ULICE	Ultraviolet LiDAR for Canopy Experiment
UNEP	United Nations Environment Program
USDA	United States Department of Agriculture
USGS	United States Geological Survey
USO	Ultra-Stable Oscillator
UTC	Coordinated Universal Time
UTM	Universal Transverse Mercator
UV	Ultra-violet
VARI	Visible Atmospherically Resistant Index
VCA	Vertex Component Analysis
VD	Virtual dimensionality
VHF	Very high frequency
VHI	Vegetation Health Index
VHSR	Very High Spatial Resolution
VIC	Variable Infiltration Capacity
VIIRS	Visible Infrared Imager Radiometer Suite
VIS	Visible
VISAT	Video, Inertial, and Satellite GPS
Vis-NIR	Visible and Near Infrared
VITO	Flemish Institute for Technological Research (Belgium)
VLA	Very Large Array
VLBI	Very Large Baseline Interferometry
VOS	Volatile Organic Compound
VPD	Vapor Pressure Deficit
VSDI	Visible and Shortwave Infrared Drought Index

VWC	Vegetation Water Content
WALID	Water LiDAR Simulation Model
WASS	Wide Area Augmentation System
WDI	Water Deficit Index
WGHM	Water GAP Global Hydrology Model
WGS	World Geodetic System
WMA	Winter Metric Anomaly
WMO	World Meteorological Organization
WMS	Wide-area Master station
WRS	Wide-area Reference Stations
WSI	Water stress index
WTC	Wet Troposphere Correction
XML	Extensible Markup Language
ZSSD	Zero-mean Sum of Squared Difference

Introduction

The social and economic importance of forest and agricultural resources in many parts of the world, along with the increasing world population, economic development, and climate change, have led to development of methods and techniques for managing these resources. Food safety-related issues with the challenge of meeting the growing needs of populations that are increasing by several million per year are of great importance for scientists and policy makers. As for forests, they are the root of the livelihoods of over a billion people worldwide. They contain more than 80 percent of the terrestrial biodiversity of the planet and help protect watersheds which are essential supplying drinking water. Forests also store carbon, both in living and dead biomass. They thus play an important role in the global carbon cycle. However, climate and anthropogenic changes present enormous challenges to the conservation of agricultural and forest lands on Earth.

This book, describing the main uses of remote sensing in agriculture and forestry, and supported by scientists that are internationally renowned in their fields, will allow to update knowledge and describe issues that are currently being researched and that will be developed for years to come. It is designed for remote sensing or geophysics research teams, engineers and students in their 2nd (engineering schools, Masters) and 3rd university cycles (PhD).

The first chapter of this volume concerns the use of optical remote sensing for mapping the primary soil properties, essential to understanding agricultural environments function. The second chapter provides different methods and pointers on how to estimate the biophysical parameters of the vegetation cover, and the third chapter examines methods for land cover mapping.

The following chapter discusses the use of remote sensing in the development of indicators and models of crop management. It discusses the use of remote sensing in monitoring agricultural uses (biomass production, yield, precision agriculture, irrigation, etc.). The assimilation of remote sensing products and data in operating crop models is also presented.

The second part of the book covers uses related to understanding and monitoring of the dynamics of vegetation. The first chapter discusses the monitoring of crops in tropical areas by radar and optical remote sensing. A second chapter presents the monitoring of the agricultural landscape by radar remote sensing.

The last three chapters analyze the properties of the forest cover (cover dynamic, height, biomass) using three different techniques (optical, LiDAR and radar).

We thank all the scientists who contributed to the preparation of this volume, the authors of the chapters of course, and also the experts of the Scientific Committee. This project was conducted with support from IRSTEA (French Research Institute of Science and Technology for the Environment and Agriculture), CNRS (French National Center for Scientific Research) and CNES (French National Center for Space Studies).

We also thank our families for their support and love, Prof. André Mariotti (Professor Emeritus at the University of Pierre and Marie Curie) and Dr. Pierrick Givone (Scientific Director at IRSTEA) for their encouragement and support in the realization of this project.

1

Mapping of Primary Soil Properties Using Optical Visible and Near Infrared (Vis-NIR) Remote Sensing

1.1. Introduction

Information regarding soils and their variability in different landscapes is increasingly sought after to improve decision-making regarding a wide range of global issues, such as agricultural production, climate change and the problems of environmental degradation. This information comes in the form of a group of soil properties, indicated by field observations or laboratory analyses (i.e. organic matter (OM), particle size, calcium carbonate, iron, pH, humidity, etc.), the list and the determination procedures were the subjects of standardization [SOI 51, BAI 95]. This group of soil properties, also called "primary" soil properties, is what makes up current soil databases [GAU 93, BAT 09] and the resulting soil maps. These primary soil properties are used as input data for pedotransfer functions [BOU 89] to estimate properties called "functional" soil properties (i.e. available water content, structural soil stability, a pesticide's adsorption coefficient, etc.). These functional properties are then used to help decision about soil management, and also as input parameters of crop models [BRI 10], carbon dynamics models [PAR 89, FAL 02], hydrological models [ARN 98, MOU 02] and erosion models [LEB 05, CER 10].

Chapter written by Cécile GOMEZ and Phillipe LAGACHERIE.

The main factor limiting the use of these models is the lack or low density and accuracy of the determination of primary soil properties needed to organize them. In fact, the existing soil databases in the world are neither sufficiently comprehensive nor sufficiently precise to meet the demands of soil data, especially to organize agri-environmental models as mentioned above [ROS 04]. Therefore, there is a major challenge to develop alternative methods for mapping soil properties over large areas, with high spatial resolution, while presenting acceptable cost implementation.

Visible and near infrared imagery (Vis-NIR, 400–2500 nm) is an emerging and powerful tool for the mapping of primary soil surface properties [BRI 16] with high spatial resolutions and over large areas, and thus a valuable tool for agronomic and environmental models. Tillage allows soil redistribution on the first cultivated horizon, especially in cultivated areas. Thus, despite the low surface penetration of electromagnetic waves in the Vis-NIR (of the order of the wavelength, i.e. μm), each primary soil surface property estimated by Vis-NIR imaging can be considered as representative of the first cultivated horizon.

Multispectral Vis-NIR imaging benefits from decades of existence, but is limited by its small number of spectral bands, allowing for the mapping of some major soil types [ESC 87]. In parallel, many studies have shown that laboratory Vis-NIR spectroscopy may be considered as an alternative to physico-chemical soil analyses for a quantitative estimation of primary soil properties [VIS 06, NOC 15] such as: clay minerals, texture, calcium carbonate, organic matter, organic carbon, total carbon, iron, pH, cation-exchange capacity. Based on these works, recent studies have successfully expanded the use of laboratory Vis-NIR spectroscopy to hyperspectral Vis-NIR imagery, for primary soil properties mapping [SEL 06, GOM 12].

In this chapter, we present the spectral responses of different primary soil properties, as well as the most commonly used methods to estimate these primary soil properties. Then, we present studies that led to mapping of soil properties through the direct use of the Vis-NIR multispectral and hyperspectral remote sensing data. Finally, we present studies that have used the Vis-NIR remote sensing products to provide a more extensive soil map and/or on a larger scale.

1.2. Spectral signatures of soils

The spectral signature of a material in the Vis-NIR spectral domain is characterized by:

– its general form (Figure 1.1(a));

– the intensity of its reflectance, also called albedo (Figure 1.1(b));

– specific absorption bands (Figure 1.1(c)).

The absorption phenomenon results from electronic and vibrational transitions because of the absorption of specific photon energy within ionic and molecular components of the observed material [BRI 16].

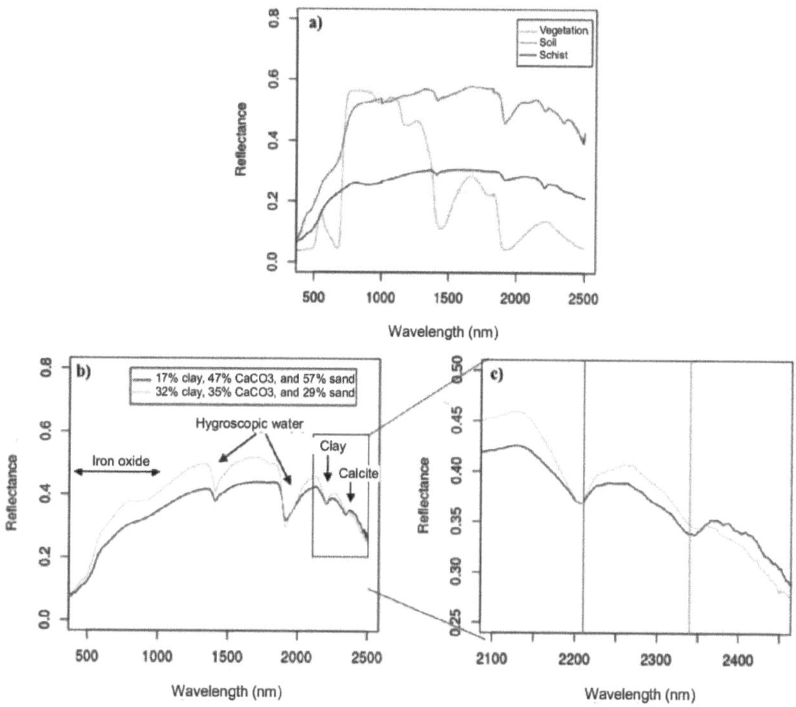

Figure 1.1. *Reflectance spectra acquired in the laboratory: a) on a sample of green vegetation (vine leaf, green), soil (brown) and schist (black); b) on two different soil samples; c) zoom on the spectral range from 2100 to 2400 nm. The red and green vertical lines indicate the center of absorption band at 2206 nm and 2340 nm, respectively. For a color version of this figure, see www.iste.co.uk/baghdadi/3.zip*

Three major chemical components affect the soil spectrum: minerals (clay, ferric and carbonate), organic matter (OM) and water [STO 81]. The presence of calcium carbonate ($CaCO_3$) in the soil causes an absorption band around 2340 nm, because of vibrations at the molecular level associated with the C–O bond [GAF 86] (Figure 1.1(c)). The presence of clay minerals, such as kaolinite, montmorillonite and illite in the soil induce an absorption band around 2200 nm because of the combination of vibrations associated with the OH bond and the OH–Al–OH bonds [HUN 71, CHA 02] (Figure 1.1(c)). Other clay minerals have specific spectral responses in NIR spectral domain, relative to vibrations associated with Mg and Fe–OH bonds. Iron in its trivalent state produces three absorption bands between 400 and 1000 nm because of electronic processes (Figure 1.1(b)). The hematite is characterized by absorption bands centered at 550, 630 and 860 nm and goethite is characterized by absorption bands centered at 480, 650 and 920 nm [HUN 71, MOR 85]. The absorption bands around 500 nm and 900 nm dominate the reflectance in the visible spectrum, whereas the absorption band around 640 nm is seldom marked.

Organic matter is an important soil component, and has a strong influence on soil reflectance characteristics throughout the entire Vis-NIR domain (i.e. Tables 2 and 3 in [BEN 97]). The vibrations of O–H and C–H bonds cause absorptions, particularly because of lignin and cellulose (i.e. Tables 2 and 3 in [BEN 97]). The physico-chemical complexity of soil's organic matter (fresh or dry) explains the difficulty in precisely identifying specific absorption bands. Nevertheless OM has low effect on spectrum when its content is below 2% [BAU 70].

Finally, increasing the water content in soil induces:

– a general reduction in reflectance along the spectrum because of light absorption by water;

– a reduction or increase of absorption bands associated to minerals (e.g. carbonate around 2340 nm and for clay around 2200 nm, respectively);

– the appearance of absorption bands at 1400 and 1900 nm because of the presence of liquid water (H_2O) or associated with the OH bond in soil [BOW 65].

The soil structure, linked to its particle size, the sample geometry, the view angle and the geometry of the light source (incident angle and azimuth angle) influence the general shape of the reflectance spectrum through a change of spectral intensity and absorption band depth [BAU 85, BEN 95]. In particular, the more the grain size increases, the more the surface of spectral distribution decreases, and so the more the reflectance spectrum intensity decreases [BAU 85]. Thus, the spectrum of clay soil will tend to have more albedo than the spectrum of sandy soil (Figure 1.1(b)). Finally, the soil structure usually doesn't affect the absorption band's position [BEN 95].

These spectral characteristics have been studied in laboratory Vis-NIR spectroscopy for decades and described, among others by Ben-Dor *et al.* [BEN 98]. In addition to the wide variety of the physico-chemical soil properties affecting the Vis-NIR spectrum, airborne and satellite imagery introduces new complex factors linked to measurements that are less well-known than in the laboratory. Absorption bands at 1400 and 1900 nm, characteristic of soil water in laboratory measurements, cannot be used for remote sensing as both spectral ranges are affected by atmospheric water vapor. On both these spectral ranges, the remote sensing signal is mainly absorbed and is too noisy to be exploitable for soil study. Some studies have shown the usefulness of water absorption band boundaries, unaffected by the atmosphere, to identify and map the soil water content using remote sensing [HAU 08]. Moreover, as explained above, sample geometry (soil surface in a natural environment), the angle of the image (influenced by flight dynamics in airborne conditions, such as the pitch and roll) and the geometry of the light source (which may vary during an airborne acquisition over several hours) influence the overall spectral shape by a change in the spectral intensity in the spectrum and spectral absorption bands.

Some primary soil properties, such as pH and cation-exchange capacity (CEC), have no particular spectral characteristics and, therefore, cannot be determined by direct spectral analysis. Despite this, in some pedological contexts, a number of studies have successfully exploited correlations between these properties and soil properties or minerals markers with spectral characteristics [BEN 95, BEN 02, MAR 13, ONG 14].

1.3. Estimation of soil properties from their spectral signatures

The recognition analysis of a Vis-NIR reflectance soil spectrum allows us to distinguish and identify soil types [DEM 02, DEM 04]. However, this method cannot provide a quantitative estimation of primary soil properties.

Such a quantitative estimation of primary soil properties can be achieved by developing estimation models called Spectrotransfer Functions (STF) linking a set of *Y* dependent (response) variables (soil property) and a set of *X* predictor variables (Vis-NIR spectrometric measurements). In this section, we will present two approaches allowing for the development of STFs (sections 1.3.1 and 1.3.2), the principle of calibration and validation of STFs (section 1.3.3) and various indicators of performance of these SFTs (section 1.3.4).

1.3.1. *The spectral indices approach*

Spectrotransfer functions for estimating primary soil properties were initially developed by analyzing absorption bands and spectrum geometry. The general principle is based on the relationship between spectral indices and soil properties. The spectral indices calculations are based on ratio of (at least) two reflectance values focusing on absorption band area, depth and asymmetry, and also slope between two reflectance values [VAN 04].

The first developments of spectral indices for the study of soils were based on the relationship between the reflectance values of three visible spectral bands (red, green and blue at 693, 556 and 477 nm, respectively). These studies have been proposed successively by Escadafal [ESC 94], Madeira *et al.* [MAD 97] and Mathieu *et al.* [MAT 98] to estimate iron oxide content from multi-spectral Vis-NIR data. These spectral color indices (brightness index, index color, etc.) are based on physico-chemical relationships between iron oxides and reflectance (section 1.2). For example, the redness index can be related to the hematite content and is calculated from Landsat TM bands, according to the following expression [MAD 97]:

$$\textit{redness index} = \frac{\mathrm{TM3}^2}{\mathrm{TM1}\,.\,\mathrm{TM2}^3} \qquad [1.1]$$

where TM3, TM2 and TM1 correspond to the red band (630–690 nm), the green band (520–600 nm) and the blue band (450–520 nm), respectively. This redness index uses the first hematite absorption band at 550 nm (section 1.2).

A spectral index based on the ratio of reflectance values around the absorption band at 2200 nm, because of the combination of OH and OH–Al vibration modes (section 1.2), has also been proposed for fine particles estimation [LEV 07]. This index, known as the SWIR Fine particles index (FI) is calculated according to the following expression:

$$FI = \frac{R_{\lambda_1}^2}{R_{\lambda_2} . R_{\lambda_3}^3} \qquad [1.2]$$

where R_{λ_1}, R_{λ_2}, and R_{λ_3} are, respectively, the reflectance values at wavelengths of $\lambda_1 = 2133$ nm, $\lambda_2 = 2225$ nm and $\lambda_3 = 2209$ nm and can be related to the textural clay content.

Different spectral indices based on absorption band area (Figure 1.2) have been proposed using the hyperspectral Vis-NIR imaging data, thereby extending the range of primary soil properties that can be determined by remote sensing. The method of Continuum Removal (CR) is a pretreatment generally used before the calculation of this area. This method consists of normalizing the spectrum by its convex spectral envelope, allowing us to remove the baseline ("continuum") that contains target measurement parameters (humidity differences between targets, illumination between shots ...) and thus highlighting the absorption bands related to physico-chemical properties of the target [CLA 84] (Figure 1.2). To estimate the textural clay content, a spectral index based on absorption band area around 2200 nm, because of the combination of vibrations associated with the OH bond and the OH–Al–OH bonds (red vertical line Figures 1.2(b) and 1.1(c)), was developed from the CR spectrum according to the following formula [LAG 08]:

$$CR_{\lambda_3} = \min\left(1, \frac{R_{\lambda_3}}{R_{\lambda_2} + \frac{\lambda_1 - \lambda_3}{\lambda_1 - \lambda_2}\left(R_{\lambda_1} - R_{\lambda_2}\right)}\right) \qquad [1.3]$$

where R_{λ_1}, R_{λ_2} and R_{λ_3} are, respectively, the reflectance values at wavelengths of $\lambda_1 = 2143$ nm, $\lambda_2 = 2269$ nm and $\lambda_3 = 2206$ nm. Similarly, Lagacherie *et al.* [LAG 08] proposed to estimate $CaCO_3$ content, using the absorption band centered around 2340 nm because of vibrations at the molecular level associated with the CO bond (green vertical line in

Figure 1.1(c)). Lagacherie *et al.* [LAG 08] used the spectral bands $\lambda_1 = 2275$ nm, $\lambda_2 = 2375$ nm and $\lambda_3 = 2348$ nm in equation [1.3]. Finally, spectral indices based on absorption bands area related to biochemical components (cellulose, lignin, etc.) were tested for soil organic carbon estimation [BAR 08].

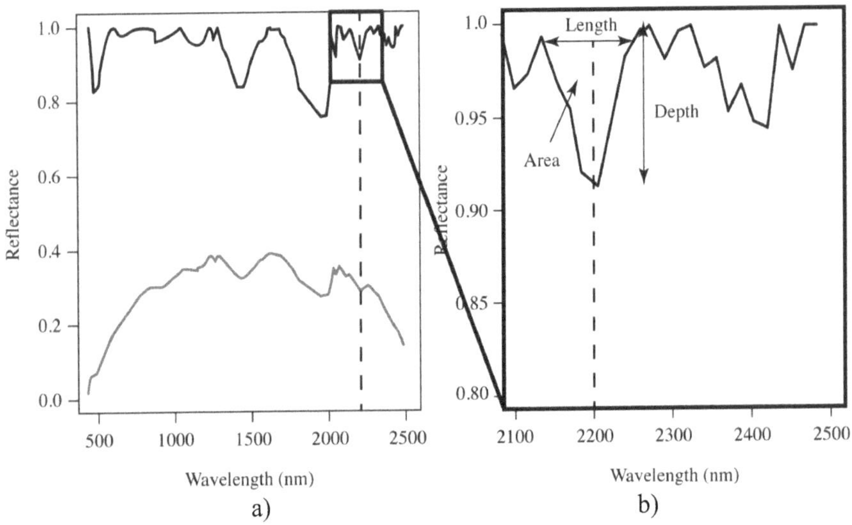

Figure 1.2. *a) Example of a soil spectrum measured by the HyMap airborne sensor (gray spectrum) and its continuum removal (black spectrum), and b) focus on an absorption band over the CR. The dotted vertical line indicates the center of the absorption band at 2200 nm. For a color version of this figure, see www.iste.co.uk/baghdadi/3.zip*

1.3.2. *The chemometric approach*

1.3.2.1. *General concept*

Given the complexity of soil spectra, the use of spectral indices may be insufficient to estimate some primary soil properties. To take the entire spectrum into account, and not just a specific part, a second approach is to use chemometric techniques. Chemometrics is defined as the set of statistical methods, charts or symbols used to better understand the information obtained in chemistry [BER 06, ROG 10]. In our case, a chemometric approach is used to search for the statistical relationships between the whole spectrum (not a specific part, as mentioned above) and soil properties in multivariate regression models.

1.3.2.2. *Possible (and recommended) pretreatments*

Before being used to calibrate a regression model, the spectrum generally undergoes pretreatments to remove information independent from the properties. To remove any multiplicative effects present in the spectrum, for example, because of particle size and shape of the target, a Standard Normal Variate (SNV) correction is usually applied to the spectra. Moreover, multiplicative effects may become additive effects by applying a logarithm transformation [ROG 10]. And, in order to remove additive effects, for example, because of the light scattering in the target, a derivative (first or second) can be carried, or a value can be subtracted, as the average of the spectrum [ROG 10].

1.3.2.3. *Regression methods*

The chemometric approach regroups many multivariate regression methods based on building statistical models to estimate the desired soil property, using a calibration base. In Vis-NIR remote sensing for soil sciences, two multivariate regression methods are commonly used:

– Multiple Linear Regression (MLR) is used for both multi-spectral and hyperspectral remote sensing. This method is useful when the Vis-NIR reflectance spectra are acquired in a small number of wavelengths (multi-spectral data case) and when there are few collinearities between reflectance values at each wavelength. The MLR regression model used to estimate $\hat{y}$ soil property is written as:

$$\hat{y} = \sum_{i=1}^{N} b_i x_i + a \qquad [1.4]$$

where N is the number of wavelength, b_i is the regression coefficient associated with the wavelength i, x_i is the reflectance value at the wavelength i and a the residue;

– Partial Least Squares Regression (PLSR) is widely used in Vis-NIR laboratory spectroscopy and hyperspectral remote sensing. This method allows the management of: (1) co-linearity between the reflectance values at different wavelengths; and (2) a number of predictors (wavelengths) that is larger than the number of samples composing the database [*X*, *Y*]. This second characteristic is particularly true in the case of the laboratory Vis-NIR spectroscopy (laboratory spectrometers acquiring reflectance of more than 2000 wavelengths), but less so in the case of hyperspectral remote sensing (airborne and satellite sensors acquiring reflectance from 120 to 350 wavelengths).

PLSR is used to minimize the covariance between *X* (descriptive variables: wavelengths) and the dependent variables *Y* (sought out soil property). The PLSR projects the variables *X* and *Y* in an area of reduced size (set of orthogonal vectors, called latent variables, maximizing the covariance between *X* and *Y*). A detailed description of the PLSR is given in Wold *et al.* [WOL 01].

Besides the MLR and PLSR methods, other regression methods were tested on hyperspectral remote sensing:

– Penalized-Spline Signal Regression (PSR) is a method that takes into account correlations between reflectance values at neighboring wavelengths. PSR forces the coefficients of the regression to vary smoothly across the wavelengths. This is done by projecting coefficients onto a set of smooth functions (B-splines). New constraints are thus introduced into the regression algorithm, preserving the entirety of the spectrum. This method thus differs from the PLSR by taking into account the order of the wavelengths. Initially proposed by Hastie and Mallows [HAS 93], the PSR was developed and put into practice by Marx and Eilers [MAR 99]. Note that the PSR provides stable and solid predictions under different measurement conditions [MAR 02];

– Stepwise Multiple Linear Regression (SMLR) is an iterative technique for selecting the wavelengths most suited to the prediction of the desired property [WIL 87, MAR 89]. At each iteration, the wavelengths are reviewed in order from the model based on an input threshold, or to eliminate them according to a release threshold. The procedure stops when no more wavelengths can be added or removed from the model according to the input and output selected thresholds;

– Support Vector Machine Regression (SVMR) is an automatic learning technique initially developed for classification, and then extended to regression and based on the reduction of a cost function [VAP 95, VAP 98]. The SVMR has the ability to handle large input spaces efficiently, to deal with noisy patterns and multi-modal class distributions, and their restriction on only a subset of training data in order to fit a (non-linear) function.

1.3.3. *Calibration/Validation*

The implementation of STFs often requires the establishment of a database [*X*, *Y*] consisting of Vis-NIR reflectance spectra acquired by the sensor on bare soil surfaces (descriptive variables *X*) associated with the soil property measured on a laboratory soil sample collected on the same bare soil surfaces (dependent variables *Y*). The quality of database [*X*, *Y*] and, therefore, the estimation model that will follow, are based on following rules:

– definition of the sampling plan. The soil diversity of the study site needs to be fully represented. When a soil map is available in the study area, sampling can be stratified by soil unit, i.e. soil samples are taken in each soil unit, proportionate to the size of these units. When no soil map is available, the sampling can be stratified by soil-forming factors (land use, topography and geology), using, for example, the Latin hypercube sampling method [BUD 06]. It is important to note that in the Vis-NIR remote sensing, the constraint of having pixels over bare soil may be higher than the constraint of having a stratified sampling map by soil unit;

– field soil samples collection. Each soil sample must be a composite of several soil sub-samples collected at different locations on a surface larger or equal to the size of the corresponding pixel of the Vis-NIR image, which will then be used to spatialize the soil property(ies). Moreover, each sub-sample should be as representative as possible of the soil depth characterized by the optical Vis-NIR remote sensing, all the while with a soil volume sufficient for laboratory analysis. A compromise is usually found by sampling from 0 to 10 cm of depth. Finally, in a context of cultivation, the soil surfaces are tilled, which greatly reduces the impact of the sampling depth, at least in the first 30 cm of soil.

There are no precise figures on the minimum number of samples needed to build a database for soil properties estimation. Nevertheless, it is accepted that a data number [*X*, *Y*] inferior to 50 does not allow to develop a robust estimation model, whatever the size of the study area.

The use of a calibration [X_{cal}, Y_{cal}] and validation [X_{val}, Y_{val}] database (sub-databases of [*X*, *Y*]) allows for the calibration of the STFs presented above and theirs validations, respectively. The data allocation in the calibration [X_{cal}, Y_{cal}] and validation of [X_{val}, Y_{val}] database is an important and complex step. If the two sets ([X_{cal}, X_{cal}] and [X_{val}, Y_{val}]) are too similar, the model may be too optimistic (underestimation of the prediction error). Inversely, if the two sets ([X_{cal}, Y_{cal}] and [X_{val}, Y_{val}]) are too different, the model will seem inaccurate (overestimation of the prediction error).

To ensure model robustness, it is necessary to build a calibration database representative of the study area in terms of either soil properties or of spectral signatures. It is also necessary that when a validation database is available, it is homogeneous regarding the calibration database, again either in terms of soil properties or of spectral signatures. In theory, calibration [X_{cal}, Y_{cal}] and validation [X_{val}, Y_{val}] databases would have to fulfill the following criteria:

– similar distribution of Y_{cal} and Y_{val} variables;

– maximum spectral variability within X_{cal}, where spectra remain in a realistic domain (avoiding outliers).

These two criteria are difficult to meet simultaneously. Different calibration [X_{cal}, Y_{cal}] and validation [X_{val}, Y_{val}] database construction methods exist, relying on either of the above criteria. A similar distribution of Y_{cal} and Y_{val} variables is frequently used. The Y variables need to be sorted in ascending order. Then, both samples having the lowest Y soil properties are assigned to the calibration database. Then, the next sample is placed in the validation set, and the procedure is continued by alternately placing both following samples in the calibration set and the next sample in the validation set. The assignment of samples in each of the bases follows this process until the last sample, thereby constructing two uniform sets containing 2/3 and 1/3 of the samples, respectively. To ensure a significant spectral variability, methods such as Kennard-Stone [KEN 69] or DUPLEX [SNE 77] can also be used.

1.3.4. *Performance indicators of these models*

Spectrotransfer function performance is usually evaluated from various parameters, using the calibration database [X_{cal}, Y_{cal}]:

– the determination coefficient of the calibration database (R^2_{cal});

– Standard Error of Calibration (SEC) defined by the estimated standard deviation of the random error between the n observed values Y_{cal} and the n predicted values $\hat{Y}_{cal}$. It is the degree to which the model is consistent with the calibration data:

$$SEC = \sqrt{\sum_{i=1}^{n} \frac{\left(\hat{y}_{cal\_i} - y_{cal\_i}\right)^2}{n}} \quad [1.5]$$

where y_{cal_i} is the observed value of the sample i and $\hat{y}_{cal\_i}$ is the estimated value of this sample i.

– Ratio of Performance to Deviation (RPD_c) is defined by the ratio between the standard deviation (sd) of the observed values Y_{cal} and the SEC [CHA 01]:

$$RPD_c = \frac{sd(Y_{cal})}{SEC} \quad [1.6]$$

When the user also has a validation database [X_{val}, Y_{val}], the quality of the estimation model can be assessed by [BEL 10]:

– the determination coefficient of the validation database (R^2_{val});

– the bias between the m observed values Y_{val} and the m predicted values $\hat{Y}_{val}$, following:

$$Biais = \sum_{i=1}^{m} \frac{\hat{y}_{val\_i} - y_{val\_i}}{m} \quad [1.7]$$

– the standard validation error or Standard Error of Prediction (SEP) is defined by the estimated standard deviation of the random error between the m observed values Y_{val} and the m predicted values $\hat{Y}_{val}$ following:

$$SEP^2 = \sum_{i=1}^{m} \frac{\left(\hat{y}_{val\_i} - y_{val\_i}\right)^2}{m} \quad [1.8]$$

where y_{val_i} is the observed value of the sample i and $\hat{y}_{val\_i}$ is the estimated value of this sample i

– Ratio of Performance to Deviation (RPD_v) is defined by the ratio between the standard deviations of the observed values Y_{cal} *and* the SEP [CHA 01]:

$$RPD_v = \frac{sd(Y_{val})}{SEP} \quad [1.9]$$

It is interesting to note that RPD_v and R^2_{val} indicators are redundant [BUD 13] as:

$$RPD_v = \frac{1}{\sqrt{\left(1 - R^2_{val}\right)}} \quad [1.10]$$

As such, the Ratio of Performance to Inter-quartile (RPIQ) index was introduced in [BEL 10] as the ratio of the *interquartile* (IQ) and the *SEP*:

$$RPIQ = \frac{IQ}{SEP} \quad [1.11]$$

The IQ is the interquartile, which is the difference between the third quartile (value below which we find 75% of the samples) and the first quartile (value below which we find 25% of the samples).

When no validation data are available, STFs can be validated using the cross-validation technique. Cross-validation involves alternately removing a small number (k) of samples, to build a model on the remaining n-k observations, then to use this model to produce an estimated value on the removed k samples. If all the samples have been removed, it is possible to determine the R^2_{cv} indicator according to the following formula:

$$R^2_{cv} = 1 - \frac{\sum_{i=1}^{n}\left(y_{cal\_i} - \hat{y}_{cal\_i/i}\right)^2}{sd(y_{cal})} \quad [1.12]$$

where y_{cal_i} is the observed value of the sample i and $\hat{y}_{cal\_i/i}$ is the estimated value of this sample i when this sample is removed from the calibration base to build the model. The numerator of this ratio is also called Predictive Error Sum of Squares (PRESS). In most spectrometric studies, the number k is set to 1 (called leave-one-out cross-validation).

1.4. Direct uses of estimation models

Since the emergence of multi-spectral Vis-NIR satellite, many studies have sought to highlight the link between soil properties and soil reflectance. Initially used for the classification of soil types, work in multispectral Vis-NIR imaging has enabled the mapping of large soil sets [ESC 87]. The first quantitative estimations of primary soil properties by remote sensing data date from the early 2000s [BEN 02]. The increasing availability of the hyperspectral Vis-NIR airborne and satellite data since the 2000s has allowed for increasing works on primary soil properties mapping (Table 1.1).

In this section, we will present examples of the multi-spectral Vis-NIR imaging data use (section 1.4.1) and the hyperspectral Vis-NIR imaging data use (section 1.4.2) for primary soil properties mapping, then we will present the current limitations of these techniques (section 1.4.3).

Reference	Sensor Used	Approach	Surface Area Mapped	Properties Studied
[SUL 05]	IKONOS (*M*)	Chemometric (MLR)	~ 100 ha	Total carbon and clay
[NAN 06]	LANDSAT TM (*M*)	Chemometric (MLR)	~ 184 km^2	10 properties including texture, organic carbon, and CEC
[JAR 10]	LANDSAT TM (*M*)	Chemometric (MLR)	~ 300 km^2	Organic carbon
[VAU 13]	SPOT (*M*)	Chemometric (MLR)	~ 220 km^2	Organic carbon
[SHA 15]	LANDSAT TM (*M*)	Spectral indices	~ 160 km^2	Clay
[BEN 02]	DAIS airborne (0.4-2.5 μm) (*H*)	Chemometric (MLR)	~ 27 km^2	Soil field moisture, soil saturated moisture, organic matter, pH, electric conductivity
[CRO 06]	DAIS airborne (0.4-2.5 μm) (*H*)	Spectral indices	~ 20 km^2	Gravel coverage
[SEL 06]	HyMap airborne (0.4-2.5 μm) (*H*)	Chemometric (MLR and PLSR)	~ 200 km^2	Sand, clay, organic carbon, and total nitrogen
[LAG 08]	HyMap airborne (0.4-2.5 μm) (*H*)	Spectral indices	~ 24 km^2	Clay and $CaCO_3$
[GOM 08a]	HyMap airborne (0.4-2.5 μm) (*H*)	Chemometric (PLSR) and spectral indices	~ 24 km^2	Clay and $CaCO_3$
[GOM 08b]	HYPERION satellite (0.4-2.5μm) (H)	Chemometric (PLSR)	~ 16 km^2	Organic carbon
[STE 10]	AHS-160 airborne (0.4-2.5 μm) (*H*)	Chemometric (PLSR, PSR, and SVMR)	~ 420 km	Organic carbon
[GER 12]	HyMap airborne (0.4-2.5 μm) (*H*)	Chemometric (PLSR)	~ 200 and 260 km^2	Clay and organic carbon
[STE 12]	AHS-160 airborne (0.4-2.5 μm) (*H*)	Chemometric (PLSR, PSR, and SVMR)	~ 420 km^2	Organic carbon
[GOM 12]	AISA-DUAL airborne (0.4-2.5 μm) (*H*)	Chemometric (PLSR)	~ 300 km^2	Clay, sand, silt, $CaCO_3$, and CEC
[LU 13]	HYPERION satellite (0.4-2.5 μm) (*H*)	Chemometric (SMLR and PLSR)	~ 176 km^2	Organic carbon, pH, total phosphorus, CEC

Table 1.1. *Examples of work on the use of Vis-NIR remote sensing for the estimation of soil properties (M: multi-spectral and H: hyperspectral)*

1.4.1. *Example of primary soil properties mapping using multispectral Vis-NIR imaging*

Two examples were chosen to describe primary soil properties mapping using multi-spectral imaging data. The first example is the work of Nanni and Demattè [NAN 06] showing the potential of the Landsat TM satellite for collecting data for the mapping of 10 primary soil properties over an area of 184 ha in a cultivated area (sugarcane plantation in Brazil). The 10 soil properties are the following: texture (clay, sand, and silt), organic matter, total iron, total SiO_2, sum of cations, Al saturation, titanium dioxide (TiO_2), and CEC. Soil samples were collected from each point of regular 100 × 100 m grids, resulting in one sample per hectare, so 184 surface soil samples were collected (Figure 1 in [NAN 06]). Of these 184 samples, only those located on bare soil surfaces at the time of the Landsat TM data acquisition were used in a calibration database. However, neither the number nor the location of these samples are shown in the article.

For each soil property, a STF linking a set of Y dependent variables (soil property) and a set of X predictor variables (Vis-NIR measurements acquired by the Landsat TM sensor) was built using the MLR model (section 1.3.2) and uses the calibration database [X_{cal}, Y_{cal}] and without validation data (section 1.3.3).

Based on R^2_{cal} analysis (section 1.3.4), the STFs developed to estimate the soil properties of clay, sand, organic matter, total iron, total SiO_2, sum of cations, TiO_2 and CEC, have satisfactory performances (Table 4 in [NAN 06]). Despite the good performance of these functions, Nanni and Demattè [NAN 06] do not show estimated soil properties mapping.

The second example is extracted from Shabou *et al.* [SHA 15]. They also studied the potential of Landsat TM satellite data for mapping clay content over the cultivated Tunisian environment. Shabou *et al.* [SHA 15] have developed a STF based on the relation between a linear spectral index ($MID_{infrared}$ = (TM5 – TM7) / (TM5 + TM7)) and the textural clay content Y according to the equation:

$$Y = a\,MID_{infrared} + b \qquad [1.13]$$

where TM5 and TM7 are the spectral band from 1.55 to 1.75 μm and from 2.08 to 2.35 μm, respectively, a is the coefficient of linear regression and b is the residue. As well as Nanni and Dematté [NAN 06], and despite a very low number of calibration data [X_{cal}, Y_{cal}] (n = 30), the developed spectrotransfer function [1.13] offers a correct clay estimation performance. The estimated clay content mapping has been performed at the scale of a 670 km^2 study area and at a spatial resolution of 30 m, by applying the spectrotransfer function [1.13] to all the pixels of bare soil (Figure 1.3). The analysis of the estimated clay content map shows that the clay content distributions are consistent with the soil units of the soil map produced in 1975 by the Tunisian Ministry of Agriculture.

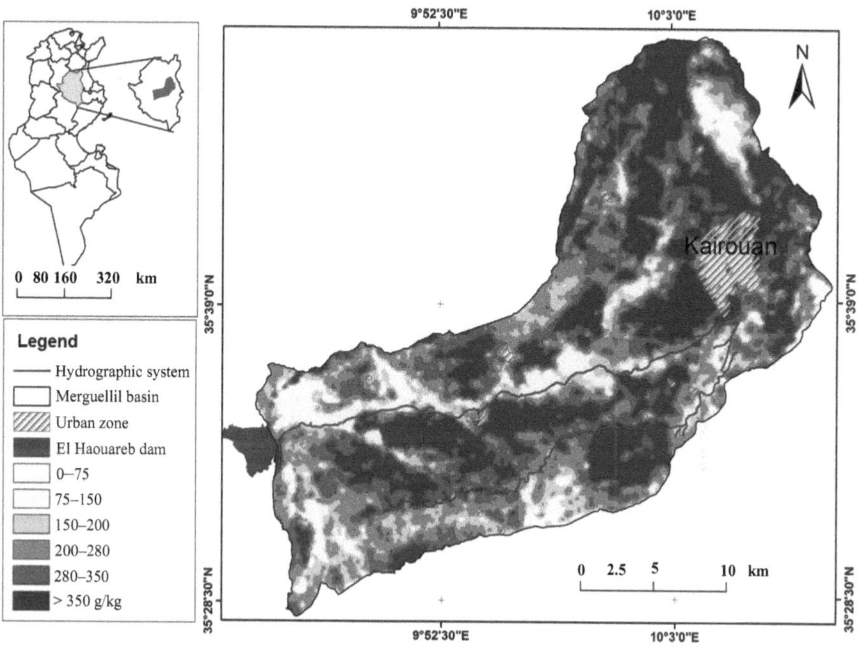

Figure 1.3. *Map of the textural clay content, obtained by using multi-spectral Landsat TM satellite data (spatial resolution of 30 m) and a spectral index [SHA 15]. For a color version of this figure, see www.iste.co.uk/baghdadi/3.zip*

1.4.2. *Example of primary soil properties mapping using hyperspectral Vis-NIR imagery*

The study used to describe the mapping of primary soil properties using Vis-NIR hyperspectral imaging was provided by Gomez *et al.* [GOM 12]. This work used airborne AISA-DUAL data acquired over a cultivated area (Cap Bon, Tunisia) of 300 km^2. Here, they focused on the estimation of eight primary soil properties: texture (clay, sand and silt), $CaCO_3$, iron, pH, organic carbon and CEC. The interest of this example is based on three points:

– the study area is one of the largest so far studied (Table 1.1) and thus covers a large variety of soil;

– the range of soil properties is one of the largest (Table 1.1);

– the spectrotransfer functions associated to successful calibration were applied to the study area allowing a detailed analysis of benefits and limits of estimated primary soil properties maps.

A sampling of 129 surface soils, each representing a 10 × 10 m square, was achieved on the study area. This sampling was stratified according to the soil groups of the study area, taking into consideration the accessibility and bare soil coverage at the time of airborne data collection. For each of the eight soil properties, a spectrotransfer function linking the variables to predict Y (soil property) to the descriptive variables X (Vis-NIR airborne data measured by AISA-DUAL sensor, 0.4–2.5 µm) was built using PLSR regression model (section 1.3.2) calibrated from the 129 soil samples [X_{cal}, Y_{cal}]. No validation database was used (section 1.3.3). Based on the analysis of R^2_{cv} and standard error of cross-validation (SECV) performance indicators (section 1.3.4), only four soil properties can be estimated by STF using airborne Vis-NIR AISA-DUAL data: clay, sand, iron and CEC (Table 1.2 and Figure 1.4). In this study, these four properties meet the two rules previously identified by Ben-Dor *et al.* [BEN 02] as necessary for their prediction:

– the soil property has a specific spectral signature because of a chemical component or an associated physical structure, or is correlated with a soil property that has a specific spectral signature because of a chemical component or associated physical structure;

– the soil property has a wide range of value (Table 1.2).

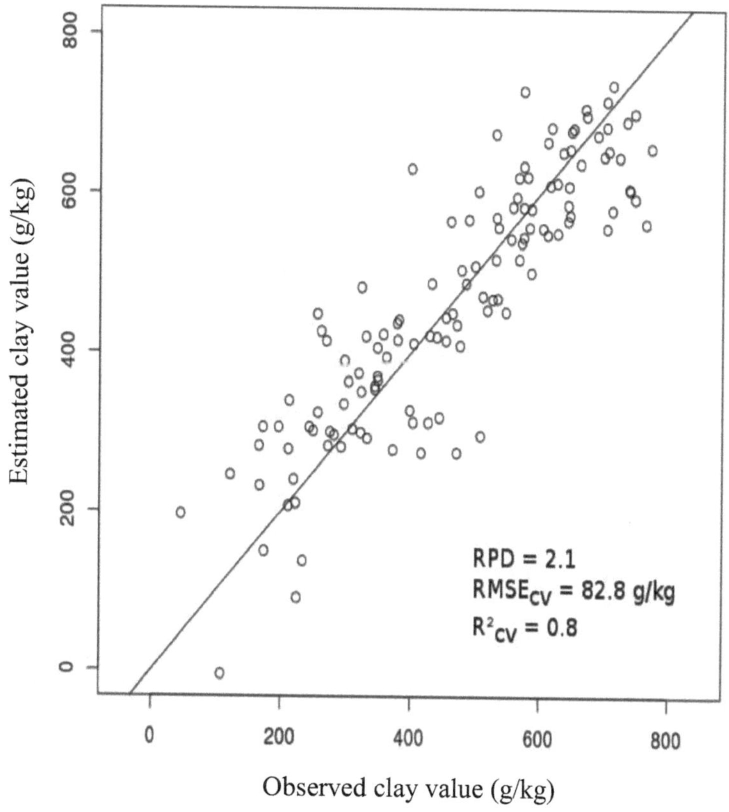

Figure 1.4. *Clay textural values measured in the laboratory versus clay textural values estimated by PLSR model (g/kg)*

A map of these four primary soil properties (clay, sand, iron, and CEC) was conducted across the 300 km^2 study area. The estimated clay map exhibits a complex regional soil pattern, mainly because of variations in lithology (Figure 1.5). Differences both in estimated clay contents and soil pattern appeared between the Pliocene area, located in the southeast corner of the image, and the Miocene area, covering the rest of the image. Pliocene area exhibits low levels of textural clay and they are weakly variable. Inversely, the Miocene area presents a wide range of textural clay content values whose spatial variations are very well revealed by the estimated clay map. They follow the geological pattern formed by the alternating sandstone and marl outcrops, yielding low and high values of clay content, respectively

(Figure 1.5). These soil patterns vary across the southeast/northwest direction, with decreasing distance between successive sandstone outcrops and the occurrence of a large sandstone outcrop in the middle of the study area. The estimated clay map of a watershed (6.6 km^2) is characterized by the alternating sandstone and marl banks. Deposition of sandy material (from the erosion of sandstone areas in the valleys) perpendicular to the outcrops, add to the complexity of the previously described clay content pattern.

Thus, an analysis of the estimated primary soil property maps, produced with a high spatial resolution (~5 m), allows us to identify the soil surface pattern, and to create new perspective for understanding of the process of soil implementation in the studied areas. The spatial resolution of these maps (5 m) allows us to consider an improvement of existing soil maps and an estimate, at the field scale, of functional surfaces properties (useful soil water reserves, soil erodibility, etc.) linked to the estimated primary soil properties. Finally, the next section (section 1.5) presents examples of estimated primary soil surface properties use, for mapping these properties (1) in subsurface and (2) over the entire study area (including semi-vegetated and urban areas).

	Property Variability	[Min; Max] [Avg; SD]	Associated Specific Spectral Response	Correlated with a property having an associated specific spectral response	R^2_{cv}
Iron (g/100g)	Strong	[0.3; 3.3] [1.6; 0.5]	Yes	Yes	0.7
CEC (cmol+/kg)	Strong	[2.8; 34.3] [19.5; 6.6]	No	Yes	0.7
Clay (g/kg)	Strong	[46; 772] [463.7; 175.9]	Yes	Yes	0.8
Sand (g/kg)	Strong	[32; 896] 326; 202.3]	No	Yes	0.7
Silt (g/kg)	Medium	[58; 429] [210.2; 58.7]	No	Yes	0.1
$CaCO_3$ (g/kg)	Medium	[1; 346] [36.9; 52.3]	Yes	No	0.1
pH	Weak	[5.8; 8.8] [8.2; 0.5]	No	No	0.2
Organic carbon (g/kg)	Weak	[2.7; 21.8] [8.9; 2.8]	Yes	No	0.3

Table 1.2. *Characteristics of soil properties studied by Gomez et al. [GOM 12] and the associated performance indicators (R^2_{cv})*

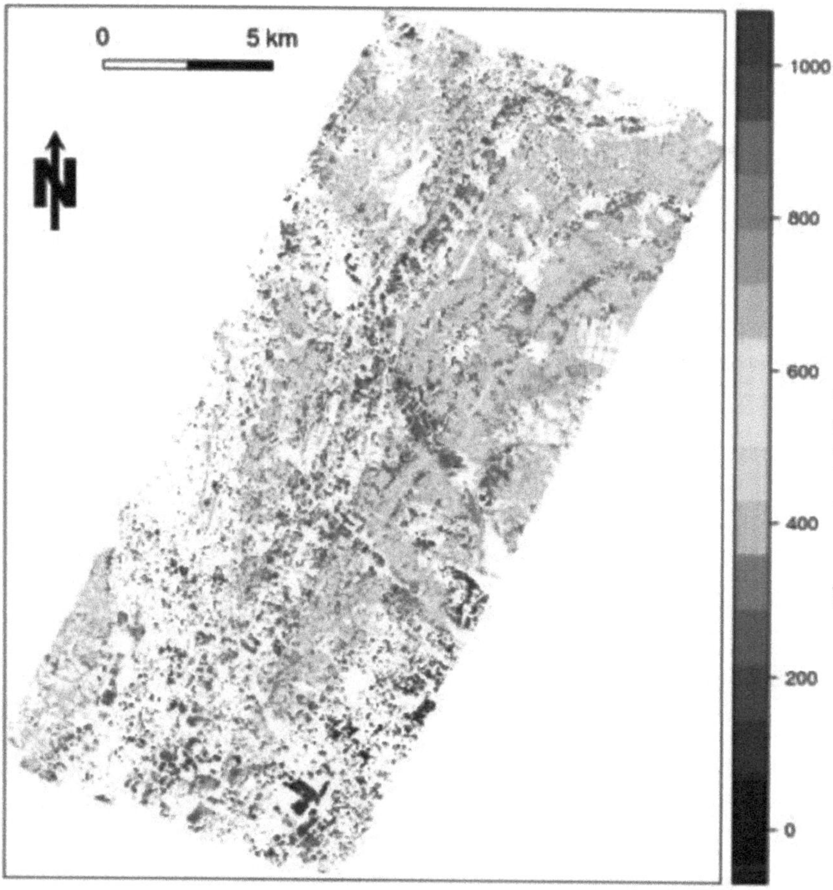

Figure 1.5. *Estimated clay content map, obtained using airborne hyperspectral AISA-DUAL data (spatial resolution of 5 m) and a PLSR model [GOM 12]. The white areas are masked pixels (vegetation and urban zones). For a color version of this figure, see www.iste.co.uk/baghdadi/3.zip*

1.4.3. *Limits*

Since the beginning of Vis-NIR remote sensing use for soil properties estimation, the superiority of the STFs developed from X variables measured in laboratory, compared with STFs developed from X variables acquired by an airborne sensor, has been demonstrated [BEN 02]. This observation was found by numerous studies, highlighting the presence of three major factors disrupting the soil spectral measurement and affecting, therefore, the quality of estimation functions [STE 08, BEN 09, LAG 08]: (1) a low signal to noise ratio of the imaging sensor; (2) the atmosphere; and (3) some disruptive factors affecting the target (soil areas). Regarding the last factor, the presence of surface elements (green vegetation, dry residues, coarse elements, etc.), the soil roughness variability (which may be because of different tillage methods in cultivated areas) and the surface moisture level variability, are major limitations to the use of Vis-NIR remote sensing data for mapping primary soil properties. In practice, to fully reduce these limits, it is recommended to only use Vis-NIR remote sensing on bare and dry soil surfaces. Finally, regardless of the sensor type, estimation model, and surface, it was highlighted that the soil property have to follow the following rules [BEN 02]:

– the soil property must have a specific spectral signature because of a chemical or physical structure, or be correlated to a soil property having a specific spectral signature because of an associated chemical or physical structure;

– the soil property must have a wide range of values. Thus, the predictability of a primary soil property depends, among other things, on the soil diversity of the study site.

1.5. Use of the Vis-NIR remote sensing products for digital soil mapping

Vis-NIR remote sensing provides useful information on the variations of some soil properties, but also has limitations (section 1.4.3) that limits its consideration as a source of sufficient data for all soil properties at all soil depths. This is why it is important to place Vis-NIR remote sensing in a wider framework, which is that Digital Soil Mapping (DSM). DSM is defined as “the creation and enrichment of spatial soil information systems referenced by numerical models inferring spatial and temporal changes in the soil and their properties using spatial data observations of the soil environment” [LAG 07].

The general principle of the DSM is summarized in Figure 1.6. It consists of predicting soil classes or soil properties (*S* to the left of the equation in Figure 1.6) using the spatial data on soil available in the area of interest and the spatial data representing the landscape features that are in relation (causal or not) with the soil or one of its properties. This spatial data, referred to hereinafter as “soil covariates”, is assumed to be readily available in digital form on the area of interest (digital terrain model, land use map, geological map, etc.). Estimates of soil classes or soil properties are produced by prediction functions induced either by statistical models or geostatistics, or by formalization of soil scientists’ expertises. These functions are calibrated by means of spatial data on soil, providing an estimate of the prediction uncertainty.

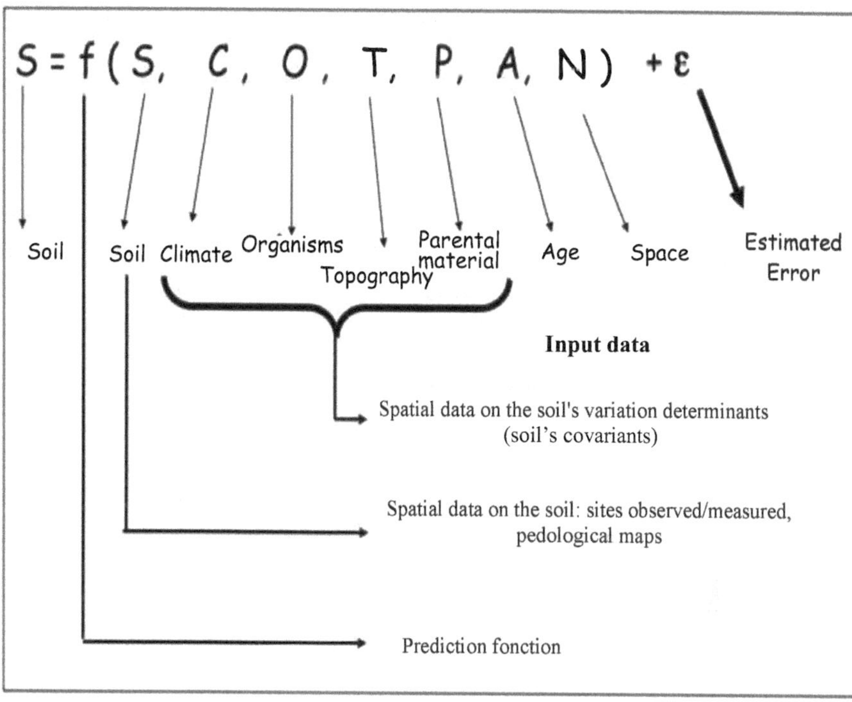

Figure 1.6. *General principle of digital soil mapping (from [MCB 03])*

In terms of the proposed use of digital soil mapping, the use of remote sensing products can be considered according to the two ways described below.

1.5.1. *Using Vis-NIR remote sensing products as a new data source for DSM*

In the case where the Vis-NIR remote sensing product used specifically relates to the property to be mapped, each pixel of the remote sensing product will be considered as a new site where a measurement of this property is available. The interest of remote sensing products compared with spatial sampling of measured sites which are conventionally used in digital soil mapping is the considerably higher spatial density which can capture in a very detailed way and over large areas the changes in the soil property (see example in Figure 1.5). However, the soil property is known with less accuracy than if a soil sample was analyzed in the laboratory, which must be taken into account in the choice of soil prediction functions.

This method of using the Vis-NIR remote sensing products has been applied in Languedoc-Roussillon (France) [LAG 12] and in the Cap Bon region (Tunisia) [CIA 12], in particular to predict the clay content in the soil surface. The method was adopted to improve the spatial interpolation between the measured sites using clay content images as auxiliary variables estimated by Vis-NIR remote sensing. The adopted spatial interpolation function was co-kriging [WAC 95]. Thanks to this co-kriging, clay content estimations were obtained over the entire study area (Figure 1.7(a)), thanks to the use of both measured sites and Vis-NIR remote sensing estimations present in the area. It is important to note that the clay content information estimated by co-kriging (Figure 1.7(b)) varies greatly depending on whether they are located on areas associated with Vis-NIR remote sensing estimates (Figure 1.5) or areas without such estimates, as they were vegetated during the Vis-NIR airborne data acquisition.

Finally, the use of a DSM approach allows for the most comprehensive analysis of the areas without direct Vis-NIR predictions, as these areas are affected by a disturbance factor (such as vegetation) which prevents the direct estimation of the soil property from Vis-NIR spectra (the "holes" in Figure 1.5 filled in Figure 1.7(a)), and as the same time the use of a DSM approach delivers an estimation error on these areas (Figure 1.7(b)).

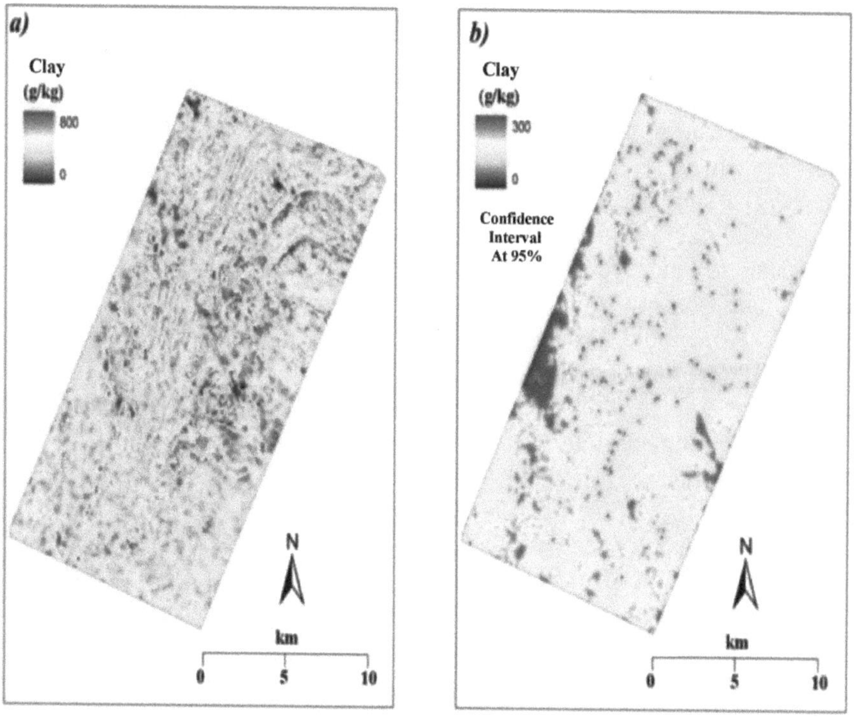

Figure 1.7. *Digital mapping of the clay content using a non-exhaustive estimate from the hyperspectral Vis-NIR remote sensoring (Figure 1.5): a) estimated clay content values; b) estimated prediction uncertainty (width of the 95% confidence interval) [CIA 12]. For a color version of this figure, see www.iste.co.uk/baghdadi/3.zip*

1.5.2. *Using Vis-NIR remote sensing products as DSM variables*

In the case where the Vis-NIR remote sensing product used relates to another soil property than those to be mapped, this product will be considered as a new soil covariate. The advantage compared with conventionally used covariates in digital soil mapping (e.g. the relief data) is that Vis-NIR remote sensing provides a direct representation of certain soil properties which are covariates *a priori* better correlated with the soil property to map. This is particularly the case with correlations between soil property surface (e.g. 0–5 cm) and the same property for a deep soil layer (e.g. 60–100 cm), seen respectively as a covariate and property to be mapped.

An example of deep layers soil properties prediction using the Vis-NIR remote sensing estimation of the soil surface property is provided by Lagacherie *et al.* [LAG 13] for the Cap Bon region (Tunisia). The approach consisted of calibrating, with a set of soil profiles and a soil analysis of the study area, linear regressions that estimate differences in soil property values between the surface (0–5 cm) and a given depth interval (0–15 cm, 15–30 cm, 30–60 cm and 60–100 cm). Covariates used to predict these differences were, first, soil surface property maps obtained from the hyperspectral Vis-NIR remote sensing and, secondly, three relief indicator derived from a digital terrain model, translating the soil erosion-redeposition process that may impact the link between surface properties and deep properties (altitude, vertical curvature of the relief, membership index to the geomorphological unit "flat-bottomed valley"). Once these "surface-depth" prediction functions were established, they were applied with adding the estimated differences to the estimates of surface properties by the hyperspectral Vis-NIR remote sensing. The results demonstrate an acceptable performance of the four tested properties (clay content, sand, iron, and CEC) up to a depth of 30–60 cm (R^2 between 0.55 and 0.81), but which degradation with depth (R^2 between 0.38 and 0.43 at 60–100 cm).

As before, the use of the digital soil mapping approach mobilizing the Vis-NIR remote sensing product and other information sources (a DTM, for example) allows us to push the limits of the Vis-NIR remote sensing making the prediction of deep soil properties possible. This, however, is only applicable in situations where the correlations between surface properties and deep properties are sufficient, which can exclude the predictions of layers that are too deep or part of very complex pedological covers.

1.6. Perspectives

The presence of surface elements (such as, rocks and dry vegetation residue) and the soil surface characteristics (e.g. moisture and roughness) can disrupt the spectral measurement [BEN 09] and, therefore, affect the quality of the soil property estimates. The non-identification of these surface elements and soil surface characteristics during the calibration and validation data selection results in the constitution of heterogeneous databases in terms of features and states. Correcting these unknown factors for each studied pixel would improve the prediction models performances.

In large quantities, surface elements can prevent us from estimating soil properties. As such, the presence of green vegetation cover is considered to be a prime disturbance factor when estimating soil properties [BAR 11]. It then seems necessary to consider methodological developments capable of using the spectral measurement on partially-vegetated surfaces frequently found in cultivated areas.

The construction of Vis-NIR remote sensing soil property estimation models is based on the use of predictive *X* variables (reflectance at different wavelengths) and variables to predict *Y* (the desired soil property). The combined use of remote sensing data of different types (and, therefore, spectrally complementary such as combining Landsat 8 and Sentinel-2 data, or Vis-NIR and thermal data) could be intended to assess the gain in precision when the resolution of predictive *X* variables is increased.

Finally, although the potential of airborne hyperspectral Vis-NIR data has been demonstrated for soil mapping, the cost (in money and time) airborne data acquisition remains, in many areas of the world, an important obstacle. The launch of two hyperspectral Vis-NIR satellite sensors is planned for 2018: the Italian sensor PRISMA (PRecursore IperSpettrale della Missione Applicativa) and the German sensor EnMAP (Environmental Mapping and Analysis Program, http://www.enmap.org). In addition, three other hyperspectral Vis-NIR satellite sensors are currently being developed: the American HyspIRI sensor (Hyperspectral InfraredImager), Italo-Israeli SHALOM (Spaceborne Hyperspectral Land and Application Management Ocean Mission) and French HYPXIM (Hyperspectral X Imagery). Each sensor is characterized by a spectral configuration (full width at half maximum number of spectral bands and spectral resolution), a spatial resolution and a signal to noise ratio. Apart from the HyspIRI sensor, which, with its low spatial resolution (60 m) does not seem suited for digital soil mapping studies [GOM 15], the other four hyperspectral satellite sensors are promising tools for large scale digital mapping of primary soil properties.

1.7. Key points

– Soils have a spectral signature in the visible and near infra-red, resulting from a set of chemical and physical components contained therein. These components are related to some soil properties studied by soil scientists (clay content, organic carbon, calcium carbonate, iron).

– Approaches by chemometric or spectral indices have been developed to estimate the primary soil properties from their spectral signature, in the laboratory, in field and by remote sensing. These approaches are calibrated by physic-chemical soil analysts performed on a set of soil samples, representative to the soil variability of the study area.

– The literature shows that the hyperspectral Vis-NIR remote sensing of soil properties provides satisfactory results for soil properties:

- having a specific associated spectral response;

- correlated to a property with a specific spectral response;

- having a sufficient range of variation.

– The results of multi-spectral remote sensing are more unequal and less predictable.

– These works of soil properties mapping using hyperspectral Vis-NIR remote sensing are nevertheless restricted to soil surfaces which are bare and uniformly dry during the Vis-NIR remote sensing acquisition.

– Using Vis-NIR remote sensing of soil in the wider context of the digital soil mapping makes it possible to go beyond certain limits (mapping areas inaccessible to measurement, mapping properties of soil deep layers) when used in combination with other spatial data (such as DEM).

1.8. Bibliography

[ARN 98] ARNOLD J.G., SRINIVASAN R., MUTTIAH R.S. *et al.*, "Large-area hydrologic modeling and assessment : Part I. Model development", *Journal of the American Water Resources Association*, vol. 34, no. 1, pp. 73–89, 1998.

[BAI 95] BAIZE D., JABIOL B., *Guide pour la description des sols*, Collection techniques et pratiques, INRA, Paris, 1995.

[BAT 09] BATJES N.H., "Harmonized soil profile data for applications at global and continental scales: updates to the WISE database", *Soil Use and Management*, vol. 25, no. 2, pp. 124–127, 2009.

[BAR 08] BARTHOLOMEUS H., SCHAEPMAN M.E., KOOISTRA L. *et al.*, "Spectral reflectance based indices for soil organic carbon quantification", *Geoderma*, vol. 145, nos. 1–2, pp. 28–36, 2008.

[BAR 11] BARTHOLOMEUS H., KOOISTRA L., STEVENS A. *et al.*, "Soil organic carbon mapping of partially vegetated agricultural fields with imaging spectroscopy", *International Journal of Applied Earth Observation and Geoinformation*, vol. 13, no. 1, pp. 81–88, 2011.

[BAU 70] BAUMGARDNER M.F., KRISTOF S.J., JOHANNSEN C.J. *et al.*, "Effects of organic matter on the multispectral properties of soils", *Proceedings of Indiana Academy of Science*, vol. 79, pp. 413–422, 1970.

[BAU 85] BAUMGARDNER M.F., SILVA L.F., BIEHL L.L. *et al.*, "Reflectance properties of soils", *Advances in Agronomy*, vol. 38, pp. 1–44, 1985.

[BEL 10] BELLON-MAUREL V., FERNANDEZ-AHUMADA E., PALAGOS B. *et al.*, "Critical review of chemometric indicators commonly used for assessing the quality of the prediction of soil attributes by NIR spectroscopy", *Trends in Analytical Chemistry*, vol. 29, no. 9, pp. 1073–1081, 2010.

[BEN 97] BEN-DOR E., INBAR Y., CHEN Y., "The reflectance spectra of organic matter in the visible near-infrared and short wave infrared region (400–2500) during a controlled decomposition process", *Remote Sensing of Environment*, vol. 61, pp. 1–15, 1997.

[BEN 95] BEN-DOR E., BANIN A., "Near infrared analysis (NIRA) as a simultaneous method to evaluate spectral featureless constituents in soils", *Soil Science*, vol. 159, pp. 259–269, 1995.

[BEN 98] BEN-DOR E., IRONS J.A., EPEMA A., "Soil spectroscopy", in RENCZ A. (ed.), *Manual of Remote Sensing*, 3rd ed. John Wiley and Sons, New York, 1998.

[BEN 02] BEN-DOR E., PATKIN K., BANIN A. *et al.*, "Mapping of several soil properties using DAIS-7915 hyperspectral scanner data – a case study over clayey soils in Israel", *International Journal of Remote Sensing*, vol. 23, pp. 1043–1062, 2002.

[BEN 09] BEN-DOR E., CHABRILLAT S., DEMATTÊ J.A.M. *et al.*, "Using imaging spectroscopy to study soil properties", *Remote Sensing of Environment*, vol. 113, no. 1, pp. S38–S55, 2009.

[BER 06] BERTRAND D., DUFOUR E., *La spectroscopie infrarouge et ses applications analytiques*, 2nd ed., Tec & Doc-Lavoisier, 2006.

[BOU 89] BOUMA J., "Using soil survey data for quantitative land evaluation", *Advances in Soil Science*, pp. 177–213, 1989.

[BOW 65] BOWERS S.A., HANKS A.J., "Reflection of radiant energy from soil", *Soil Science*, vol. 100, pp. 130–138, 1965.

[BUD 06] BUDIMAN M., MCBRATNEY A.B., "A conditioned Latin hypercube method for sampling in the presence of ancillary information", *Computers & Geosciences*, vol. 32, no. 9, pp. 1378–1388, 2006.

[BUD 13] BUDIMAN M., MC BRATNEY A.B., "Why you don't need to use RPD?", *Pedometron*, vol. 33, pp. 14–15, 2013.

[BRI 10] BRISSON N., GATE P., GOUACHE D. *et al.*, "Why are wheat yields stagnating in Europe? A comprehensive data analysis for France", *Field Crops Research*, vol. 119, pp. 201–212, 2010.

[BRI 16] BRIOTTET X., "Radiometry in the optical domain", in BAGHDADI N., ZRIBI M. (eds), *Optical Remote Sensing of Land Surfaces*, ISTE Press, London and Elsevier, Oxford, 2016.

[CER 10] CERDAN O., GOVERS G., LE BISSONNAIS Y. *et al.*, "Rates and spatial variations of soil erosion in Europe: a study based on erosion plot data", *Geomorphology,* vol. 122, pp. 167–177, 2010.

[CIA 12] CIAMPALINI R., LAGACHERIE P., MONESTIEZ P. *et al.*, "Co-kriging of soil properties with Vis-NIR hyperspectral covariates in the Cap Bon region (Tunisia)", in *Digital Soil Assessments and Beyond*, CRC Press, 2012.

[CHA 02] CHABRILLAT S., GOETZ A.F.H., KROSLEY S. *et al.*, "Use of hyperspectral images in the identification and mapping of expansive clay soils and the role of spatial resolution", *Remote Sensing of Environment*, vol. 82, pp. 431–445, 2002.

[CHA 01] CHANG C.-W., LAIRD D.A., MAUSBACH M.J. *et al.*, "Near-infrared reflectance spectroscopy–principal components regression analysis of soil properties", *Soil Science Society of America Journal,* vol. 65, pp. 480–490, 2001.

[CLA 84] CLARK R.N., ROUSH T.L., "Reflectance spectroscopy: quantitative analysis techniques for remote sensing applications", *Journal of Geophysics Residence*, vol. 89, pp. 6329–6340, 1984.

[CRO 06] CROUVI O., BEN-DOR E., BEYTH M. *et al.*, "Quantitative mapping of arid alluvial fan surfaces using field spectrometer and hyperspectral remote sensing", *Remote Sensing of Environment*, vol. 104, no. 1, pp. 103–117, 2006.

[DEM 02] DEMATTÊ J.A.M., "Characterization and discrimination of soils by their reflected electromagnetic energy", *Pesquisa Agropecuária Brasileira,* vol. 37, no. 10, pp. 1445–1458, 2002.

[DEM 04] DEMATTÊ J.A.M., CAMPOS R.C., ALVES M.C. *et al.*, "Visible-NIR reflectance: a new approach on soil evaluation", *Geoderma,* vol. 121, pp. 95–112, 2004.

[ESC 87] ESCADAFAL R., POUGET M., "Cartographie des formations superficielles en zone aride (Tunisie méridionale) avec Landsat TM", *Photo-Interprétation*, vol. 4, 1987.

[ESC 94] ESCADAFAL R., "Soil spectral properties and their relationships with environmental parameters – Examples from arid regions", in HILL J., MÉGIER J. (eds), *Imaging Spectrometry – A Tool for Environmental Observations*, Springer, 1994.

[FAL 02] FALLOON P., SMITH P., "Simulating SOC changes in long-term experiments with RothC and CENTURY: model evaluation for a regional scale application", *Soil Use and Management*, vol. 18, pp. 101–111, 2002.

[GAF 86] GAFFEY S.J., "Spectral reflectance of carbonate minerals in the visible and near infrared (0.35-2.55 pm): calcite, aragonite and dolomite", *Journal of geophysical Research, Solid Earth*, vol. 71, pp. 151–162, 1986.

[GAU 93] GAULTIER J.P., LEGROS J.P., BORNAND M. *et al.*, "L'organisation et la gestion des données pédologiques spatialisées : Le projet DONESOL", *Revue de Géomatique*, vol. 3, no. 3, pp. 235–253, 1993.

[GER 12] GERIGHAUSEN H., MENZ G., KAUFMANN H., "Spatially explicit estimation of clay and organic carbon content in agricultural soils using multi-annual imaging spectroscopy data", *Applied and Environmental Soil Science*, 2012.

[GOM 08a] GOMEZ C., LAGACHERIE P., COULOUMA G., "Continuum removal versus PLSR method for clay and calcium carbonate content estimation from laboratory and airborne hyperspectral measurements", *Geoderma*, vol. 148, no. 2, pp. 141–148, 2008.

[GOM 08b] GOMEZ C., VISCARRA ROSSEL R.A., MCBRATNEY A.B., "Soil organic carbon prediction by hyperspectral remote sensing and field vis-NIR spectroscopy : an Australian case study", *Geoderma*, vol. 146, nos. 3–4, pp. 403–411, 2008.

[GOM 12] GOMEZ C., COULOUMA G., LAGACHERIE P., "Regional predictions of eight common soil properties and their spatial structures from hyperspectral Vis–NIR data", *Geoderma*, p. 176–185, 2012.

[GOM 15] GOMEZ C., OLTRA CARRIO R., LAGACHERIE P. *et al.*, "Sensitivity of soil property prediction obtained from Hyperspectral Vis-NIR imagery to atmospheric effects and degradation in image spatial resolutions", *Remote Sensing of Environment*, vol. 164, pp. 1–15, 2015.

[HAS 93] HASTIE T., MALLOWS C., "A discussion of 'a statistical view of some chemometric regression tools', by Frank IE and Friedman JH", *Technometrics*, vol. 35, pp. 140–143, 1993.

[HAU 08] HAUBROCK S.N., CHABRILLAT S., KUHNERT M. *et al.*, "Surface soil moisture quantification and validation based on hyperspectral data and field measurements", *Journal of Applied Remote Sensing*, vol. 2, p. 023552, 2008.

[HUN 71] HUNT G.R., SALISBURY J.W., LENHOFF C.J., "Visible and near-infrared spectra of minerals and rocks: III. Oxides and hydroxides", *Modern Geology*, vol. 2, p.195–205, 1971.

[JAR 10] JARMER T., HILL J., LAVÉE H. *et al.*, "Mapping soil organic carbon in semi-arid and arid ecosystems of Israel", *Photogrammetric Engineering and Remote Sensing*, vol. 75, no. 1, pp. 85–94, 2010.

[KEN 69] KENNARD R.W., STONE L.A., "Computer aided design of experiments", *Technometrics,* vol. 11, no. 1, pp. 137–148, 1969.

[LAG 07] LAGACHERIE P., MCBRATNEY A.B., "Spatial soil information systems and spatial soil inference systems: perspectives for digital soil mapping", in LAGACHERIE P., MCBRATNEY A.B., VOLTZ M. (eds.), *Digital Soil Mapping: an Introductory perspective*, Elsevier, 2007.

[LAG 08] LAGACHERIE P., BARET F., FERET J.B. *et al.*, "Estimation of soil clay and calcium carbonate using laboratory, field and airborne hyperspectral measurements", *Remote Sensing of Environment*, vol. 112, no. 3, pp. 825–835, 2008.

[LAG 12] LAGACHERIE P., BAILLY J.S., MONESTIEZ P. *et al.*, "Using scattered hyperspectral imagery data to map the soil properties of a region", *European Journal of Soil Science*, vol. 63, no. 1, pp. 110–119, 2012.

[LAG 13] LAGACHERIE P., SNEEP A.-R., GOMEZ C. *et al.*, "Combining Vis–NIR hyperspectral imagery and legacy measured soil profiles to map subsurface soil properties in a Mediterranean area (Cap-Bon, Tunisia)", *Geoderma*, pp. 168–176, 2013.

[LEB 05] LE BISSONNAIS Y., DAROUSSIN J., JAMAGNE M. *et al.*, "Pan-European soil crusting and erodibility assessment from the European Soil Geographical Database using pedotransfer rules", *Advances in Environmental Monitoring and Modelling*, vol. 2, no. 1, pp. 1–15, 2005.

[LEV 07] LEVIN N., TSOAR H., MAIA L.P. *et al.*, "Dune whitening and inter-dune freshwater ponds in NE Brazil", *CATENA*, vol. 70, no. 1, pp. 1–15, 2007.

[LU 13] LU P., WANG L., NIU Z. *et al.*, "Prediction of soil properties using laboratory VIS–NIR spectroscopy and Hyperion imagery", *Journal of Geochemical Exploration*, vol. 132, pp. 26–33, 2013.

[MCB 03] MCBRATNEY A.B., MENDONCA SANTOS M.L., MINASNY B., "On digital soil mapping", *Geoderma*, vol. 117, nos. 1–2, pp. 3–52, 2003.

[MAD 97] MADEIRA NETTO J.S., BEDIDI A., POUGET M. *et al.*, "Visible spectrometric indices of hematite (Hm) and goethite (Gt) content in lateritic soils: the application of a Thematic Mapper (TM) image for soil-mapping in Brasilia, Brazil", *International Journal of Remote Sensing*, vol. 18, pp. 2835–2852, 1997.

[MAR 99] MARX BD, EILERS P.H.C., "Generalized linear regression on sampled signals and curves: a P-spline approach", *Technometrics*, vol. 41, pp. 1–13, 1999.

[MAR 02] MARX B.D., EILERS P.H.C., "Multivariate calibration stability: A comparison of methods", *Journal of Chemometry*, vol. 16, pp. 129–140, 2002.

[MAR 89] MARTEN G. C., SHENK J. S., BARTON F.E. II, Near Infrared Reflectance Spectroscopy (NIRS): Analysis of Forage Quality, U.S. Dept. of Agric. Handbook 643, 1989.

[MAR 13] MARIN-GONZALEZ O., KUANG B., QURAISHI M.Z. *et al.*, "Online measurement of soil properties without (Fe3O4), direct spectral response in near infrared spectral range", *Soil & Tillage Research,* vol. 132, pp. 21–29, 2013.

[MAT 98] MATHIEU R., POUGET M., CERVELLE B. *et al.*, "Relationships between satellite based radiometric indices simulated using laboratory reflectance data and typic soil color of an arid environment", *Remote Sensing of Environment*, vol. 66, pp. 17–28, 1998.

[MOR 85] MORRIS R.V., LAUER H.V., LAWSON C.A. *et al.*, "Spectral and other physicochemical properties of submicron powders of Hematite (α-Fe2O3), Maghematite (γ-Fe2O3), Magnetite Goethite (α-FeOOH), and Lepidocrocite (γ-FeOOH)", *Journal of Geophysical Research*, vol. 90, pp. 3126–3144, 1985.

[MOU 02] MOUSSA R., VOLTZ M. ANDRIEUX P., "Effects of the spatial organization of agricultural management on the hydrological behaviour of a farmed catchment during flood events", *Hydrological Processes*, vol. 16, pp. 393–412, 2002.

[NAN 06] NANNI M.R., DEMATTÊ J.A.M., "Spectral reflectance methodology in comparison to traditional soil analysis", *Soil Science Society of America Journal,* vol. 70, no. 2, pp. 393–407, 2006.

[NOC 15] NOCITA M., STEVENS A., VAN WESEMAEL B. *et al.*, "Soil spectroscopy: an alternative to wet chemistry for soil monitoring", *Advances in Agronomy*, vol. 132, pp. 139–159, 2015.

[ONG 14] ONG C.C.H., CUDAHY T.J., "Mapping contaminated soils: using remotely-sensed hyperspectral data to predict pH", *European Journal of Soil Science*, vol. 65, no. 6, p. 1365–2389, 2014.

[PAR 89] PARTON W.J., SANFORD R.L., SANCHEZ P.A. *et al.*, "Modeling soil organic matter dynamics in tropical soil," in COLEMAN D.C., OADES J.M., UEHARA G. (eds), *Dynamics of Soil Organic Matter in Tropical Ecosystems*", Univ. of Hawaii Press, Honolulu, 1989.

[ROG 10] ROGER J.M., "La robustesse des étalonnages multidimensionnels, application aux données spectrales", *Techniques de l'Ingénieur*, SL 265, vol. 11 2010.

[ROS 04] ROSSITER D.G., "Digital soil resource inventories: status and prospects", *Soil Use & Management,* vol. 20, no. 3, pp. 296–301, 2004.

[SEL 06] SELIGE T., BOHNER J., SCHMIDHALTER U., "High resolution topsoil mapping using hyperspectral image and field data in multivariate regression modeling procedures", *Geoderma*, vol. 136, nos. 1–2, pp. 235–244, 2006.

[SHA 15] SHABOU M., MOUGENOT B, CHABAANE Z.L. *et al.*, "Soil clay content mapping using a time series of Landsat TM data in semi-arid lands", *Remote Sensing*, vol. 7, pp. 6059–6078, 2015.

[SNE 77] SNEE R.D., "Validation of regression models: methods and examples", *Technometrics*, vol. 19, no. 4, pp. 415–428, 1977.

[SOI 51] SOIL SURVEY STAFF, "Soil survey manual", Agricultural Handbook 18, USDA, U.S. Govt. Print. Off. Washington, DC, vol. 503, 1951.

[STE 08] STEVENS A., WESEMAEL V.B., BARTHOLOMEUS B. *et al.*, "Laboratory, field and airborne spectroscopy for monitoring organic carbon content in agricultural soils", *Geoderma*, vol. 144, nos. 1–2, pp. 395–404, 2008.

[STE 10] STEVENS A., UDELHOVEN T., DENIS A. *et al.*, "Measuring soil organic carbon in croplands at regional scale using airborne imaging spectroscopy", *Geoderma*, vol. 158, nos. 1–2, pp. 32–45, 2010.

[STE 12] STEVENS A., MIRALLES I., WESEMAEL B., "Soil organic carbon predictions by airborne imaging spectroscopy: comparing cross-validation and validation", *Soil Science Society of America Journal*, vol. 76, no. 6, pp. 2174–2183, 2012.

[STO 81] STONER, E.R., BAUMGARDNER M.F., "Characteristic variations in reflectance of surface soils", *Soil Science Society of America Journal*, vol. 45, pp. 1161–1165, 1981.

[SUL 05] SULLIVAN D.G., SHAW J.N., RICKMAN D., "IKONOS imagery to estimate surface soil property variability in two Alabama physiographies", So*il Science Society of America Journal,* vol. 69, no. 6, pp. 1789–1798, 2005.

[VAN 04] VAN DER MEER F., "Analysis of spectral absorption features in hyperspectral imagery", *International Journal of Applied Earth Observation and Geoinformation*, vol. 5, no. 1, pp. 55–68, 2004.

[VAP 95] VAPNIK V., *The Nature of Statistical Learning Theory*, Springer-Verlag, New York, 1995.

[VAP 98] VAPNIK V., *Statistical Learning Theory*, Wiley, 1998.

[VAU 13] VAUDOUR E., BEL L., GILLIOT J. M. *et al.*, "Potential of SPOT multispectral satellite images for mapping topsoil organic carbon content over Peri-Urban croplands", *Soil Science Society of America Journal*, vol. 77, no. 6, pp. 2122–2139, 2013.

[VIS 06] VISCARRA ROSSEL R.A., WALVOORT D.J.J., MCBRATNEY A.B. *et al.*, "Visible, near-infrared, mid-infrared or combined diffuse reflectance spectroscopy for simultaneous assessment of various soil properties", *Geoderma*, vol. 131, pp. 59–75, 2006.

[WAC 95] WACKERNAGEL H., *Multivariate Geostatistics*, Springer-Verlag, vol. 256, 1995.

[WIL 87] WILLIAMS P.C., NORRIS K.H., "Near-infrared technology in the agricultural and food Industries", *American Association of Cereal Chemists*, St. Paul, MN, vol. 330, 1987.

[WOL 01] WOLD S., SJOSTROM M., ERIKSSON L., "PLS-regression: a basic tool of chemometrics", *Chemometrics and Intelligent Laboratory Systems*, vol. 58, pp. 109–130, 2001.

2

Estimation of Biophysical Variables from Satellite Observations

2.1. Introduction

The supply of biophysical variables derived from satellite observations serves several applications related to agriculture, the environment, resource management and the climate. Moreover, these same variables may be acquired at a range of scales:

– at the very local level, the use of precision agriculture allows adapting cultural practices to the spatial variability of the plot;

– across the landscape or territory, to manage the environment and natural resources;

– at the continental or global scale for the study of climate, biogeochemical cycles and vegetation dynamics.

The observations offered by current satellites are well suited to these needs, with spatial resolution ranging from a kilometer to tens of meters according to the classification proposed by Morisette [MOR 10]. In addition, repeated observations at high frequency allows tracking the dynamics of vegetation, which provides insight into how it functions.

Some of the vegetation's biophysical variables listed include Essential Climate Variables (ECV), recognized for their essential role in

Chapter written by Fred BARET.

describing the interactions between the surface and the atmosphere in relation to climate studies [GCO 11]. These ECVs include the Leaf Area Index (LAI) and the Fraction of Absorbed Photosynthetically Active Radiation (FAPAR). Significant progress has been made in estimating these variables from satellite observations, in particular by improving the embedded instruments' measurement capabilities [LIA 04]. However, the satellites' onboard sensors measure only the radiation reflected or emitted by the surface and therefore do not directly provide the values of LAI or FAPAR. It is therefore necessary to transform the radiance measurements recorded by sensors into LAI and FAPAR estimates. The estimation algorithms should be as specific as possible and efficient in computing given the enormous volumes of data they must process. Moreover, they should require minimal calibration, since they are intended to be applied in a variety situations [WAL 04]. Biophysical variables of interest estimated from satellite data become "satellite products" as processing chains automatically generate ready to use data fulfilling the needs of the user community.

Many methods have been proposed to find surface properties using satellite observations [BAR 07a, GOE 89, KIM 00, MYN 88, VER 11a]. They include empirical approaches requiring calibration on experimental datasets. If these approaches are very simple, they are often limited by the size and diversity of the calibration dataset, as well as by the uncertainties associated with the ground measurements. Alternatively, more complex approaches using radiative transfer models (RTM) have been proposed, requiring no experimental calibration datasets. The RTMs describe physical processes associated with the paths of photons in the canopy. They simulate the radiation field reflected from a surface according to the geometric configuration of observation and illumination given the characteristics of the vegetation and soil. It is therefore necessary to invert these models to estimate the characteristics of the canopy corresponding to the radiation measured by the satellite's sensor. The inversion therefore attempts to find few RTM input variables of interest based on its output variables.

Using the available literature, this chapter aims to analyze how the biophysical variables can be estimated from satellite observations in the optical domain with kilometric and decametric spatial resolutions. We will focus on the ECV variables LAI and FAPAR, which will be defined further on. We will then describe the principles of RTM inversion implementation to estimate LAI and FAPAR. The theoretical performances of these methods will be analyzed and several techniques that may improve the estimates will be discussed in detail. Finally, the possible combination of methods, products and sensors will be presented. To conclude, we will highlight the main difficulties while suggesting some lines of inquiry to resolve these difficulties.

2.2. Definition of the canopy biophysical variables accessible from remote sensing observations

Four main biophysical variables of vegetation, which are LAI, the coverage fraction, the fraction of absorbed photosynthetically active radiation absorbed by vegetation, and the chlorophyll content, are accessible from the measured reflectance. The definition of these variables is given in the following sections.

2.2.1. *Leaf Area index: LAI, GLAI, PAI, GAI, apparent and effective values*

The LAI is defined as half of the total developed area of vegetation by horizontal ground surface unit (m^2/m^2) [CHE 92]. It is a structure variable that quantifies the interface between the vegetation and the atmosphere to exchange mass and energy. The LAI controls the processes of photosynthesis, transpiration and interception of rain by the canopy. For photosynthesis and transpiration, the LAI should be restricted to green surfaces, leading to define the Green Leaf Area Index (GLAI). In addition, the surface of green photosynthetically active organs other than the leaves must also be included, leading to define the Green Area Index (GAI).

LAI, GLAI and GAI can be measured by very tedious and expensive destructive methods. Indirect methods that are based on the measurement of the gap fraction (Po) have been developed [JON 04a]. Measuring the gap fraction does not distinguish between green elements and non-vegetation green elements, leading to define the Plant Area Index (PAI), which includes all the vegetation elements regardless of potential photosynthetic activity. However, color photographs taken from the top of the canopy can also be used to measure the Green Fraction (GF), defined as the fraction of green vegetation covering the soil in a particular viewing direction. Assuming that the green leaves are found primarily in the upper layers of the canopy, which is generally the case, this technique provides an estimate of the GAI [BAR 10]. Similarly, remote sensing observations are mainly sensitive to the green elements of the canopy: this is because of the chlorophyll's absorption which shows a very strong and characteristic spectral variation. Remote sensing observations are therefore closely linked to the GAI [DUV 11]. Table 2.1 shows that indirect methods provide access only to the GAI or PAI based on their ability or not to distinguish green elements.

		Only Green	Green + Non-Green	Only leaves	All elements	Methods
LAI	*Leaf Area Index*		✓	✓		Only destructive methods
GLAI	*Green Leaf Area Index*	✓		✓		Only destructive methods
GAI	*Green Area Index*	✓			✓	Green Fraction (GF) measurments from the top of the field: photography, remote sensing
PAI	*Plant Area Index*		✓		✓	Gap fraction (Po) measurements from the bottom of the field (photography, transmittance)

Table 2.1. *Definitions of LAI, GLAI, GAI, and PAI and the associated measurement methods. All amounts are expressed in $m^2.m^{-2}$*

Indirect PAI and GAI estimation methods require assumptions about the structure of the canopy. The one that corresponds to the turbid environment is most commonly used: the elements of the canopy are then assumed to be an infinitesimal size and to be randomly distributed in the volume of vegetation. However, this strong assumption is rarely verified, because the leaves have finite dimensions and can be aggregated at different scales from that of the branch (leaves grouped around a twig), to that of the plant (leaves and branches clustered around the plant) and of the field (plants grouped in bunches), up to the landscape (vegetation groups into patches in the landscape). This may lead to biases in the PAI estimation from the gap fraction [WAL 03], or GAI from the reflectances [CHE 05]. Consequently, it is necessary to introduce additional concepts to specify how the estimate is made. "Effective" PAI or GAI values will be defined as the estimated quantities from the directional measurement of the gap fraction (Po) or the GF using the Miller formula [MIL 67] which assumes that the plant elements are randomly distributed in the canopy. However, applying the Miller formula is based on the measurement of Po or GF in all directions of the hemisphere, which is rarely possible in practice. Similarly, the GAI estimate from satellite observations correspond to "apparent" values [MAR 94] that depend on the viewing configuration and inversion technique, including assumptions about the structure of the canopy often associated with the RTM as we shall see in the following sections.

2.2.2. *FAPAR: illumination geometry and distinction between green and non-green elements*

The FAPAR is defined as the fraction of absorbed photosynthetically active radiation (PAR: solar radiation in the spectral range between 400 and 700 nm) absorbed by vegetation [MOT 11]. The FAPAR is widely used in simple models of primary production corresponding to biomass accumulation of photosynthesis [MCC 09]. It is therefore necessary to consider only the photosynthetically active parts corresponding to the green elements of the canopy, similarly to the GAI definition. The radiation balance in the spectral range of the PAR spectral domain allows computing the FAPAR [MOT 11]. The value of FAPAR can be approximated by the fraction of radiation intercepted (the FIPAR) corresponding to the complement to the unity of the gap fraction [BEG 91]. However, these measurement techniques cannot separate the absorption or interception of green elements from the non-green elements. Only the GF measurement taken from the top of the canopy in the direction of the incoming light provides a direct measure of FIPAR.

The FAPAR and FIPAR are not intrinsic characteristics of the canopy and result from the interaction of the light with the vegetation. The FAPAR and FIPAR express radiative fluxes that are dependent upon illumination conditions. The lighting conditions can be described by a single component from the solar direction (called black-sky), dependent on the position of the sun, and a diffuse component from the sky (called white-sky) [MAR 94]. Most FAPAR products available are defined by the black-sky component corresponding to the position of the sun at the time of the satellite overpass

[WEI 14], usually around 10:30 solar time. Baret *et al.* [BAR 04] propose to use the value at 10:00 solar time, as it is a very good approximation of the daily integrated value used by primary production models that typically operate at a daily time step.

2.3. Inversion methods of radiative transfer models

The light reflected by the canopy is the result of radiative transfer processes in vegetation [BRI 16]. It depends on the variables of the canopy state (vegetation structure, the optical properties of the elements) and illumination conditions. The observation layout then defines the reflectance field sampled at the time of measurement: wavelengths, direction(s) concerned and the spatial sampling. The state variables include the variables of interest for uses such as LAI (Figure 2.1). Other variables such as the FAPAR can only be calculated from the knowledge of the state variables of vegetation and even the RTM.

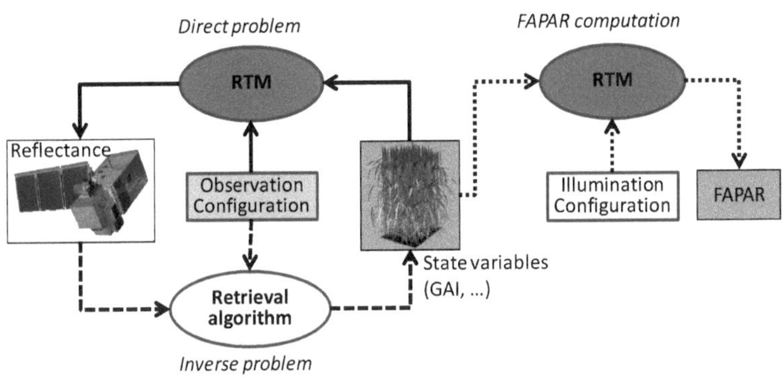

Figure 2.1. *Direct (solid lines) and inverse problems (dotted lines) considered for state variable remote sensing estimation. The calculation of the FAPAR from the estimated state variables is also shown on the right*

The estimation techniques can be classified into two broad categories:

– techniques focusing on the biophysical variables of interest corresponding to certain RTM input variables: they minimize the distance between measured and simulated reflectance (Figure 2.2(a));

– those focusing on radiometric observations corresponding to the model's outputs: they minimize the distance between the GAI learning base and the inverse model's estimated GAI (Figure 2.2(b)). Once the inverse model is calibrated, it can be directly used.

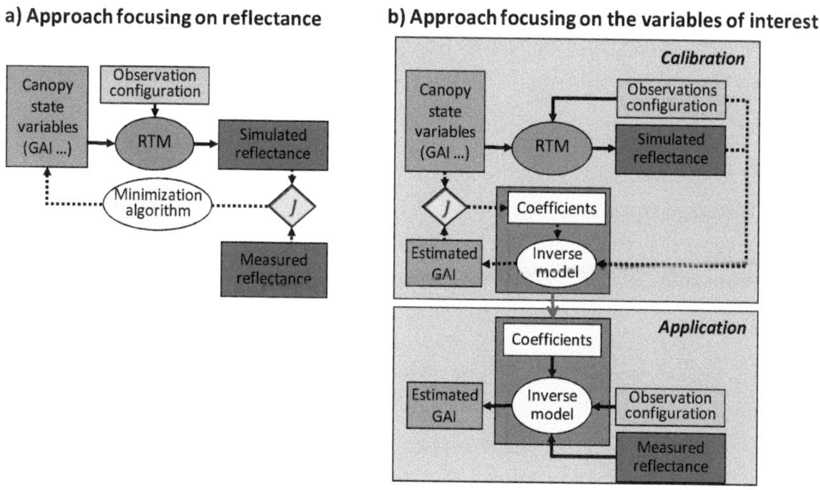

Figure 2.2. *The two main approaches that are used to estimate canopy characteristics from the reflectance measurements. In panel a), the approach focusing on reflectance. In panel b) the approach focusing on the vriables of interest. J represents the cost function to minimize the mismatch on reflectance (a) or variable of interest (b)*

2.3.1. *Approaches focusing on radiometric variables*

Approaches focusing on the radiometric variables adjust the canopy state variables by minimizing the distance between the observed reflectance and those simulated by the RTM (Figure 2.2(b)). The distance is quantified by a cost function (J) that takes into account the uncertainties associated with the model and the measurements. It can be established theoretically from the maximum likelihood assuming that all uncertainties are independent and normally distributed [TAR 87]:

$$J = (R - \hat{R})^t \cdot W^{-1} \cdot (R - \hat{R}) \quad [2.1]$$

where R and $\hat{R}$ are the observed and simulated reflectance vectors respectively, and W is the covariance matrix for uncertainties. However, the description of the W covariance matrix constitutes a major difficulty of this approach. In most cases, only the diagonal terms of the matrix corresponding to the variance (σ^2) associated to the uncertainties are described, assuming that errors are uncorrelated. Under these conditions, equation [2.1] simplifies into a simple Euclidean distance:

$$J = \sum_{n=1}^{N} \frac{(R_n - \hat{R}_n)^2}{\sigma_n^2} \qquad [2.2]$$

where N is the number of degrees of freedom equal to the number of used configurations (bands, viewing directions or illumination...). More sophisticated cost functions have been proposed including a regularization term that maintains the solution close to the expected values. This will be discussed later.

Minimization algorithms are iterative. From an initial solution (vector of state variables), the cost function is evaluated and the state variables are then iteratively adjusted to minimize the cost function. Several minimization techniques are available, such as the simplex algorithm [NEL 65], the gradient descent algorithm [BAC 02a, BIC 99, GOE 85, JAC 95, LAU 08, VOB 10], Markov Monte Carlo's chains [ZHA 03], simulated annealing [BAC 01] and genetic algorithms [FAN 03a]. The possibility of obtaining a sub-optimal solution corresponding to a local minimum of the cost function is a major challenge associated with this type of algorithm. This problem can be mitigated by using multiple initial solutions distributed in the space of the canopy state variables. Others propose disrupting the trajectory of the cost function to better explore the possibilities. However, these improvements require more power and computing time. Computing performance can be improved by using an analytical expression of the gradients of the cost function through the adjoint model [LAU 08]. In addition, the physical RTM can be replaced by a metamodel which consists of an analytic function that accurately describes the relationship between the canopy state variables and the simulated reflectance. In addition to increasing the speed of calculation, the meta model has the advantage of simplifying the calculation of the adjoint model [JAM 05].

Look Up Tables (LUT) are also an interesting alternative to iterative approaches. These are pre-computed tables containing simulations of reflectance and the corresponding state variables [GAN 12, KNY 98, WEI 00]. This technique is easier to implement computationally and avoids the problem of a local minimum because the search is done for all the simulated cases contained in the tables, without going through a specific path of the vegetation's state variables. The LUT should be constructed so that the response surface is described as best as possible, that is to say with a finer sampling where the reflectance is highly sensitive to the canopy state variables [WEI 00]. This sampling method is different from that used for the approaches focusing on the biophysical variables, as we shall see later. The different cases contained in the LUT are sorted according to the value of the cost function (*J*). The solution is then considered as corresponding to the lowest value of the cost function, similarly to what is done with the iterative minimization techniques. It may also be defined by the distribution of the best solutions corresponding to the set of cases with the lowest values of the cost function. This set of best solutions may correspond to a given fraction of all simulated cases [COM 02] or can be defined using a threshold computed from uncertainties associated to the measurements and to the model [KNY 98]. A more rigorous approach to exploit the LUT is to weigh each case by its likelihood as done in the GLUE method [MAK 02].

2.3.2. *Approaches focusing on the variables of interest: machine learning*

This approach (Figure 2.2(b)) consists of calibrating the coefficients of an inverse analytical model on a set of training cases using an iterative minimization technique. For this purpose, a cost function that measures the distance between the variables of interest contained in the learning base and the values estimated from the inverse model is minimized. Once the inverse model is calibrated, it can be used to estimate the variables of interest using the measured reflectance.

The learning database can be constructed from experimental data [ASR 84, CHE 02, DEN 06, HUE 88,VERR 12]. However, as mentioned in the introduction, the generally small sampling of all possible cases offered by the learning database (observation geometry, lighting conditions, vegetation type, stage of development, state of stress, sub-underlying soil properties) are a major limitation of the performances of these methods. In addition, measurement errors in the variables of interest and in the corresponding reflectances can generate uncertainties and biases in the estimates [FER 05]. Finally, *in situ* measurements necessary to characterize the vegetation variables of interest have footprints ranging from a few meters to tens of meters, requiring the development of sometimes dedicated spatial sampling strategies to represent pixels observed by the satellite [WEI 07]. In the case of sensors with a kilometric to hectometric resolution, this task is obviously even more difficult [MOR 06].

These are the reasons why it is often better to use the RTM to generate learning bases through computer simulation, which can represent a wide range of situations and conditions [BAC 06, BAR 91, BAR 07b, GAN 12, HUE 97, KNY 99, SEL 85].

2.3.2.1. *Approaches based on vegetation indices (VI)*

The simplest approaches are based on multiple linear or polynomial regressions between the variables of interest and canopy reflectances or various combinations of these reflectances corresponding to the vegetation index (VI). These vegetation indices are designed to minimize the influence of disturbing factors such as the reflectance of the underlying soil [BAR 91, RIC 77] or atmospheric effects [HUE 94]. The strong non-linearity exsting between the reflectance and canopy state variables can also be reduced by using reflectance transformations (usually power, exponential or logarithmic) or through combinations of spectral bands, allowing the use of simple linear regression with the canopy variables. Operational algorithms based on these principles were developed: MGVI from MERIS data which is then extended to other sensors [GOB 08], the "back-up" MODIS algorithm based on NDVI [MYN 02] the POLDER algorithm using the DVI index calculated from corrected reflectance directional effects [ROU 02]. Although capable of performing well, vegetation indices are inherently limited by the empiricism of their formulation and the very small number of spectral bands used (usually 2–3).

2.3.2.2. *Machine learning approaches*

Alternatively, machine learning has been proposed since the 1990s. Among these methods, neural networks (Figure 2.3) were used intensively [GON 99]. Baret *et al.* [BAR 95] and Verger *et al.* [VER 11a] have shown that neural networks used with individual spectral reflectance were more efficient than approaches based on vegetation indices. Fang and Liang [FAN 05] found that neural networks are as efficient as the multiple regression projection pursuit. Neuronal approaches were applied to MERIS [LAC 06] and VEGETATION [BAR 07b] at a kilometric spatial resolution. Neural networks have also been applied to decametric resolution airborne data POLDER [WEI 02] and Landsat satellite sensors [FAN 03b], CHRIS

[VER 11a] and FORMOSAT [CLA 13]. Other automatic learning methods were also assessed, such as Support Vector Machines (SVM) or Gaussian Process Regression (GPR) [VER 12]. Methods like GPR appear interesting when learning is performed on experimental data. However, when applied to large learning datasets, they appear limited by the computing capacity [MAC 03]. However, one advantage of this approach applied to experimental data is to provide an estimation of the associated uncertainties. In the case of the use of simulated learning database from the RTM, the uncertainties associated with simulated reflectance must be specified, which is not always easy.

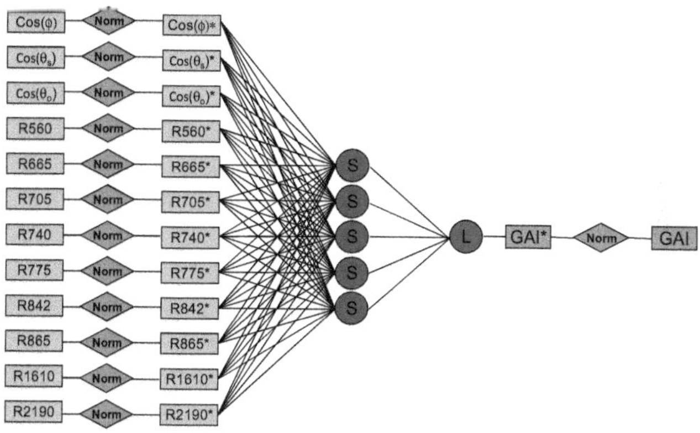

Figure 2.3. *An example of a neural network used to estimate the GAI using canopy reflectances from the MSI Sentinel-2 sensor. Norm is the standardization of inputs and outputs. S and L represent the tangent sigmoid and linear transfer functions associated with each neuron. φ, θs and θ_o represent respectively the relative azimuth between the viewing and illumination directions, zenith angles of the sun and that of the view direction. R560–R2190 represent the canopy reflectances in the bands of the MSI Sentinel-2 sensor. From [BAR 09]*

The learning database is a major component in the success of machine learning approaches. It should represent the distributions and co-distributions of the variables in actual conditions. It is at this level that the '*a priori*' knowledge of these distributions is introduced into the machine learning approaches. They can therefore be considered as part of the family of Bayesian approaches. The density of the possible cases will decrease rapidly with the required number of the vegetation variables. The experimental designs can then be exploited to ensure that the space of canopy realization is roughly evenly populated [BAC 02b]. Machine learning approaches can also be used as smoothing methods: they interpolate between the different cases of the learning database. Consequently, estimates made outside the definition domain (corresponding to the convex hull of the learning database reflectance) should be considered with caution. To simplify the learning database and make it a more robust method, it is possible to eliminate cases that have been simulated but never observed [BAR 07a]. However, this procedure requires having a large database representative of all possibly encountered situations.

2.3.3. *Advantages and limitations of several approaches*

The advantages and limitations of the several approaches mentioned above will be discussed by distinguishing different aspects:

–the computing power required: Machine learning, once trained, are very efficient: the inverse model is usually relatively simple and can be run very quickly. However, the learning or calibration process may require very high computing power, especially when the inverse model is complex with many coefficients and when the learning base is very large. The implementation of the LUT-based techniques in operational processing lines is very easy and effective. In contrast, the iterative minimization methods require large computing capabilities, even though improvements are possible using a meta model as seen previously. The use of the adjoint model will also limit the number of RTM calls. Finally, a segmentation of the reflectance space also reduces the execution time [PIN11];

– the flexibility of observation layouts: The iterative minimization methods allow for the estimation of canopy state variables using varying combinations of observation layouts. It is also possible to inverse an RTM simultaneously on several pixels. This creates interesting possibilities to exploit spatial or temporal constraints as we shall see later on. Basic LUT techniques theoretically allow an exploitation of observation layout configurations at the cost of an increase in the size of tables, making them more difficult to handle. In contrast, machine learning requires a fixed number of entries. The characteristics of the observation must be explicitly entered by the inputs of the inverse model as illustrated in Figure 2.3. However, increasing the number of entries makes the learning process more difficult and time consuming. An alternative is to develop an inverse model for each individual observation layout considered, which increases learning and operational implementation;

– *the integration of prior information*: The information is used to restrict the number of possible solutions by limiting the differences between the values of estimated variables and the expected values. Approaches focusing on radiometric data incorporate previous information directly in the cost function through the regularization term as discussed in section 2.5.3. However, in the case of the LUTs, it is also possible to restrict the simulations in the range of expected situations, like what is done for the MODIS LAI and FAPAR algorithms, which depend on the observed vegetation type [SHA 05]. For machine learning techniques, the information is introduced through the distribution and co-distribution of the RTM entries: when the variables of interest must be estimated in tested situations (species, stage of development, type of underlying soil), it is more effective to calibrate a model specific to each situation. Q u *et al.* [QU 08] propose the use of Bayesian networks where the RTM simulations are operated in parallel with a description of the distribution of canopy state variables, which may depend on the stages of development and the type of vegetation;

– *the associated uncertainties*: Approaches focusing on the radiometric data used to calculate the uncertainty associated with the estimates by propagating the uncertainties of reflectance measurement and the RTM, using the derivatives of the cost function to the measured point [LAU 05]. When using the LUTs, uncertainties can be estimated by Monte Carlo methods, or approximated by the standard deviation of the set of solutions corresponding to measurement uncertainties [KNY 98]. In machine learning, uncertainty about the measures can be assessed as proposed by Aires *et al.* [AIR 04], and may also include the errors associated with the estimation process. A simpler alternative is proposed by Baret *et al.* [BAR 13], based on the learning database. Although estimates of the uncertainties are possible in theory, they are usually limited by poor knowledge of uncertainties associated with the measured reflectance and the RTM used. Knyazikhin *et al.* [KNY 98] use an uncertainty of 20% applied to the MODIS reflectance data at the top of the canopy. Based on a thorough evaluation of satellite reflectance measurements, Baret *et al.* [BAR 07b] propose the use of an additive uncertainty of the order of 0.05 and multiplicative of around 3%. This example shows that the uncertainties associated with reflectance in each of the spectral bands are very poorly known. In addition, the structure of uncertainties also plays an important role and is unfortunately very difficult to describe;

– *the robustness of the estimates and the quality assessment*: A quality index must be associated with estimated values for information on how the inversion process was working. For iterative minimization techniques, we can use the criteria to stop the iterations [GIL 02]: algorithms may encounter numerical problems that are often associated with small values of the cost function. Few numerical problems are expected when using the LUTs and machine learning, we then focus mainly on the definition domain for the entries, and on the expected physical range for the estimates [BAR 13]. The robustness of the estimates will depend on the algorithm itself and also on the degree to which the inverse problem is ill-posed, according to the measurement layout, the RTM and measurement uncertainties.

2.4. Theoretical performances in estimating the different variables of interest

Several biophysical variables are potentially accessible. However, the performances of the estimation of the apparent values from satellite-derived observations depend on the assumptions on the structure of the vegetation and the observation layout. Furthermore, the different definitions of GAI and FAPAR must also be discussed in terms of the associated estimation uncertainty. We will analyze the theoretical performances in estimating the different characteristics of the canopy, using a numerical experiment. A large canopy reflectance data set was simulated in the measurement configurations of the Sentinel-2 sensor [MAL 12] from various combinations of input entries of the RTM SLC [VER 07] coupled with the PROSPECT model describing the optical properties of leaves [JAC 90]. Sven spectral bands (560, 670, 705, 740, 865, 1610 and 2190 nm) were chosen to sample the main absorption features of chlorophyll and water. The SLC model [VER 07] simulates leaf clumping on a plant scale: plants are randomly distributed and are represented by revolution ellipsoids filled with a turbid medium. The leaf clumping is mainly described by the soil fraction covered by the crowns of the plants in the vertical direction. The SLC model makes it possible to also simulate a turbid environment when plants crowns completely cover the ground. Three typical solar directions and five reference directions are considered here.

Besides the knowledge of GAI used as input for simulations, GAI_{true}, the black-sky FAPAR ($FAPAR_{bs}$) the white-sky FAPAR ($FAPAR_{ws}$), the GF and the effective GAI, GAI_{eff}, are equally calculated. All the simulations are stored in the LUT used to estimate the five variables of interest $FAPAR_{bs}, FAPAR_{ws}, GF, GAI_{eff}$ and GAI_{true}: a subset of simulated cases extracted from the LUT will be reserved to test the theoretical performance of the estimate. Measurement noise is associated with the corresponding reflectance to create more realistic simulations. The solution is finally set in the LUT as the case which corresponds to the minimum of the cost function shown in equation [2.2], where σ^2 is the reflectance variance calculated from the measurement uncertainties introduced. No constraint or previous information is therefore used in the cost function. The estimates were made using either the LUT corresponding to turbid medium cases, or the LUT corresponding to the set of turbid medium and clumped cases. The reader is referred to Kandasamy *et al.* [KAN 10] for more details on this numerical experiment.

The results presented in Figure 2.4 show that the GF and $FAPAR_{ws}$ are the best estimated variables. In addition, their performances are dependent on the assumptions made on the canopy structure. $FAPAR_{bs}$, is also well estimated, but with a slight performance degradation, especially when the canopy tested are clumped. The GAI values are more difficult to estimate, especially for the actual GAI, GAI_{true}, in the case of clumped canopies. In contrast, relatively stable performances are observed for the effective GAI, GAI_{eff}, fairly independent from assumptions about the canopy structure.

This simple numerical experiment shows that the estimated true GAI from a unidirectional reflectance measurement provides estimates of very variable quality, particularly in the case of clumped canopies. The presence of clumped canopies in the LUT does not improve the estimates. Additional constraints or previous information need to be used to compensate for the additional variables introduced in the radiative transfer model to describe the clumping. This is the subject of the next section.

2.5. How to manage the under-determined and ill-posed nature of the inverse problem?

2.5.1. *The under-determination and ill-posed nature of the inverse problem*

RTMs depend on a set of input variables that characterize the components of the vegetation: the underlying soil, leaves and canopy structure

(Figure 2.5). Several models have been proposed to describe the reflectance of the soil. They are either based on physical foundations focusing mainly on the directional variability of reflectance [JAC 92, LIA 96] or are empirical and mainly describing the spectral variation [LIU 02, PRI 90]. A minimum of six parameters are required to describe both the spectral and directional variability of the reflectance of the soil. Leaf properties (transmittance and reflectance) are simulated from the content in biochemical compounds that absorb radiation (chlorophyll, water and dry matter mainly), the mesophyll structure and the surface characteristics of the leaves [DAW 98, JAC 90]. A minimum of four parameters is necessary at the leaf level. The simplest description of the canopy structure is made with two parameters: the LAI and leaf inclination. Accordingly, the description of the spectral and directional reflectance of the canopy will require a minimum of 12 parameters that are generally unknown and should be estimated by inverting the RTM. The inversion process in parallel must rest on at least the same number of independent observations, that is to say the combinations of spectral bands and viewing directions (or illumination).

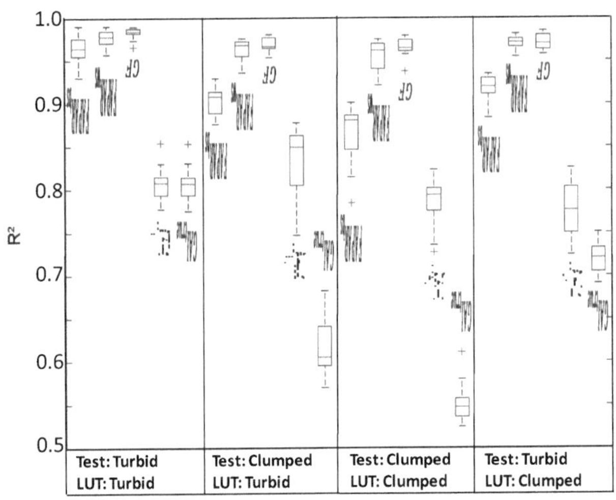

Figure 2.4. *Theoretical performances (R^2 of the regression between the reference value of the variable and its estimated value) for $[FAPAR_{bs}, FAPAR_{ws}, GF, GAI_{eff}, GAI_{true}]$ based on the considered cases (turbid or clumped) and assumptions on the canopy structure used in the LUT (turbid or clumped). Box and whiskers plots represent the results for the 15 geometric observation layouts (three positions of the sun and five viewing directions); the median is the red line and the blue box contains 50% of the data; whiskers correspond to other values outside of the extreme values recorded by a red +. For a color version of this figure, see www.iste.co.uk/baghdadi/3.zip*

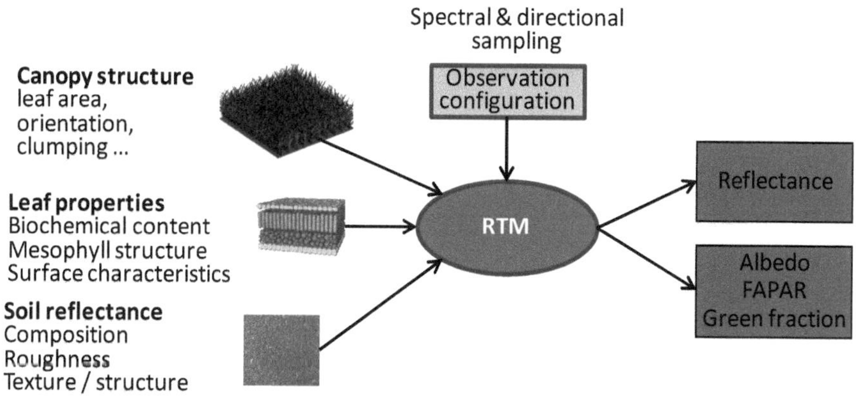

Figure 2.5. *The RTMs used to simulate the canopy reflectance based on field structure, leaf properties and soil. These models can also simulate other radiative properties such as albedo, FAPAR and the green fraction*

The characteristic dimensionality of reflectance measurements was assessed in different ways, usually by independently considering the spectral and directional dimensions. Several studies show that the function describing the directional distribution of the reflectance can be decomposed on the basis of empirical or semi-empirical orthogonal functions, generally having two to four kernels [BRE 02, ROU 92]. Other studies report a high level of redundancy between spectral bands [GRE 01, LIU 02, PRI 94, THE 04] with a dimensionality ranging from 5 to 60 depending on the data reviewed and the method used to quantify the dimensionality. More recently, Laurent *et al.* [LAU 11] found a dimensionality of 3 or 4 for the data captured by the CHRIS sensor with high spectral resolution and with multiple viewing directions. These results confirm those of Settle [SET 04], which show a high degree of redundancy between spectral bands and directions. It is therefore clear that in most situations, the inversion of RTM is an underdetermined problem: the number of unknown variables is greater than the dimensionality of observations.

Because of the underestimation and uncertainties associated with the model and the measurements, the inverse problem is generally ill-posed: the solution is not unique and does not depend continuously on observations [GAR 64]. As a result, very similar RTM simulated values of spectral and directional reflectance (Figure 2.2(a)) may correspond to very different solutions. This may be because of two main factors:

– *the low sensitivity of the reflectance to a particular variable*: This is the case for high values of GAI because of the well-known saturation problem: a small variation in the measured reflectance may result in a very large variation of the estimated GAI value. Similarly, when the GAI is very high, the estimated soil reflectance values will be very inaccurate because the canopy reflectance will not depend on the soil properties;

– *links between variables*: This is obviously the case when the variables always appear to be combined together in the same way, in the form of a product for example: it is impossible to separately estimate each variable. This phenomenon is also observed for some variables which do not formally appear as products in the model as shown by various studies [BAR 07a].

The ill-posed nature of RTM inversion must be countered by leveraging additional information [KNY 99]. This can be achieved either by introducing information in the inversion process, or by exploiting constraints on some variables. Moreover, the choice of a more realistic RTM compared with the considered targets will reduce the uncertainties associated with it. These aspects are described separately in the following sections.

2.5.2. *Reducing RTM uncertainties*

The realism of RTM has a strong influence on the estimation performance. The RTM is to be built on physical bases and a description of the canopy structure, leaf properties and soil should be consistent with the canopy considered. The soil is relatively well described, mainly by empirical models as we have seen previously. The leaf properties are relatively well described by the PROSPECT [JAC 09] or LIBERTY [DAW 98, MOO 08] models, at least if the directional effects are not considered [COM 14]. The canopy structure is the main limiting factor of realistic reflectance RTM simulations. To take into account the architectural features of a particular canopy, you must know the type of vegetation observed. Depending on the spatial resolution of observations and the heterogeneity of the scene, this information is not always available. The kilometric spatial resolution observations generally correspond to a mixture of different types of vegetation, and the use of a specific RTM is not possible. In contrast, with the decametric resolution, pixels have a high probability of containing only one type of vegetation that can be easily identified. Under these circumstances, the use of RTM to observe a specific vegetation type must provide more realistic and accurate simulations of reflectance. Lopez-Lozano [LOP 08] compared the inversion of an RTM using either a turbid medium or a more realistic 3D description of the canopy structure (Figure 2.6). The results clearly show that the estimated GAI is much more accurate with a 3D description than with the assumption of turbid medium, especially for vines. The improvement is less important for the corn crop for which the assumption of a turbid medium is not very far from reality. The difference between the apparent accessible GAI by RTM inversion and the true GAI will depend on the RTM used and the type of vegetation considered. Moreover, the apparent GAI value will depend on the observation layout illustrated, for example, by the strong effect of the azimuthal orientation of the vine rows. Using a more realistic RTM requires a description of the vertical and horizontal variation of the green area density. Realism gain is then obtained at the cost of an increase in the number of the canopy unknown variables that must be offset by integrating information on the distributions and co-distributions of these additional variables of the canopy structure.

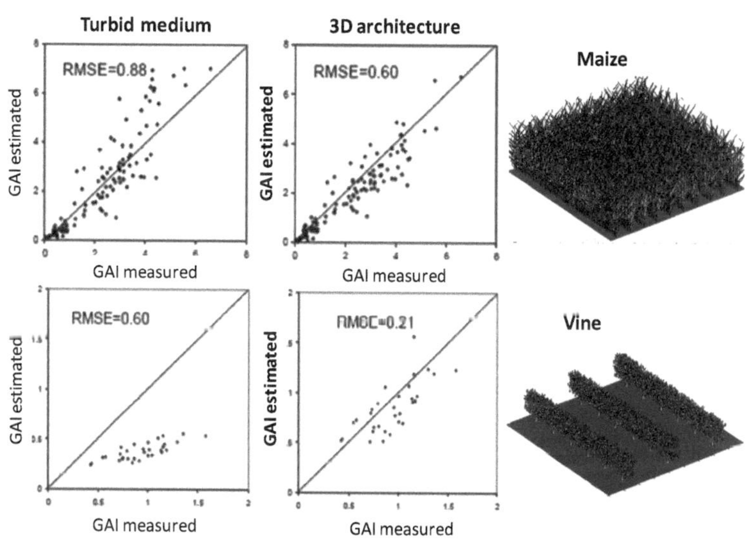

Figure 2.6. *Quality comparison of the GAI estimation when using a "turbid medium" RTM (left) and a realistic 3D description of the structure (right). Examples of maize fields (top) and vines (bottom). RMSE is the square root of the average error [LOP 08]*

2.5.3. *Using prior information*

The prior information is the knowledge available on distributions and co-distributions of RTM input variables. It is used directly in the machine learning approaches to generate the calibration database of the inverse model. For approaches focusing on radiometric variables (LUT and iterative minimization), the prior information is introduced in a cost function regularization term:

$$J = \underbrace{(R-\hat{R})^t \cdot W^{-1} \cdot (R-\hat{R})}_{\text{Radiometric information}} + \underbrace{(\hat{V}-V_p)^t \cdot C^{-1} \cdot (\hat{V}-V_p)}_{\text{Prior information}} \qquad [2.3]$$

where $\hat{V}$ and V_p are the estimated value and the RTM input variables value vectors, and C is the covariance matrix characterizing the prior information. The first term of that equation [2.3] corresponds to the distance between the observed and simulated reflectance (equation [2.2]) and the second term corresponds to the distance between the estimated values and RTM input variables value. The equation [2.3] is based on the Bayes theorem [BAY 63] which is widely used for estimation problems [TAR 05]. However, even if the theory is well known, it is still used very little by remote sensing in the optical domain [COM 02, LAU 08, LEW 12, PIN 11].

The implementation of the cost function corresponding to the equation [2.3] requires a good description of the W and C covariance matrices, as well as V_p values. W coefficients should reflect both measurement uncertainties and RTM-related errors. The measurement uncertainties can be deduced from the sensor specifications, whereas RTM-related errors are much less well known and may depend on the situations encountered, such as the type of vegetation and development considered. In addition, the terms of the W covariance matrix are generally very poorly known, and measurement errors, particularly those related to the model, are very structured, that is to say with the terms of significant covariance. When many input measurements are used as in the case of hyperspectral observations, the covariance terms will weight the contribution of each wavelength in the cost function by taking into account the correlations, generally strong, observed between bands. The difficulty of measuring W covariance terms explains why a small subset of spectral bands and viewing directions is typically used when more configurations are available such as in the case of hyperspectral and multi-directional observations. Parsimony corresponding to a reduction of the dimensionality of the measurements is a highly desired property in most estimation problems [TEN 00]. In the case of machine learning, the reduction of the dimensionality is also often beneficial: it reduces the number of coefficients of the inverse model facilitating the learning process.

The introduction of previous information in the inversion process is used to restrict the posterior distributions of the estimated variables. However, the reduction in the dispersion of the solutions is achieved at the cost of a loss of accuracy of the estimates: solutions are biased toward the previous values of the variables of interest, as shown in Figure 2.7.

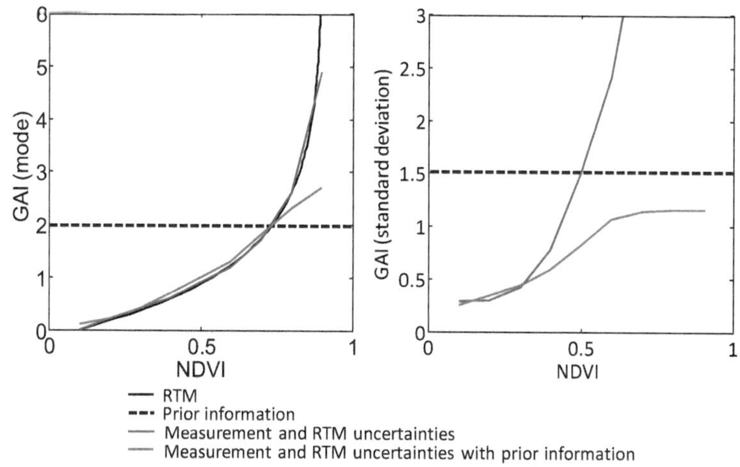

Figure 2.7. *Mode (left panel) for distributions of the solutions (GAI) of the inverse problem as a function of the measured value. In this case, GAI is estimated using a simple empirical radiative transfer model: NDVI = RT (GAI) where NDVI is the normalized difference vegetation index calculated as NDVI = $(R_{nir}-R_r) / (R_{nir}+ R_r)$, with R_{nir} and R_r being respectively the reflectance in the near infrared and red bands respectively. For a color version of this figure, see www.iste.co.uk/baghdadi/3.zip*

The mode corresponds to the maximum frequency of the distribution of the solutions corresponding to the maximum likelihood. Four cases are considered:

– using only the previous information (no observations used: black dashed line);

– using only the RTM (GAI = RT^{-1} (NDVI)) the model and measurements being considered as perfect (black solid line);

– inverting the RTM and taking into account the measurement and the RTM errors (red line);

– by inverting the RTM taking into account measurement and RTM errors as well as the previous information (green line). On the right panel, the standard deviation of the distribution of the solutions is presented for the four cases. The standard deviation of the case of the perfect RTM and measurements is not shown, as the standard deviation of the solution will be zero by definition [BAR 07a].

2.5.4. *Use of additional constraints*

2.5.4.1. *Temporal constraints*

The change of the canopy structure and leaf optical properties results in incremental processes under the climate and soil dependency and genetic characteristics of the plants. Brutal or chaotic temporal variations are not expected outside accidental situations such as those caused by fire, flood, clear cutting, or harvesting. The smooth nature of the dynamics of the state variables of the canopy across a growing season can be used as an additional constraint in the inversion process as proposed by Lewis *et al.* [LEW 12]. The use of models that describe the temporal evolution of some variables shows improvement in the GAI estimates of corn, especially for large GAI values sensitive to saturation [KOT 05]. The semi-empirical nature of the GAI evolution model with parameters that possess a biological meaning allows for easier association of the corresponding previous distribution. However, the results show that the improved estimates come mainly from the temporal smoothing effect: the adjustment of the dynamic GAI model on the individual GAI estimates [KOT 05]. This explains the success of the methods of temporal compositing applied to kilometric spatial resolution observations: very little information is available on the dynamics of the surface outside the smooth and continuous temporal profiles [ATK 12, CHE 04, KAN 13, LEW 12, REF 13, VER 10 ZHU 13]. Classic sigmoidal shape growth phases and vegetation senescence [JON 04b, ZHA 03] and knowledge of climatology considered as a temporal profile [SAM 08, VER 12a] were also used.

2.5.4.2. *Spatial constraints*

Most algorithms are applied independently to each pixel of an image, neglecting the possible use of spatial structures observed on images. Some authors have tried to exploit these spatial structures observed at the decametric resolution. The "object" approach proposed by Atzberger [ATZ 04] is based on the use of the covariance between different state variables of vegetation or soil observed on a limited spatial window representing the same type of vegetation as found in an agricultural field. The results show a significant improvement of GAI estimates, the chlorophyll and water content, explained by a possible limitation of compensation between the different variables estimated in the RTM inversion process [ATZ 04]. These principles were then operated using a few simple rules that can be applied across an agricultural field, or more broadly on a similar set of connected pixels representing an object [ATZ 10]. It is therefore necessary to identify different objects in an image, which has now become a popular approach, especially for image segmentation [PEN 11].

2.5.4.3. *Holistic estimation using coupled models: from inversion to assimilation*

Estimating the characteristics of an element of the system without solving the whole system is suboptimal: each element of the system imposes constraints on other items throughout the radiative transfer process and temporal or spatial constraints, as previously seen. This is clearly demonstrated in the case of radiative coupling between the leaves and the canopy: estimating the structural characteristics of the canopy from the reflectance measured in several spectral bands can be achieved by inverting the RTM model for each band. Leaf characteristics represented by their reflectance and transmittance, sometimes grouped in the single scattering albedo, must be estimated for each band considered [PIN 91]. This generally leads to inconsistencies between the estimated structural characteristics of the canopy in each spectral band. This can also lead to spectrally inconsistent optical properties of the leaves or soils when no model imposes physical constraints on the estimates. The resolution of the whole system at once, using a model of leaf optical properties coupled with a canopy RTM, will improve estimates by imposing spectral constraints coming from the optical leaf property model. In addition, the resolution of the whole system reduces the number of variables to estimate: smaller number of parameters to represent the leaf properties in several spectral bands and no replication of the canopy characteristics which are by definition spectrally invariant. The advantage of such an holistic approach was also demonstrated by [LAU 11] when using the RTM coupled between the atmosphere and the vegetation cover.

Lauvernet *et al.* [LAU 08] proposed a "multi-temporal object" approach to take into account the spatial and temporal constraints. The reflectance data are seen here at the top of the atmosphere. The RTM of the atmosphere, the canopy, the leaves, and the soil are coupled to simulate the reflectance at the top of the atmosphere from a set of input variables of each sub-model. The spatial and temporal constraints are based on the assumption that the atmosphere is stable over a limited window space of a few kilometers, but can vary from one date to another. Conversely, the leaf and canopy characteristics change little from one date to another ("smooth" nature of these characteristics), but can vary greatly from one pixel to another [HAG 08]. The resolution of the system on a set of pixels and dates reduces the number of unknowns, the atmospheric characteristics shared by the pixels, and the leaf and canopy characteristics being shared between dates. The results show improved performance for most variables, particularly for the atmosphere and canopy characteristics such as the GAI.

The improved performance from these holistic approaches is achieved at the cost of a significant increase in the complexity of the problem, especially in terms of the number of variables to be estimated simultaneously and the computing power needed to operate the coupled models. Machine learning type approaches are generally ineffective for such dimensionality. The iterative minimization implemented with the adjoint model [LAU 08, LEW 12, VOB 10] provides a well suited and efficient solution. This inversion technique may also be applied to the RTM-coupled models that represent both the structure and function of the canopy, as proposed by Weiss *et al.* [WEI 01]. However, considerable efforts are needed to achieve a realistic description of the dynamics of the canopy structure, while promoting a consistent radiative transfer model in the canopy and a vegetation functioning model.

2.6. Combination of methods and sensors to improve estimates

2.6.1. *Hybrid methods and ensemble products*

Verger *et al.* [VER 08] show that neural networks could be used effectively to replace the MODIS algorithm [SHA 05] based on the LUT approach: neural networks being calibrated on a data set containing the MODIS sensor reflectance and the corresponding values of the MODIS LAI products. This method is called "hybrid" because it uses the approach focusing on the biophysical variable based on machine learning (in this case, a neural network) over a set of cases in which the biophysical variables have been previously calculated from an approach focusing on the radiometric data (in this case, the LUT technique). This principle is now used to connect the long-time series of NDVI from AVHRR [TUC 05] to the LAI MODIS products by exploiting the period 2000–2009, during which the AVHRR and MODIS sensors made simultaneous observations [ZHU 13]. This hybrid approach appears different from direct machine learning on the RTM simulations as proposed by Bacour *et al.* [BAC 06] or Baret *et al.* [BAR 07b].

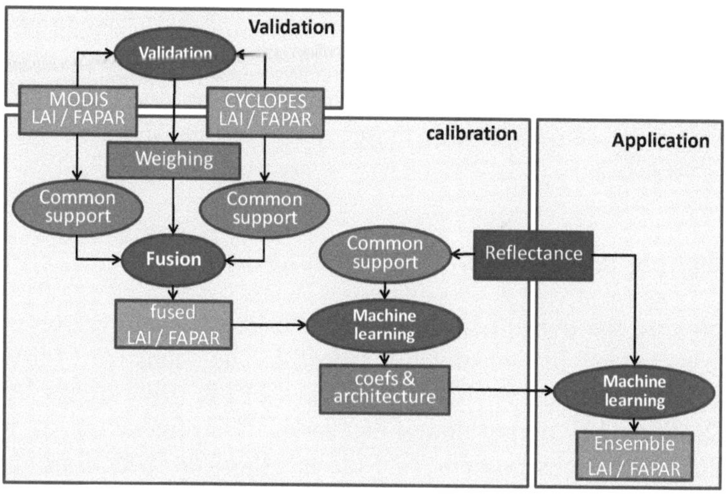

Figure 2.8. *GEO V1 [BAR 13], OV2 [VER 14], GLASS [XIA 14] algorithm principles using several existing products to develop a comprehensive product*

The results of the kilometric product validation exercises have helped to better quantify the performance of the currently available products [GAR 08]. The most realistic products have been selected and eventually combined to create a new product that capitalizes all past developments (Figure 2.8). For GEOV1 products [BAR 13], GEOV2 [VER 14], GLASS [XIA 14], a learning database is first created. It contains a representative sample across the globe of MODIS and CYCLOPES LAI and FAPAR products from which the fused product will be calculated as a weighed sum of the two original products. The original biophysical products (LAI and FAPAR) in the training set must have the same spatial and temporal support to be combined. This is achieved by simple interpolation which will also perform the first smoothing. A weighted average is then calculated for the fused product (Figure 2.8). The weighting is derived from validation exercises that provide either uncertainties [XIA 14] or heuristics [BAR 13]. The merged products and the corresponding reflectances are then used to calibrate a machine learning algorithm. The algorithm, once calibrated, will eventually be operationally used to transform the reflectance observed into estimates of biophysical variables of interest (Figure 2.8). For GEOV1 [BAR 13], a neural network is used to estimate the instantaneous values of LAI and FAPAR for each observation date available for the respective sensor. A smoothing and gap filling algorithm (gaps correspond to the missing data) is then applied to the flash estimates [VER 11b]. In the case of the GLASS [XIA 14] product, observations of a full year are used to obtain an annual temporal profile calculated using a generalized regression neural [SPE 91]. The results show that in general, these ensemble products have improved performances as compared with those of the original products [CAM 13, FAN 13, XIA 14].

2.6.2. *Sensor combination to build long-time series*

Monitoring the dynamics of vegetation throughout the season or years can characterize its phenology, identify its function [GAN 10, JON 04b], detect anomalies [CIA 05], breaks [VER 10] or trends [ALC 10, DEJ 12, FEN 12] during long-time series. The frequency of revisits, consistency of the observations, and the length of the period are the main limiting factors to consider when working with time series. Kilometric observations provided by polar orbiting satellites can cover the globe every day over more than three decades (Figure 2.9). However, the theoretical daily frequency is reduced mainly because of the occurrence of clouds. The combination of different sensors over the same period will slightly improve the availability of observations exploitable because of the strong spatio-temporal correlations of cloud cover: Yang *et al.* [YAN 06] have shown no improvement in the availability of usable data by combining MODIS sensors aboard TERRA and AQUA satellites. Inversely, Hagolle *et al.* [HAG 05] show a significant improvement using the observations of both VEGETATION instruments on SPOT 4 and SPOT 5. Verger *et al.* [VER 11b] merge data from MODIS and VEGETATION sensors and achieve a significant reduction of gaps in the time series and improved the quality of LAI estimates. These contrasting results are because of very different principles of temporal fusion algorithms, emphasizing the importance of the temporal composition process which consists primarily of smoothing and possibly completing missing observations in the time series [KAN 13].

Interest in using data from different satellites becomes obvious for decametric observations for which the time between two observations is several days or even weeks. Merging estimates from several decametric sensors do not pose major difficulties, as illustrated in Figure 2.10: a very high temporal consistency of the GAI estimates is generally observed when the same algorithmic principles are applied to the different sensors. This confirms the results of Verger *et al.* [VER 08] and Gobron *et al.* [GOB 08] who obtained kilometric resolution and who have demonstrated that applying the same algorithm to different sensors provides very consistent estimates if the differences between the specific observations to each sensor configurations are taken into account. If all decametric satellites currently in orbit represent a strong potential for monitoring the vegetation, their concurrent use has been little exploited, except in the case of homogeneous constellations such as those of rapid-eye or DMC [SUN 02]. Difficulties in their access and the costs of the images explain why they have rarely been combined.

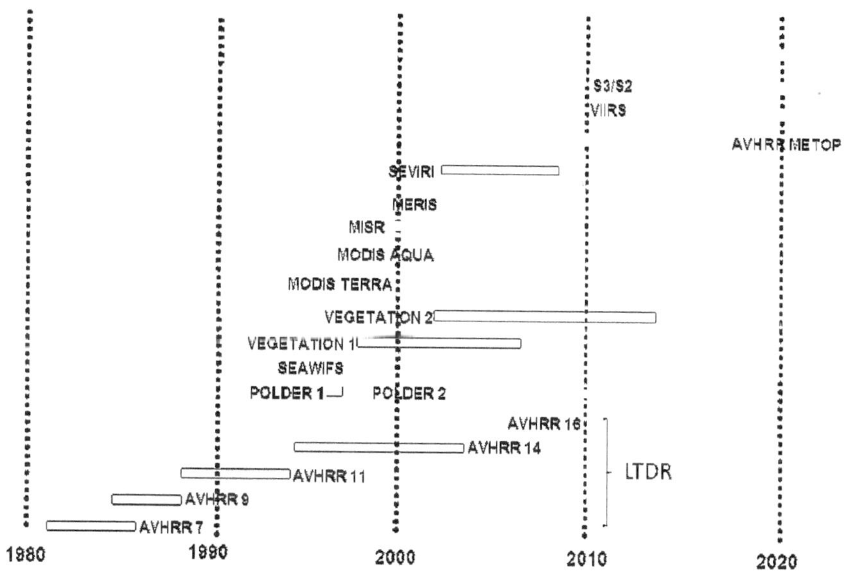

Figure 2.9. *Series sensors at the kilometric spatial resolution available from 1980 to the present day*

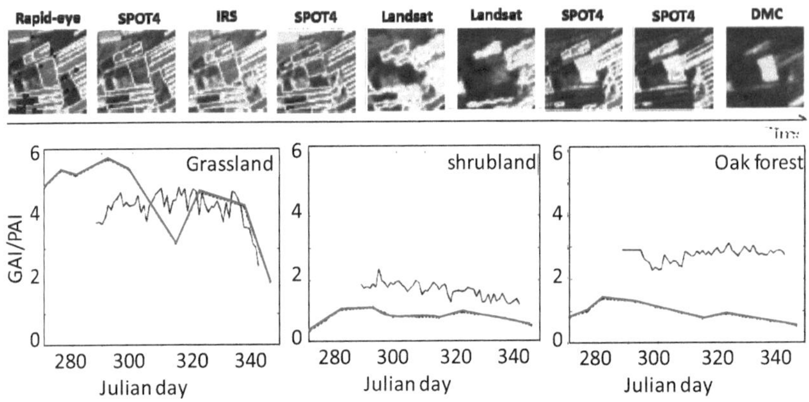

Figure 2.10. *Operating a heterogeneous constellation of satellites to describe the seasonal variation of GAI. Above, images of GAI estimated from different sensor observations during the season. Below, the red curve shows an example of the seasonal variation of GAI estimated by satellites for three types of vegetation (grassland, scrubland, oak forest). The black curve corresponds to PAI (Plant Area Index) measurements performed on the ground. Results obtained on the website of La Crau (43.5° latitude, 4.9° longitude). The algorithm used for all the sensors is similar to that described in Verger et al. [VER 11a]. No smoothing is applied to the estimates of GAI or PAI. The differences between the estimates made from satellite and ground measurements are mainly explained by differences in definition between the GAI derived from satellite and the PAI measurement of the ground, which includes the non-green elements of significant importance for the garrigue and the green oak forest. The rapid fall of GAI and PAI is noted after day 340 for the Grassland, because of the senescence induced by low winter temperatures. For a color version of this figure, see www.iste.co.uk/baghdadi/3.zip*

The fusion of infrequent decametric observations and almost daily observations in kilometric or hectometric resolution is very promising because it can provide very frequent decametric observations. However, such combination of sensors has been little used. It was primarily used for classification [KAR 09], including pan-sharpening [FAI 97, FAS 08] and vegetation indices [GAO 06]. Research efforts are needed to better exploit the complementarity of frequent coarse spatial resolution with the less frequent decametric resolution observations.

The succession of several kilometric sensors has been exploited to build long-time series since 1981 (Figure 2.9). The consistency of biophysical variables estimations using various sensors to build the time-series must be very strong to identify possible trends, which can be of very low amplitude [BEC 11]. This is currently achieved by the using a unique algorithm for the sequence of available sensors. Zhu *et al.* [ZHU 13] transform the long series of observations of NDVI calculated from the AVHRR sensors (Figure 2.11) into a chronic LAI and FAPAR by calibrating a neural network on the MODIS products during a common period between AVHRR and MODIS. The temporal consistency between sensors is provided here in the calculation of the NDVI from the GIMMS algorithm [TUC 05]. Verger *et al.* [VER 12b] have also built a long-time series of observations from AVHRR up to 2000 supplemented with data from the VEGETATION sensor after 2000. The reflectance data used has been carefully processed to ensure the best possible consistency [NAG 09]. A neural network was then calibrated on the overlapping period between AVHRR, MODIS and VEGETATION. LAI and FAPAR products from MODIS and VEGETATION were merged as previously seen and used as a basis for learning. A temporal composition algorithm [VER 12a] is finally applied to flash estimates to eliminate outliers, smoothing the temporal profiles and filling the gaps corresponding to missing data.

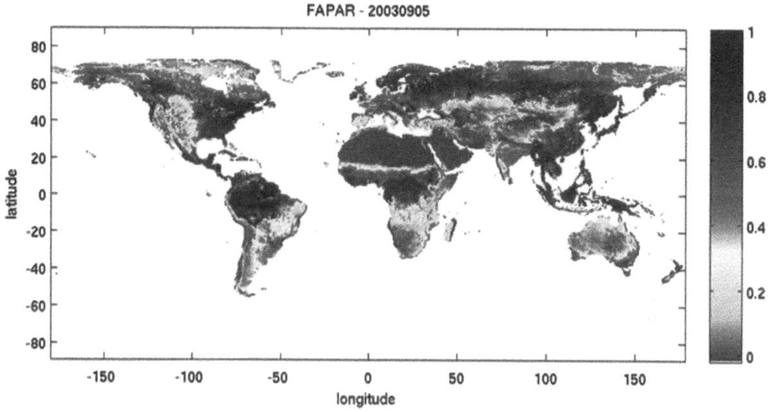

Figure 2.11. *Global mapping of FAPAR GEOV1 as estimated by the algorithm applied to the VEGETATION sensor for September 5, 2003. For a color version of this figure, see www.iste.co.uk/baghdadi/3.zip*

2.7. Conclusion

This review of techniques for estimating vegetation biophysical variables from satellite observations shows the significant progress made in recent years on the maturity of the algorithms. Several products are now available to the community, most of them calculated from observations at kilometric spatial resolution as shown in Table 2.2. The multiplicity of products ensures sufficient confidence in their performances from the observed agreement among themselves and with ground observations. The validation exercise is essential to identify problems, improve products and quantify uncertainties. The RMSE (square error) associated with the FAPAR is in the order of 0.10–0.15 in absolute value [WEI 14] and the LAI is estimated with an RMSE that is slightly less than 1.0. However, the relatively small number of validated ground measurements available limits the evaluation of the precision of these products.

Products	Sensors	LAI	FAPAR	Resolutionspatiale	Revisit frequency (days)	Observation period	Reference
MODIS C5	MODIS	✓	✓	1 km	8	2000-	[MYN 02]
CYCLOPES V3	VEGETATION	✓	✓	0.009°	10	1999-2007	[BAR 07b]
GLOBCARBON	VEGETATION	✓	✓	0.009°	30	1999-2007	[DEN 06]
JRC-FAPAR	SEAWIFS		✓	2 km	1	1997-2006	[GOB 06]
JRC-TIP	MODIS	✓	✓	0.01°	16	2000-	[PIN 11]
GIMMS_3g	AVHRR	✓	✓	8 km	30	1981-2013	[GAN 10]
GLASS	MODIS/AVHRR	✓	✓	1 km	10	1981-2014	[XIA 12]
GEOV1_VEG	VEGETATION	✓	✓	0.009°	10	1999-	[BAR 13]
GEOV2_VEG	VEGETATION	✓	✓	0.009°	10	1999-	[VER 14]

Table 2.2. *The various global LAI and FAPAR products currently available*

When very little independent information or constraints are available, as in the case of kilometric observations, the best estimated variables are the GF, the FAPAR and the FIPAR. The apparent values of the estimated LAI or GAI are then very close to the effective values, LAI_{eff} or GAI_{eff}, while the actual LAI or GAI values are poorly estimated, with errors that rely heavily on the measurement configuration. It also depends on the type of vegetation in connection with the leaf clumping. Xiao *et al.* [XIA 12] proposed a correction of this effect, but the performance validation is still very partial. All of these results on validation and uncertainties associated with the definition of estimated variables should be widely disseminated to the users of these LAI or GAI products. Alternatively, the use of the GF is expected to achieve a greater consistency with the vegetation functioning models used to describe the interactions with the climate. The integration of satellite observations into such models should thus make the assumptions on canopy structure more explicit.

This review has shown that two types of approaches can be distinguished when estimating the characteristics of the vegetation from satellite observations.The approaches focusing on biophysical variables calibrate an inverse model by machine learning. Using ground observations in place of RTM simulations will allow integration of the measurement error of the reflectance and would not include the RTM error, since no RTM will be used in this purely empirical approach. The machine learning methods are very effective in this case and can be integrated easily into operational processing chains. However, the difficulty of sampling all possible cases in addition to the errors made on the ground measurements of the variables of interest justifies the use of RTM. The RTM should be well suited to the type of vegetation considered. Unfortunately, the use of realistic RTM specific to each type of vegetation requires more variables to describe the spatial heterogeneity of leaf area density. The use of previous information and specific constraints will be imperative to solving this ill-posed inverse problem. Approaches focusing on the radiometric data such as the iterative minimization therefore appear then very attractive when exploiting an ensemble of constraints that may be available at the decametric resolution. This approach should lead to the assimilation of the satellite reflectance measurement in models describing the optical and structural properties and the functionig of the vegetation. The expected increase in the supply of frequent decametric observations, especially with the Sentinel-2 mission, will certainly guide research and development in this direction, to exploit all available information including the measurements and the knowledge on physical and biological processes.

2.8. Key points

The estimate of vegetation characteristics from satellite observations is now well understood by the scientific community. The provision of operational products, currently at the kilometric resolution for the most part, demonstrates its relative maturity. Using radiative transfer models can explicitly take into account the particular configuration of observation. However their inversion to estimate the characteristics of the vegetation is generally an ill-posed problem. The integration of information and constraints in the inversion process improves the accuracy of estimates. Holistic approaches made possible by the availability of frequent decametric observations are associated with a better description of the type of vegetation and therefore its structure and its functioning should lead to significant improvements in the estimation of the biophysical variables. It will also be possible to access a more detailed and realistic description of the vegetation funtionning.

2.9. Bibliography

[AIR 04] AIRES F., PRIGENT C., ROSSOW W.B., "Neural network uncertainty assessment using Bayesian statistics: a remote sensing application", *Neural Computation*, vol. 16, pp. 2415–2458, 2004.

[ALC 10] ALCARAZ-SEGURA D., LIRAS E., TABIK S. *et al.* "Evaluating the consistency of the 1982-1999 NDVI trends in the Iberian Peninsula across Four time-series derived from the AVHRR sensor: LTDR, GIMMS, FASIR, and PAL-II", *Sensors*, vol. 10, pp. 1291–1314, 2010.

[ASR 84] ASRAR G., FUCHS M., KANEMASU E.T. *et al.*, "Estimating absorbed photosynthetic radiation and leaf area index from spectral reflectance in wheat", *Agronomy Journal*, vol. 76, pp. 300–306, 1984.

[ATK 12] ATKINSON P.M., JEGANATHAN C., DASH J. *et al.*, "Inter-comparison of four models for smoothing satellite sensor time-series data to estimate vegetation phenology", *Remote Sensing of Environment*, vol. 123, pp. 400–417, 2012.

[ATZ 04] ATZBERGER C., "Object-based retrieval of biophysical canopy variables using artificial neural nets and radiative transfer models", *Remote Sensing of Environment*, vol. 93, pp. 53–67, 2004.

[ATZ 10] ATZBERGER C., RICHTER K., "Vegetation biophysical variable retrieval using object-based inversion of hyperspectral CHRIS/PROBA data", *Hyperspectral Workshop 2010: From CHRIS/Proba to PRISMA & EnMAP and Beyond*,17–19 March 2010, Frascati, Italy, 2010.

[BAC 01] BACOUR C., Contribtion à la détermination des parameters biophysiques des couverts végétaux par inversion de modèles de réflectance: analyse de sensibilité et configurations optimales, PhD Thesis, University Paris 7, Paris, 2001.

[BAC 02a] BACOUR C., JACQUEMOUD S., LEROY M. *et al.*, "Reliability of the estimation of vegetation characteristics by inversion of three canopy reflectance models on airborne POLDER data", *Agronomie*, vol. 22, pp. 555–566, 2002.

[BAC 02b] BACOUR C., JACQUEMOUD S., TOURBIER Y. *et al.*, "Design and analysis of numerical experiments to compare four canopy reflectance models", *Remote Sensing of Environment*, vol. 79, pp. 72–83, 2002.

[BAC 06] BACOUR C., BARET F., BÉAL D. *et al.*, "Neural network estimation of LAI, fAPAR, fCover and LAIxCab, from top of canopy MERIS reflectance data: principles and validation", *Remote Sensing of Environment*, vol. 105, pp. 313–325, 2006.

[BAR 04] BARET F., BACOUR C., WEISS M. *et al.*, "Canopy biphysical variables estimation from MERIS observations based on neural networks and radiative transfer modelling: principles and validation", in ESA (ed.), *ENVISAT conference*, Salzburg (Austria), ESA, 2004.

[BAR 07a] BARET F., BUIS S., "Estimating canopy characteristics from remote sensing observations. Review of methods and associated problems", in LIANG S. (ed.), *Advances in Land Remote Sensing: System, Modeling, Inversion and Application*, Springer, pp. 171–200, 2007.

[BAR 07b] BARET F., HAGOLLE O., GEIGER B. *et al.*, "LAI, fAPAR and fCover CYCLOPES global products derived from VEGETATION: Part 1: principles of the algorithm", *Remote Sensing of Environment*, vol. 110, pp. 275–286, 2007.

[BAR 09] BARET F., WEISS M., BICHERON P. *et al.*, "S2PAD – Sentinel-2 MSI - Level 2B Products Algorithm Theoretical Basis Document", in: Vega, GMBH, 2009.

[BAR 10] BARET F., DE SOLAN B., LOPEZ-LOZANO R. *et al.*, "GAI estimates of row crops from downward looking digital photos taken perpendicular to rows at 57.5° zenith angle. Theoretical considerations based on 3D architecture models and application to wheat crops", *Agricultural and Forest Meteorology*, vol. 150, pp. 1393–1401, 2010.

[BAR 13] BARET F., WEISS M., LACAZE R. *et al.*, "GEOV1: LAI and FAPAR essential climate variables and FCOVER global time series capitalizing over existing products. Part1: Principles of development and production", *Remote Sensing of Environment*, 137, p. 299-309, 2013.

[BAR 91] BARET F., GUYOT G., "Potentials and limits of vegetation indices for LAI and APAR assessment", *Remote Sensing of Environment*, vol. 35, pp. 161–173, 1991.

[BAR 95] BARET F., CLEVERS J.G.P.W., STEVEN M.D., "The robustness of canopy gap fraction estimates from red and near infrared reflectances: a comparison of approaches", *Remote Sensing of Environment*, vol. 54, pp. 141–151, 1995.

[BAY 63] BAYES T., PRICE R., "An Essay towards solving a Problem in the Doctrine of Chance. By the late Rev. Mr. Bayes, communicated by Mr. Price, in a letter to John Canton, A. M. F. R. S", *Philosophical Transactions of the Royal Society of London*, vol. 53, pp. 370–418, 1763.

[BEC 11] BECK H.E., MCVICAR T.R., VAN DIJK A.I. *et al.*, "Global evaluation of four AVHRR–NDVI data sets: intercomparisonand assessment against landsatimagery", *Remote Sensing of Environment*, vol. 115, pp. 2547–2563, 2011.

[BEG 91] BEGUÉ A., DESPRAT J.F., IMBERNON J. *et al.*, Radiation use efficiency of pearl millet in the Sahelian zone", *Agricultural and Forest Meteorology*, vol. 56, pp. 93–110, 1991.

[BIC 99] BICHERON P., LEROY M., "A method of biophysical parameter retrieval at global scale by inversion of a vegetation reflectance model", *Remote Sensing of Environment*, vol. 67, pp. 251–266, 1999.

[BRE 02] BRÉON F.M., MAIGNAN F., LEROY M. *et al.*, "Analysis of hot spot directional signatures measured from space", *Journal of Geophysical Research*, vol. 107, no. D16, pp. AAC 1-1–AAC 1-15, 2002.

[BRI 16] BRIOTTET X., "Radiometry in the optical domain", in BAGHDADI N., ZRIBI M. (eds), *Optical Remote Sensing of Land Surfaces*, ISTE Press, London and Elsevier, Oxford, 2016.

[CAM 13] CAMACHO F., CERNICHARO J., LACAZE R. *et al.*, "GEOV1: LAI, FAPAR essential climate variables and FCOVER global time series capitalizing over existing products. Part 2: Validation and intercomparison with reference products", *Remote Sensing of Environment*, vol. 137, pp. 310–329, 2013.

[CHE 02] CHEN J.M., PAVLIC G., BROWN L. *et al.*, "Derivation and validation of Canada wide coarse resolution leaf area index maps using high resolution satellite imagery and ground masurements", *Remote Sensing of Environment*, vol. 80, pp. 165–184, 2002.

[CHE 04] CHEN J., JÖNSSON P., TAMURA M. *et al.*, "A simple method for reconstructing a high quality NDVI time series data set based on the Savitzky-Golay filter", *Remote Sensing of Environment*, vol. 91, pp. 332–344, 2004.

[CHE 05] CHEN J.M., MENGES C.H., LEBLANC S.G., "Global mapping of foliage clumping index using multi-angular satellite data", *Remote Sensing of Environment*, vol. 97, pp. 447–457, 2005.

[CHE 92] CHEN J.M., BLACK T.A., "Defining leaf area index for non-flat leaves", *Plant, Cell and Environment*, vol. 15, pp. 421–429, 1992.

[CIA 05] CIAIS P., REICHSTEIN M., VIOVY N. *et al.*, "Europe wide reduction in primary productivity caused by the heat and draught in 2003", *Nature*, vol. 437, pp. 529–533, 2005.

[CLA 13] CLAVERIE M., VERMOTE E., WEISS M. *et al.*, "Validation of coarse spatial resolution LAI and FAPAR time series over cropland in southwest France", *Remote Sensing of Environment*, vol. 139, pp. 216–230, 2013.

[COM 02] COMBAL B., BARET F., WEISS M. *et al.*, "Retrieval of canopy biophysical variables from bi-directional reflectance data using prior information to solve the ill-posed inverse problem", *Remote Sensing of Environment*, vol. 84, pp. 1–15, 2002.

[COM 14] COMAR A., BARET F., OBEIN G. *et al.*, "ACT: A leaf BRDF model taking into account the azimuthal anisotropy of monocotyledonous leaf surface", *Remote Sensing of Environment*, vol. 143, pp. 112–121, 2014.

[DAW 98] DAWSON T.P., CURRAN P.J., PLUMMER S.E., "LIBERTY – Modeling the effects of leaf biochemical concentration on reflectance spectra", *Remote Sensing of Environment*, vol. 65, pp. 50–60, 1998.

[DEJ 12] DE JONG R., VERBESSELT J., SCHAEPMAN M. *et al.*, "Trend changes in global greening and browning: contribution of short-term trends to longer-term change", *Global Change Biology*, vol. 18, pp. 642–655, 2012.

[DEN 06] DENG F., CHEN J.M., CHEN M. *et al.*, "Algorithm for global leaf area index retrieval using satellite imagery", *IEEE Transactions on Geoscience and Remote Sensing*, vol. 44, pp. 2219–2229, 2006.

[DUV 11] DUVEILLER G., WEISS M., BARET F. *et al.*, "Retrieving wheat Green Area Index during the growing season from optical time series measurements based on neural network radiative transfer inversion", *Remote Sensing of Environment*, vol. 115, pp. 887–896, 2011.

[FAI 97] FAIVRE R., FISCHER A., "Predicting crop reflectances using satellite data observing mixed pixels", *Journal of Agricultural, Biological and Environmental Statistics*, vol. 2, pp. 87–107, 1997.

[FAN 03a] FANG H., LIANG S., KUUSK A., "Retrieving leaf area index using a genetic algorithm with a canopy radiative transfer model", *Remote Sensing of Enviroment*, vol. 85, pp. 257–270, 2003.

[FAN 03b] FANG H., LIANG S., "Retrieving leaf area index with a neural network method: simulation and validation", *IEEE Transactions on Geoscience and Remote Sensing*, vol. 41, pp. 2052–2062, 2003.

[FAN 05] FANG H., LIANG S., "A hybrid inversion method for mapping leaf area index from MODIS data: experiments and application to broadleaf and needleleaf canopies", *Remote Sensing of Enviroment*, vol. 94, pp. 405–424, 2005.

[FAN 13] FANG H., JIANG C., LI W. *et al.*, "Characterization and intercomparison of global moderate resolution leaf area index (LAI) products: analysis of climatologies and theoretical uncertainties", *Journal of Geophysical Research: Biogeosciences*, vol. 118, pp. 529–548, 2013.

[FAS 08] FASBENDER D., RADOUX J., BOGAERT P., "Bayesian data fusion for adaptable image pansharpening", *IEEE Transactions on Geoscience and Remote Sensing*, vol. 46, pp. 1847–1857, 2008.

[FEN 12] FENSHOLT R., LANGANKE T., RASMUSSEN K. *et al.*, "Greenness in semi-arid areas across the globe 1981-2007 – an earth observing satellite based analysis of trends and drivers", *Remote Sensing of Environment*, vol. 212, pp. 144–158, 2012.

[FER 05] FERNANDES R., LEBLANC S.G., Parametric (modified least squares) and non-parametric (Theil-Sen) linear regressions for predicting biophysical parameters in the presence of measurements errors", *Remote Sensing of Environment*, vol. 95, pp. 303–316, 2005.

[GAN 10] GANGULY S., FRIEDL M.A., TAN B. *et al.*, "Land surface phenology from MODIS: characterization of the collection five global land cover dynamics product", *Remote Sensing of Environment*, vol. 114, pp. 1805–1816, 2010.

[GAN 12] GANGULY S., NEMANI R.R., ZHANG G. *et al.*, "Generating global Leaf Area Index from Landsat: algorithm formulation and demonstration", *Remote Sensing of Environment*, vol. 122, pp. 185–202, 2012.

[GAO 06] GAO F., MASEK J.G., SCHWALLER M. *et al.*, "On the blending of the Landsat and MODIS surface reflectance: predicting daily Landsat surface reflectance", *IEEE Transactions on Geoscience and Remote Sensing*, vol. 44, pp. 2207–2219, 2006.

[GAR 08] GARRIGUES S., LACAZE R., BARET F. *et al.*, "Validation and intercomparison of global leaf area index products derived from remote sensing data", *Journal of Geophysical Research*, vol. 113, p. G02028, 2008.

[GAR 64] GARABEDIAN P., *Partial Differential Equations*, John Willey and Sons, New York, 1964.

[GCO 11] GCOS, "Global climate observing system - systematic observation requirements for satellite-based products for climate", Update, Supplemental Details to the Satellite Based Component of the Implementation Plan for the Global Observing System for Climate in Support of the UNFCCC (2010 Update), In World Meteorological Organization, Geneva, Switzerland, p. 138, 2011.

[GIL 02] GILBERT J.C., Optimisation Différentiable: Théorie et algorithmes, Technical report, INRIA Rocquencourt, France, p. 821, 2002.

[GOB 06] GOBRON N., PINTY B., AUSSEDAT O. *et al.*, "Evaluation of fraction of absorbed photosynthetically active radiation products for different canopy radiation transfer regimes: methodology and results using Joint Research Center products derived from SeaWiFS against ground-based estimations", *Journal of Geophysical Research*, vol. 111, D13110, 2006.

[GOB 08] GOBRON N., PINTY B., AUSSEDAT O. *et al.*, "Uncertainty estimates for the FAPAR operational products derived from MERIS -Impact of top-of-atmosphere radiance uncertainties and validation with field data", *Remote Sensing of Environment*, vol. 112, pp. 1871–1883, 2008.

[GOE 85] GOEL N.S., DEERING D.W., "Evaluation of a canopy reflectance model for LAI estimation through its inversion", *IEEE Transactions on Geoscience and Remote Sensing*, vol. GE-23, pp. 674–684, 1985.

[GOE 89] GOEL N.S., "Inversion of canopy reflectance models for estimation of biophysical parameters from reflectance data", in ASRAR G. (ed.), *Theory and Applications of Optical Remote Sensing,* Wiley Interscience, 1989.

[GON 99] GONG P., WANG D.X., LIANG S., "Inverting a canopy reflectance model using a neural network", *International Journal of Remote Sensing*, vol. 20, pp. 111–122, 1999.

[GRE 01] GREEN R.O., BOARDMAN J., "Exploration of the relationship between information content and signal to noise ratio and spatial resolution in AVIRIS spectral data", in GREEN R.O. (ed.), *AVIRIS Workshop*, Pasadena, USA, 2001.

[HAG 05] HAGOLLE O., LOBO A., MAISONGRANDE P. *et al.*, "Quality assessment and improvement of temporally composited products of remotely sensed imagery by combination of VEGETATION 1 and 2 images", *Remote Sensing of Environment*, vol. 94, pp. 172–186, 2005.

[HAG 08] HAGOLLE O., DEDIEU G., MOUGENOT B. *et al.*, "Correction of aerosol effects on multi-temporal images acquired with constant viewing angles: application to Formosat-2 images", *Remote Sensing of Environment*, vol. 112, p. 1689, 2008.

[HUE 88] HUETE A.R., "A soil adjusted vegetation index (SAVI)", *Remote sensing of Environment*, vol. 25, pp. 295–309, 1988.

[HUE 97] HUETE A.R., LIU H.Q., BATCHILY K. *et al.*, "A comparison of vegetation indices over a global set of TM images for EOS-MODIS", *Remote Sensing of Environment*, vol. 59, pp. 440–451, 1997.

[HUE 94] HUETE A.R., LUI H.Q., "An error and sensitivity analysis of the atmospheric and soil correcting variants of the NDVI for the MODIS-EOS", *IEEE Transactions on Geoscience and Remote Sensing*, vol. 32, pp. 897–905, 1994.

[JAC 09] JACQUEMOUD S., VERHOEF W., BARET F. *et al.*, "PROSPECT + SAIL models: a review of use for vegetation characterization", *Remote Sensing of Environment*, vol. 113, pp. S56–S66, 2009.

[JAC 90] JACQUEMOUD S., BARET F., "PROSPECT: A model of leaf optical properties spectra", *Remote Sensing of Environment*, vol. 34, pp. 75–91, 1990.

[JAC 92] JACQUEMOUD S., BARET F., HANOCQ J.F., "Modeling spectral and directional soil reflectance", *Remote sensing of Environment*, vol. 41, pp. 123–132, 1992.

[JAC 95] JACQUEMOUD S., BARET F., ANDRIEU B. *et al.*, "Extraction of vegetation biophysical parameters by inversion of the PROSPECT+SAIL model on sugar beet canopy reflectance data. Application to TM and AVIRIS sensors", *Remote Sensing of Environment*, vol. 52, pp. 163–172, 1995.

[JAM 05] JAMET C., THIRIA S., MOULIN C. *et al.*, "Use of a neurovariational inversion for retrieving oceanic and atmospheric constituents from ocean color imagery: a feasibility study", *Journal of Atmospheric and Oceanic Technology*, vol. 22, pp. 460–475, 2005.

[JON 04a] JONCKHEERE I., FLECK S., NACKAERTS K. *et al.*, "Review of methods for *in situ* leaf area index determination. Part I: Theories, sensors and hemispherical photography", *Agricultural and Forest Meteorology*, vol. 121, pp. 19–35, 2004.

[JON 04b] JÖNSSON P., EKLUNDH L., "TIMESAT –a program for analyzing time series of satellite sensor data", *Computers and Geosciences*, vol. 30, pp. 833–845, 2004.

[KAN 10] KANDASAMY S., LOPEZ-LOZANO R., BARET F. *et al.*, "The effective nature of LAI as measured from remote sensing observations", *Proceedings of the IEEE International Geoscience and Remote Sensing Symposium, Honolulu*, pp. 789–792, 2010.

[KAN 13] KANDASAMY S., BARET F., VERGER A.et al., "A comparison of methods for smoothing and gap filling time series of remote sensing observations: application to MODIS LAI products", *Biogeosciences*, vol. 10, pp. 4055–4071, 2013.

[KAR 09] KARKEE M., STEWARD B.L., TANG L. *et al.*, "Quantifying sub-pixel signature of paddy rice field using an artificial neural network", *Computers and Electronics in Agriculture*, vol. 65, pp.65–76, 2009.

[KIM 00] KIMES D.S., KNYAZIKHIN Y., PRIVETTE J.L. *et al.*, "Inversion methods for physically-based models", *Remote Sensing Reviews*, vol. 18, pp.381–439, 2000.

[KNY 98] KNYAZIKHIN Y., MARTONCHIK J.V., MYNENI R.B. *et al.*, "Synergistic algorithm for estimating vegetation canopy leaf area index and fraction of absorbed photosynthetically active radiation from MODIS and MISR data", *Journal of Geophysical Research*, vol. 103, pp. 32257–32275, 1998.

[KNY 99] KNYAZIKHIN Y., GLASSY J., PRIVETTE J.L. *et al.*, "MODIS Leaf area index (LAi) and fraction of photosynthetically active radiation absorbed by vegetation (FPAR) product (MOD15) algorithm theoretical basis document", In: http://eospso.gsfc.nasa.gov/atbd/modistables.html, 1999.

[KOT 05] KÖTZ B., BARET F., POILVÉ H. *et al.*, "Use of coupled canopy structure dynamic and radiative transfer models to estimate biophysical canopy characteristics", *Remote Sensing of Environment*, vol. 95, pp. 115–124, 2005.

[LAU 11] LAURENT V.C.E., VERHOEF W., CLEVERS J.G.P.W. *et al.*, "Inversion of a coupled canopy-atmosphere model using multi-angular top-of-atmosphere radiance data: aforest case study", *Remote Sensing of Environment*, vol. 115, pp. 2603–2612, 2011.

[LAU 05] LAUVERNET C., Assimilation variationnelle d'observations de télédétection dans les modèles de fonctionnement de la végétation: utilisation du modèle en compte de contraintes spatiales, PhD Thesis, University Joseph Fourier, Grenoble, France, 2005.

[LAU 08] LAUVERNET C., BARET F., HASCOËT L. *et al.*, "Multitemporal-patch ensemble inversion of coupled surface-atmosphere radiative transfer models for land surface characterization", *Remote Sensing of Environment*, vol. 112, pp. 851–861, 2008.

[LEW 12] LEWIS P., GÓMEZ-DANS J., KAMINSKI T. *et al.*, "An earth observation land data assimilation system (EO-LDAS)", *Remote Sensing of Environment*, vol. 120, pp.219–235, 2012.

[LIA 04] LIANG S., *Quantitative Remote Sensing of Land Surfaces*, Wiley, New York, 2004.

[LIA 96] LIANG S., Townshend J.R.G., "A parametric soil BRDF model: a four stream approximation for multiple scattering", *International Journal of Remote Sensing*, vol. 17, pp. 1303–1315, 1996.

[LIU 02] LIU W., BARET F., GU X. *et al.*, "Relating soil surface moisture to reflectance", *Remote Sensing of Environment*, vol. 81, pp. 238–246, 2002.

[LOP 08] LOPEZ-LOZANO R., Tecnologias de informacion geografica en la cartografia de parametros biofisicos de parcelas de maiz y vina para agricultura de precision, PhD Thesis, University of Zaragoza, 2008.

[MAC 03] MACKAY D.J.C. (ed.), *Information Theory, Inference and Learning Algorithms*, Cambridge University Press, 2003.

[MAL 12] MALENOVSKÝ Z., ROTT H., CIHLAR J. *et al.*, "Sentinels for science: Potential of Sentinel-1, -2, and -3 missions for scientific observations of ocean, cryosphere, and land", *Remote Sensing of Environment*, vol. 120, pp. 91–101, 2012.

[MAK 02] MAKOWSKI D., WALLACH D., TREMBLAY M., "Using Bayesian approach to parameter estimation; comparison of the GLUE and MCMC methods", *Agronomie*, vol. 22, pp.191–203, 2002.

[MAR 94] MARTONCHIK J.V., "Retrieval of surface directional reflectance properties using ground level multiangle measurements", *Remote Sensing of Environment*, vol. 50, pp. 303–316, 1994.

[MCC 09] MCCALLUM I., WAGNER W., SCHMULLIUS C. *et al.*, "Satellite-based terrestrial production efficiency modeling", *Carbon Balance and Management*, vol. 4, no. 8, 2009.

[MIL 67] MILLER J.B., "A formula for average foliage density", *Australian Journal of Botany*, vol. 15, pp.141–144, 1967.

[MOO 08] MOORTHY I., MILLER J.R., NOLAND T.L., "Estimating chlorophyll concentration in conifer needles with hyperspectral data: an assessment at the needle and canopy level", *Remote Sensing of Environment*, vol. 112, pp. 2824–2838, 2008.

[MOR 06] MORISETTE J., BARET F., PRIVETTE J.L. *et al.*, "Validation of global moderate resolution LAI products: a framework proposed within the CEOS Land Product Validation subgroup", *IEEE Transactions on Geoscience and Remote Sensing*, vol. 44, pp. 1804–1817, 2006.

[MOR 10] MORISETTE J.T., "Toward a standard nomenclature for imagery spatial resolution", *International Journal of Remote Sensing*, vol. 31, pp. 2347–2349, 2010.

[MOT 11] MÕTTUS M., SULEV M., BARET F. *et al.*, "Photosyntheticallyactive radiation: measurement and modeling", in MEYERS R. (ed.), *Encyclopedia of Sustainability Science and Technology*, Springer, 2011.

[MYN 02] MYNENI R.B., HOFFMAN S., KNYAZIKHIN Y. *et al.*, "Global products of vegetation leaf area and absorbed PAR from year one of MODIS data", *Remote Sensing of Environment*, vol. 83, pp. 214–231, 2002.

[MYN 88] MYNENI R.B., GUTSCHICK V.P., ASRAR G. *et al.*, "Photon transport in vegetation canopies with anisotropic scattering part II. Discrete-ordinates/exact-kernel technique for one-angle photon transport in slab geometry", *Agricultural and Forest Meteorology*, vol. 42, pp. 17–40, 1988.

[NAG 09] NAGOL J.R., VERMOTE E.F., PRINCE S.D., "Effects of atmospheric variation on AVHRR NDVI data", *Remote Sensing of Environment*, vol. 113, pp. 392–397, 2009.

[NEL 65] NELDER J.A., MEAD R.A., "A simplex method for function optimization", *Computer Journal*, vol. 7, pp.308–313, 1965.

[PEN 11] PEÑA-BARRAGÁN J.M., NGUGI M.K., PLANT R.E. *et al.*, "Object-based crop identification using multiple vegetation indices, textural features and crop phenology", *Remote Sensing of Environment*, vol. 115, pp. 1301–1316, 2011.

[PIN 11] PINTY B., JUNG M., KAMINSKI T. *et al.*, "Evaluation of the JRC-TIP 0.01° products over a mid-latitude deciduous forest site", *Remote Sensing of Environment*, vol. 115, pp. 3567–3581, 2011.

[PIN 91] PINTY B., VERSTRAETE M.M., "Extracting information on surface properties from bidirectional reflectance measurements", *Journal of Geophysical Research*, vol. 96, pp. 2865–2874, 1991.

[PRI 90] PRICE J.C., "On the information content of soil reflectance spectra", *Remote Sensing of Environment*, vol. 33, pp. 113–121, 1990.

[PRI 94] PRICE J., "How unique are spectral signatures?", *Remote sensing of Environment*, vol. 49, pp.181–186, 1994.

[QU 08] QU Y., WANG J., WAN H. *et al.*, "A Bayesian network algorithm for retrieving the characterization of land surface vegetation", *Remote Sensing of Environment*, vol. 112, pp. 613–622, 2008.

[REF 13] REFSLUND J., DELLWIK E., HAHMANN A. *et al.*, "Development of satellite green vegetation fraction time series for use in mesoscale modeling: application to the European heat wave 2006", *Theoretical and Applied Climatology*, pp. 1–16, 2013.

[RIC 77] RICHARDSON A.J., WIEGAND C.L., "Distinguishing vegetation from soil background information", *Photogrammetric Engineering and Remote Sensing*, vol. 43, pp. 1541–1552, 1977.

[ROU 02] ROUJEAN J.L., LACAZE R., "Global mapping of vegetation parameters from POLDER multiangular measurements for studies of surface-atmosphere interactions: a pragmatic method and validtion", *Journal of Geophysical Research*, vol. 107, ACL 6 1–14, 2002.

[ROU 92] ROUJEAN J.L., LEROY M., DESCHAMPS P.Y., "A bidirectional reflectance model of the Earth's surface for the correction of remote sensing data", *Journal of Geophysical Research*, vol. 97, pp. 20455–20468, 1992.

[SAM 08] SAMAIN O., ROUJEAN J.L., GEIGER B., "Use of Kalman filter for the retrieval of surface BRDF coefficients with time-evolving model based on ECOCLIMAP land cover classification", *Remote Sensing of Environment*, vol. 112, pp. 1337–1346, 2008.

[SEL 85] SELLERS P.J., "Canopy reflectance photosynthesis and transpiration", *International Journal of Remote Sensing*, vol. 3, pp. 1335–1372, 1985.

[SET 04] SETTLE J., "On the dimensionality of multi-view hyperspectral measurements of vegetation", *Remote Sensing of Environment*, vol. 90, pp. 235–242, 2004.

[SHA 05] SHABANOV N.V., HUANG D., YANG W. *et al.*, "Analysis and optimization of the MODIS Leaf Area Index algorithm retrievals over broadleaf forests", *IEEE Transactions on Geoscience and Remote Sensing*, vol. 43, pp. 1855–1865, 2005.

[SPE 91] SPECHT D.F., "A general regression neural network", *IEEE Transactions on Neural Networks*, vol. 2, pp. 568–576, 1991.

[SUN 02] SUN W., STEPHENS J., SWEETING M., "Micro-mini-satellites for affordable EO constellations: rapid-eye & DMC", *Photogrammetrie Fernerkundung Geoinformation*, pp. 31–36, 2002.

[TAR 87] TARANTOLA A., *Inverse Problem Theory. Methods for Data Fitting and Model Parameter Estimation*, Elsevier B.V., Amsterdam, 1987.

[TAR 05] TARANTOLA A., *Inverse Problem Theory Problem and Methods for Model Parameter Estimation*, Society for Industrial and Applied Mathematics, Philadelphia, 2005.

[TEN 00] TENENBAUM J.B., De Silva V., Langford J.C., "A global geometric framework for nonlinear dimensionality reduction", *Science*, vol. 290, pp. 2319–2323, 2000.

[THE 04] THENKABAIL P.S., ENCLONA E.A., ASHTON M.S. *et al.*, "Accuracy assessments of hyperspectral waveband performance for vegetation analysis applications", *Remote Sensing of Environment*, vol. 91, pp. 354–376, 2004.

[TUC 05] TUCKER C.J., PINZÓN J.E., BROWN M.E. *et al.*, "An extended AVHRR 8-km NDVI dataset compatible with MODIS and SPOT vegetation NDVI data", *International Journal of Remote Sensing*, vol. 26, pp. 4485–4498, 2005.

[VER 07] VERHOEF W., BACH H., "Coupled soil-leaf-canopy and atmosphere radiative transfer modeling to simulate hyperspectral multi-angular surface reflectance and TOA radiance data", *Remote Sensing of Environment*, vol. 109, pp. 166–182, 2007.

[VER 08] VERGER A., BARET F., WEISS M., "Performances of neural networks for deriving LAI estimates from existing CYCLOPES and MODIS products", *Remote Sensing of Environment*, vol. 112, pp. 2789–2803, 2008.

[VER 10] VERBESSELT J., HYNDMAN R., ZEILEIS A. *et al.*, "Phenological change detection while accounting for abrupt and gradual trends in satellite image time series", *Remote Sensing of Environment*, vol. 114, no.12, pp. 2970–2980, 2010.

[VER 11a] VERGER A., BARET F., CAMACHO DE COCA F., "Optimal modalities for radiative transfer-neural network estimation of canopy biophysical characteristics: evaluation over an agricultural area with CHRIS/PROBA observations", *Remote Sensing of Environment*, vol. 115, pp. 415–426, 2011.

[VER 11b] VERGER A., BARET F., WEISS M., "A multisensor fusion approach to improve LAI time series", *Remote Sensing of Environment*, vol. 115, pp. 2460–2470, 2011.

[VER 12a] VERGER A., BARET F., WEISS M. *et al.*, "The CACAO method for smoothing, gap filling and characterizing anomalies in satellite time series", *IEEE Transactions on Geoscience and Remote Sensing*, vol. 51, pp. 1963–1972, 2012.

[VER 12b] VERGER A., BARET F., WEISS M. *et al.*, "Long term consistent global GEOV1 AVHRR biophysical products", *1st EARSeL Workshop on Temporal Analysis of Satellite Images*, Mykonos (Greece), pp. 1–6, 2012.

[VER 14] VERGER A., BARET F., WEISS M. "Near real-time vegetation monitoring at global scale", *IEEE Journal of Selected Topics in Applied Earth Observations and Remote Sensing* (JSTARS), vol. 7, pp.3473–3481, 2014.

[VER 12] VERRELST J., MUÑOZ J., ALONSO L. *et al.*, "Machine learning regression algorithms for biophysical parameter retrieval: opportunities for Sentinel-2 and -3", *Remote Sensing of Environment*, vol. 118, pp. 127–139, 2012.

[VOB 10] VOBBECK M., CLERICI M., KAMINSKI T. *et al.*, "An inverse radiative transfer model of the vegetation canopy based on automatic differentiation", *Inverse Problems*, vol. 26, no. 6, p. 15, 2010.

[WAL 03] WALTER J.M.N., FOURNIER R.A., SOUDANI K. *et al..*, "Integrating clumping effects in forest canopy structure: an assessment through hemispherical photographs", *Canadian Journal of Remote Sensing*, vol. 29, pp. 388–410, 2003.

[WAL 04] WALTHALL C.L., DULANEY W.P., ANDERSON M.C. *et al.*, "A comparison of empirical and neural network approaches for estimating corn and soybean leaf area index from Landsat ETM+ imagery", *Remote Sensing of Environment*, vol. 92, pp. 465–474, 2004.

[WEI 00] WEISS M., BARET F., MYNENI R. *et al.*, "Investigation of a model inversion technique for the estimation of crop charcteristics from spectral and directional reflectance data", *Agronomie*, vol. 20, pp. 3–22, 2000.

[WEI 01] WEISS M., TROUFLEAU D., BARET F. *et al.*, "Coupling canopy functioning and canopy radiative transfer models for remote sensing data assimilation", *Agricultural and Forest Meteorology*, vol. 108, pp. 113–128, 2001.

[WEI 02] WEISS M., BARET F., LEROY M. *et al.* "Validation of neural net techniques to estimate canopy biophysical variables from remote sensing data", *Agronomie*, vol. 22, pp. 547–554, 2002.

[WEI 07] WEISS M., BARET F., GARRIGUES S. *et al.*, "LAI, fAPAR and fCover CYCLOPES global products derived from VEGETATION, Part 2: validation and comparison with MODIS collection 4 products", *Remote Sensing of Environment*, vol. 110, pp. 317–331, 2007.

[WEI 14] WEISS M., BARET F., BLOCK T. *et al.*, "On line validation exercise (OLIVE): a web based service for the validation of medium resolution land products, application to FAPAR products" *Remote Sensing*, vol. 6, pp. 4190–4216, 2014.

[XIA 12] XIAO Z., LIANG S., WANG J. *et al.*, "GLASS leaf area index product derived from MODIS time series remote sensing data", *IGARSS'12*, Munich, Germany, 22 27 July, 2012.

[XIA 14] XIAO Z., LIANG S., WANG J. *et al.*, "Use of general regression neural networks for generating the GLASS leaf area index product from time-series MODIS surface reflectance", *IEEE Transactions on Geoscience and Remote Sensing*, vol. 52, pp. 209–223, 2014.

[YAN 06] YANG W., SHABANOV N.V., HUANG D. *et al.*, "Analysis of leaf area index products from combination of MODIS Terra and Aqua data", *Remote Sensing of Environment*, vol. 104, pp. 297–312, 2006.

[ZHA 05] ZHANG Q., XIAO X., BRASWELL B. *et al.*, "Estimating light absorption by chlorophyll, leaf and canopy in a deciduous broadleaf forest using MODIS data and a radiative transfer model", *Remote Sensing of Environment*, vol. 99, no. 3, pp. 357–371, 2005.

[ZHA 03] ZHANG X., FRIEDL M.A., SCHAAF C.B. *et al.*, "Monitoring vegetation phenology using MODIS", *Remote Sensing of Environment*, vol. 84, pp.471–475, 2003.

[ZHU 13] ZHU Z., BI J., PAN Y.*et al.*, "Global data sets of vegetation leaf area index (LAI)3g and fraction of photo synthetically active radiation (FPAR)3g derived from global inventory modeling and mapping studies (GIMMS) normalized difference vegetation index (NDVI3g) for the period 1981 to 2011", *Remote Sensing*, vol. 5, pp. 927–948, 2013.

3

Land Cover Mapping from Optical Images

3.1. Introduction

This chapter deals with the production of land cover maps from optical imaging. After an introduction to the subject, the different types of data used will be presented. Imaging types are described in terms of their spatial, spectral and temporal resolution. In addition to the reference data that land cover mapping requires for calibration methods, processing approaches that transform the image pixels into map information and the inevitable stage of their thematic and spatial validation are explained. At the heart of this chapter are feature extraction methods and the most common classification algorithms. Some examples are illustrated using the land cover maps that have been produced.

3.1.1. *Remote sensing imagery mapping*

Map production has been, for many years, the primary objective of remote sensing techniques, from its photographic debut by airborne means, which sought the visual interpretation of aerial photographs by experienced cartographers [GIR 10]. The rise of satellite observation of the Earth, with their systematic revisits and coverage of large areas, has opened the door to the possibility of automatically and systematically analyzing images and even the production of maps from those images.

Chapter written by Jordi INGLADA.

We should start by defining what a map is. The purpose of a map is to allow a summarized and relevant understanding of a geographical area. The map is thus a concise (or simplified), as well as effective (including the essential elements and respecting the rules of graphic semiology) representation of the physical, political or social nature of the selected space.

Remote Sensing Imaging allows us to understand the physical, biological and morphological objects that result from physical as well as human factors. These objects are identified by the analysis of land cover use, which means the physical coverage (including its biological aspects, such as vegetation) of the land surface, as described in terms of type of land use by human societies. We will also discuss landscape, through which we aim to identify and characterize homogeneous types of environments. We will also aim to distinguish artificialized areas (buildings, infrastructure), agricultural areas, forests, moors, wetlands, etc.

Unlike topographic maps, which are used to navigate in space while delivering some of the information on land cover uses, the main issue of land cover maps is to delineate, to inventory and to understand changes and typology trends in the area over time (e.g. forest change in an artificialised area) to explain the determinisms and to predict the future. Land cover maps are therefore an essential tool for planning and analyzing climate change. They are also used in models that describe environmental processes (hydrology, climate, water and carbon cycles).

3.1.2. *Land cover and land use*

The term "land cover" often encompasses two different aspects: the actual cover, that is to say, the presence of physical and biological objects, and land use, that is to say, the uses and functions of these objects.

On one hand, in some cases, one is interested in a physionomic view of the field. This is known as land cover. The aim is to distinguish the elements that form the landscape structure without trying to identify their function. In this case, for example, an agricultural area is not identified as such, but rather as herbaceous vegetation or bare soil depending on its condition at the time of production of the map.

On the other hand, it is often necessary to adopt an anthropic point of view of the landscape, to take into account the function or the type of use of the space. This is known as land use. In these cases, even if the physical cover is the same, we will distinguish different zones, such as industrial zones from commercial areas, or a sports field from a meadow.

This distinction, which is very important from the perspective of the user of the maps, does not have much impact on the methodologies used for their production. We will therefore leave that aside in the remainder of the chapter.

3.1.3. *Nomenclatures*

The legend of a land cover map is a list of mapped categories (thematic classes). The key is specific to a cartographic restitution scale in printed format (ratio between the distance on the map and the actual distance on the ground): the built-up class can only appear at very fine scales (that is to say large, such as 1/10,000), but the urban category has less meaning at these scales. Landscape objects can be grouped into different categories based on the use, on the scale, on the geographical coverage of the map and the data (imagery) used to produce it, and also on the domain specialist producer of the map, his field of expertise, budget and time allocated to collecting field references, or to the user or stakeholder of the map [LEG 96]. However, the physical reality is independent of the map and it can be organized in a nomenclature or classification. Thus, while the key of the map is determined by the scale and type of data used, the nomenclature is independent of the scale and resources used to produce the map.

Most nomenclatures are hierarchical, allowing them to be thematically exhaustive so as to consistently derive specific keys. These class hierarchies can be made *a priori* or *a posteriori*. We speak of *a priori* nomenclature when starting from abstract concepts that are detailed as we move down the class hierarchy. In the case of *a posteriori* classification, we start from concrete classes that are grouped in ascending order. Table 3.1 shows an example of hierarchical nomenclature. From this, we can define a legend for a map by choosing the desired level of detail for each branch of the hierarchy. For example, one could choose a simple hierarchy for forestry applications (Table 3.2).

Hierarchy level						
1	2	3	4	5	6	7
Vegetation	Land	Agriculture	Annual	Winter	Wheat, Barley, rapeseed	
				Summer	Corn, Sunflower, Rice	
				Mixed	Other cereal Other oleaginous Protein rich crops Seeds Vegetables Fodder Vegetable flowers	
			Multiannual	Fallows Temporary prairies Sugarcane		
			Perennial	Herbaceous Wood	Prairies Orchard Vines Nuts Olive tree Other arboriculture	
		semi-natural	Moors			
		Naturel	Forests	Decideous	Beech tree Oak tree Other trees	
				Persistent	Conifers	Scotch pine Other pines Silver fir Other resinous
					Deciduous	Evergreen oak
				Mix		
			Shrubs grasslands			
	Aquatic					
Non-vegetation	Land	Natural	Mineral surfaces			
		Artificial	Built-up Roads			
	Aquatic	Water	Streams Open water			
		Snow Ice				

Table 3.1. *Example of hierarchical nomenclature*

Vegetation	Land	Agriculture				
		semi-natural				
		natural	Forests	Decideous	Beech tree Oak tree Other trees	
				Persistent	Conifers	Scotch pine Other pines Silver fir Other resinous
					Deciduous	Evergreen oak
				Mixed		
			Shrubs Grasslands			
	Aquatic					
Non-vegetation						

Table 3.2. *Example of simplified a hierarchical nomenclature*

This hierarchy can then be flattened to develop the map key by choosing the most detailed level for each branch of the nomenclature:

1) Non-forest

2) Beech

3) Oak

4) Other trees

5) SCOTCH pine

6) Other pines

7) Silver fir

8) Other pines

9) Evergreen oak

10) Mixed forests

One of the problems frequently encountered by land cover map users is the inability to easily compare maps using different nomenclatures. To improve this situation, standardized nomenclatures were proposed. In Europe, the Corine Land-Cover nomenclature was designed in 1985. It offers a hierarchical nomenclature in three levels (5 classes, 15 classes and

44 classes) for mapping at the 1/100,000 scale built from satellite data with a minimum mapping unit of 15 or 25 ha according to the themes.

Another such initiative is the Land Cover Classification System (LCCS) proposed by the United Nations Food and Agriculture Organization (UNFAO). This is a system that allows the construction of hierarchical nomenclatures that may be specific to a certain use, but can be compared at certain levels of detail.

3.1.4. *Detection of land cover change*

Satellite imagery, combined with data processing techniques, allows for the quick production of land cover maps. Through frequent satellite revisits these maps can be produced frequently to allow changes of the observed surfaces to be studied.

Having maps that show previous states facilitates the production of new maps. One can thus work incrementally, analyzing differences and even making projections about future developments. These approaches are based on specific techniques.

3.2. The input data

We will present in this section the types of remote sensing imagery data sets that can be used for the production of land cover maps. The availability of one data type or another depends on several factors:

– economical: certain types of images have a high financial cost;

– the specific problem to be addressed: mapping some of the phenomena requires specific data;

– technological: the processing of certain types of data requires significant computing resources (long time-series covering large areas);

– skills: some imaging modalities need specialist knowledge in sensor physics and thematic knowledge of each mapped environment (urban, agricultural, forest, water).

On top of the images, reference data will be needed. The term reference data is used to refer to the land cover knowledge of a point in the territory at a given time. We often use the term ground truth, which is easy to understand but which is inappropriate because the "reference" data are sometimes obtained without going to the field, and it may not be true (it may contain errors or bias).

The reference data are required, at least, to evaluate the quality of the produced map. They are often used to calibrate automatic production methods.

3.2.1. *Types of imagery*

Depending on the mapping needs, the choice of images to be used may vary. We will present three scenarios to illustrate these choices.

3.2.1.1. *Mono-date*

Optical satellite images with very high spatial resolution (VHRS) like those provided by the Pleiades HR or SPOT6 satellites can sometimes be enough to make an accurate map for certain needs. Indeed, optical imaging with a spatial resolution close to one meter or finer, allows recognition of the main structures of the landscape (roads, buildings, rivers, etc.). For data availability reasons, the mono-date approaches have been the most frequently used.

The availability of at least one spectral band in the near infrared (NIR) also allows detection of vegetation in a relatively robust way if it is present during the acquisition, but a single date is insufficient to characterize vegetation likely to grow at a later date in the study area.

This type of imagery may be useful for mapping infrastructures, urban sprawls and urban morphologies, that are mainly characterized by their shape and relative arrangement. Therefore, a single image covering the area of interest is enough. Figure 3.1 shows an excerpt of an image with very high spatial resolution (50 cm per pixel, the result of the fusion of the panchromatic band sampled at 50 cm and multi-spectral bands sampled at 2 m).

In the airborne domain, hyperspectral data with dozens of spectral bands in the visible, infrared and thermal spectrum are also frequently used.

Figure 3.1. *Extract of an Pleiades HR image of the Rangueil district in Toulouse (France), latitude 43.5779°N, longitude 1.4666°E, natural color display RGB (red, green, blue)*

3.2.1.2. *Multi-temporal*

Unlike infrastructures and most anthropic areas, cultivated and "natural" areas are often characterized by their regular temporal behavior. Natural vegetation has regular cycles with the seasons. In the case of agricultural areas, disruptions of this behavior (crop sequences, cultural operations) overlap with these cycles.

To go beyond the vegetation–non-vegetation distinction, images of a single date are not enough. The multi-temporal imagery domain ranges from the use of two acquisitions per year (e.g. a summer and winter image to distinguish evergreen forests from deciduous forests), to the so-called hyper-temporal, where a daily acquisition frequency is achieved.

Observation systems like SPOT Vegetation (1 km of global coverage resolution each day) or MODIS (resolution from 250 m to 1 km with daily revisits) have been successfully used for mapping at continental scales, even global. However, their low spatial resolution does not allow the production of sufficiently detailed maps, even if disaggregation approaches of mixed pixels have been proposed [BEN 12].

Systems such as Landsat and recently Sentinel-2 offer revisits every few days, but with decametric spatial resolutions, which is a very good compromise for highlighting temporal behavior while having access to details of the landscape.

If the spatial resolution of the data are hardly suitable for mapping road networks or for the characterization of urban morphologies, these data are, in contrast, very useful for mapping ecosystems (e.g. wetlands), types of crops in agricultural land, or forest species with high levels of thematic detail.

Figure 3.2 shows a time series of images acquired by the Taiwanese satellite Formosat-2 (theoretical revisit 2 days, spatial resolution of 8 m and 4 spectral bands in the visible and near infrared). With this sensor, we can accurately observe changes of the agricultural landscape and therefore distinguish different crop types through their phenology and the associated agricultural practices. For example, we observe the emergence of winter crops starting on the months of January and especially in February: these are fields in red, because the display uses a color composition type RGB (near infrared, red, green).

3.2.1.3. *Multi-modalities*

In some applications, land cover classes are difficult to characterize with a single type of imaging. For example, in addition to the frequency of temporal revisits presented in the previous section, it may be helpful to have very high spatial resolution (VHSR) to remove ambiguities on certain types of objects. Another example corresponds to classes which can only be distinguished with a very fine time revisit which can only be achieved with sensors at medium or low spatial resolution (wheat/barley distinction), but this low spatial resolution does not allow mapping at plot scale.

In these cases, the mapping can be performed using several imaging modalities jointly, for example coupling optical/radar to discriminate the soil work in arable lands [VAU 14a,VAU 14b]. Currently, these uses are rare and often highly specific to particular scientific problems.

For the interested reader, it will be helpful to consult Florence Laporterie's PhD thesis about the fusion of images acquired at different spatial resolutions [LAP 03] and Amandine Robin's on the classification and change detection by jointly using images acquired at high spatial resolution and high revisit time [ROB 10].

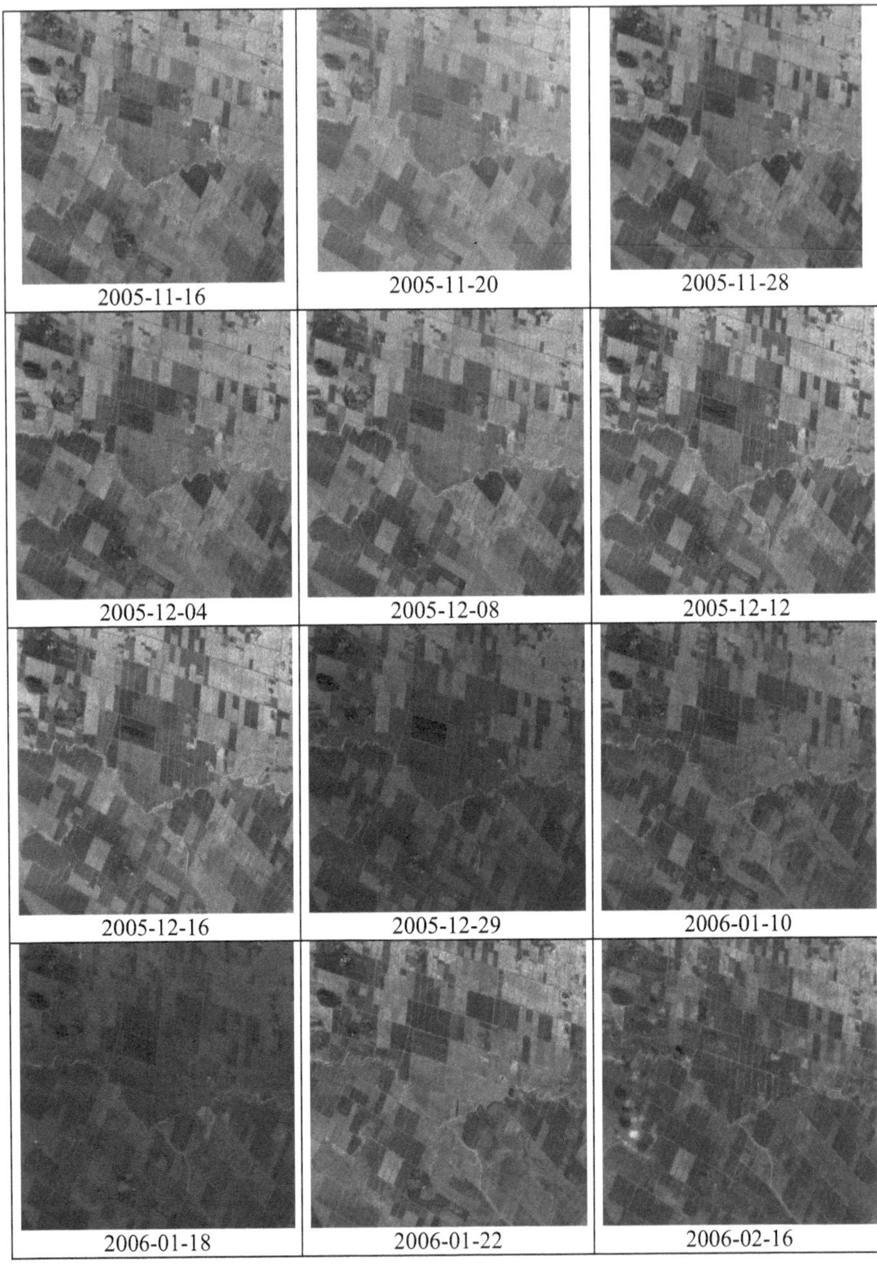

Figure 3.2. *Time series of Formosat-2 images on an agricultural site east of Marrakech (Morocco), latitude 31.65748°N, longitude 7.60126°W, RGB = (near infrared, red, green). For a color version of this figure, see www.iste.co.uk/baghdadi/3.zip*

3.2.2. *Reference data for calibration and validation*

In order for the land cover maps to be disseminated to users, it is necessary to know their quality. This quality must be quantified using standardized indicators accepted by users (see section 3.3.1).

It is therefore necessary to have reference data whose quality has been verified. These data correspond, for example, to specific locations for which we know the land cover classes at the time of the acquisition of images used to produce the land cover map. These type of data are essential, not only in output processing, to calculate statistical indicators of the validity of the processing results, but also as input for the calibration or the parameterization of the algorithm. Indeed, most classification algorithms have parameters that should be chosen wisely, and this choice is often done by automatic learning. So, we use reference data to perform this learning. In this case, it is important to separate the reference data into two sets: the training set and the validation set. This is necessary in order to eliminate any bias in the validation. The ratio between the amount of samples for validation and those for learning is often half, but if the data are limited, we can use more training samples to achieve better results.

If the origin of the reference data can be diverse, their accuracy must nevertheless be greater than that targeted by the map. Otherwise, it is not possible to produce a reliable validation. In the best case, it is preferable to resort to observation field surveys to build these reference databases, useful not only for the calibration of the model, but also for its validation. Over large areas, field surveys with their corresponding precise geographic position, must be made as close to the image acquisitions as possible and involve several people [VAU 14b]. The sampling of the field surveys should be designed so as to ensure the observation data are spatially and thematically representative.

Because of these budgetary and logistic constraints, and the sampling strategy, the use of field observation data is unfortunately not always implemented. The reference data are sometimes based solely on photo-interpretation. This procedure is based on the visual interpretation of images (possibly assisted by some algorithmic techniques). This approach is preferred in cases where the field surveys are costly or impossible. However, photo-interpretation, except in specific cases, leads to reference data that may contain labeling errors. However, the photo-interpretation makes it possible to cover much larger areas than trips to the field. Finally, in some cases, identification of certain land cover classes by photo-interpretation, such as annual crops, is simply impossible.

Beyond the use of these reference data to validate the resulting maps, we can use them for the calibration of algorithms of land cover maps themselves.

Depending on the choice of the classification method (section 3.3.3), the need for reference data for the learning process will vary in importance. It is therefore essential to anticipate these needs when organizing field surveys. The strategy and the sampling protocol (number of samples, sampling method) are a key step to anticipating the learning process and validation, but these issues are beyond the scope of the chapter [GIR 10].

3.3. Land cover map production approaches

In this section, the different steps that constitute a land cover map production chain are developed. Before addressing the processing algorithms, we will present map validation techniques.

3.3.1. *Map validation*

3.3.1.1. *The four steps of validation*

Land cover map validation has evolved significantly since the 1980s to achieve a set of practices currently accepted by the remote sensing community.

The first land cover maps were often validated in a qualitative and visual way, because there were no associated operational uses, and also because they had simplified nomenclatures and very coarse spatial resolutions. Since the resolution and nomenclatures have been refined, it was necessary to establish quantitative criteria. In the 1980s, it was difficult to obtain spatially located data for the validation and one relied mainly on aggregated statistics for large areas. The only applicable quantitative validation was then the comparison of percentages of areas covered by each class. Therefore, the maps were not validated at the pixel level.

When geolocated validation data became more easily available and its importance was understood by users, we started using categorized pixel percentages of either all combined classes, or by class of land cover.

At present, it is common to go further in the validation and to analyze the most frequently confused classes of land cover. To do this, the confusion matrix as well as indices derived from it are used.

3.3.1.2. *The confusion matrix and derived metrics*

In classification, we call confusion matrix or contingency table a double-entry table where the rows represent the true classes of reference samples and the columns represent the classes assigned in the map for these same samples. Thus, the cell corresponding to the line i and column j contains the number of samples of group i classified as belonging to group j in the map.

For example, in the confusion matrix, shown in Table 3.3, 25 urban pixels are correctly classified and there are 10 that are incorrectly classified as bare soil.

	Urban	Forest	Water	Grasslands	Bare soil	Reminder
Urban	25	1	4	5	10	0.55
Forest	2	50	3	10	2	0.74
Water	4	2	47	0	3	0.83
Grasslands	2	8	3	33	3	0.67
Bare soil	9	4	5	5	38	0.62
Precision	0.59	0.76	0.75	0.62	0.67	

Table 3.3. *Example of a confusion matrix*

From the confusion matrix, it is useful to calculate performance measures to make a summary. The most common measure is the Overall Accuracy (OA) which is the sum of the diagonal divided by the total sum:

$$OA = \frac{\sum_{i=1}^{r} n_{ii}}{\sum_{i=1}^{r}\sum_{j=1}^{r} n_{ij}} \qquad [3.1]$$

where r represents the number of classes, n_{ij} is the number of pixels. There are also measures that are calculated by class of land cover. The most common is User's Accuracy (UA), which is a fraction of the pixels classified in a given class and that really belong there:

$$UA_i = \sum_{j=1}^{r} \frac{n_{ii}}{n_{ij}} \quad [3.2]$$

and the Producer's Accuracy (PA) or recall that is the fraction of the reference pixels of a class that are correctly classified in the image:

$$PA_i = \sum_{j=1}^{r} \frac{n_{ii}}{n_{ji}} \quad [3.3]$$

The information given by these two metrics is complementary and allows the detection of over-classifications or under-classifications. These metrics can also be calculated globally by averaging over all classes:

$$PA = \frac{1}{r} \sum_{i=1}^{r} PA_i \quad [3.4]$$

$$UA = \frac{1}{r} \sum_{i=1}^{r} UA_i \quad [3.5]$$

Another frequently used metric the FScore, which is the harmonic mean between the two previous metrics:

$$\text{FScore} = \frac{2\times\text{UA}\times\text{PA}}{\text{UA}+\text{PA}} \quad [3.6]$$

It is difficult to give a general interpretation of the numerical values of these indices, but in general, we aim for overall accuracy greater than 80% and overall FScore values greater than 0.75. At the level of individual classes, we often accept that some minority classes are classified with lower accuracies, but this is the user's choice depending on the use context.

3.3.1.3. *Pixel-based or object-based validation*

Usually, land cover maps are validated by calculating metrics on pixels. Thus, metrics presented in the previous section are based on the outcome of a confusion matrix of pixel counts.

For imaging at very high spatial resolution (VHSR) and for certain land cover nomenclature classes (those containing composite objects, for example), it may be more appropriate to use specific metrics. Indeed, with these fine resolutions, there may be geometric shifts between the reference data used for validation and imaging used for processing. These discrepancies are often caused by items that have an elevation from the soil (parallax effect). This is also the case for objects which are simplified in the reference datum (the roads represented by their central axis). Several strategies exist to account for the presence of objects in the images:

1) instead of conducting random sampling of reference pixels available for validation, sampling can be done by taking into account the size of the objects, so that the confusion matrix represents the mapped scene;

2) one can also use metrics based on shape comparisons or ensemble distances, such as Haussdorf or Fréchet distances;

3) finally, we can use specific measures to compare segmentations, such as Hoover metrics [LAP 03].

However, at present, there is no commonly accepted approach in the community for validation based on objects as is the case for pixel-based approaches.

3.3.2. *Feature extraction*

Here, we are interested in the transformation of image data into relevant information for nomenclature algorithms.

Classification algorithms are used to assign a land cover class for each object (pixel or region) present in the study area. We could simply give as input to these algorithms pixel values (or sequences of values in the case of time series images). However, this approach proves ineffective once the land cover classes to distinguish are complex or very similar to each other. It is therefore often necessary to convert the raw image data so as to facilitate the work of the classification algorithm. This transformation is called feature extraction.

Depending on the type of images available and of the nomenclature of target classes, different types of features can be extracted.

3.3.2.1. *Radiometric indices*

The radiometric indices are quantities computed using combinations of the values of different channels for the same object (pixel or region). In the case of optical imaging, reflectances in different spectral bands (blue, green, red, near infrared, etc.) are used. Calculating radiometric indices therefore first requires the atmospheric correction of the images, that is to say the the transformation of raw digital numbers into surface reflectance values. This can be done by means of radiative transfer models (such as MODTRAN, ATCOR) or *in situ* reflectance measurements synchronous to image acquisition.

These radiometric indices are used to highlight specific properties of the observed surfaces. For example, in the case of multi-spectral optical imaging, the presence of vegetation may be highlighted using the red and near-infrared bands, because the vegetation has high reflectance values in NIR and low in red. Thus, vegetation indices such as the Normalized Difference Vegetation Index (NDVI) have been proposed:

$$NDVI = \frac{NIR-R}{NIR+R} \quad [3.7]$$

where NIR is reflectance in the near infrared; R, reflectance in the red.

The NDVI is close to 1 for highly vegetated areas while it is close to 0 or negative for areas without vegetation. It is thus clear that if the aim is to differentiate vegetated surfaces from the rest, the work of the classification algorithm will be easier if the input information is expressed in NDVI values than if it is in reflectance values in different spectral bands.

The literature offers many vegetation indices (see Chapter 2) using other combinations of spectral bands to correct certain NDVI limitations (saturation phenomenon related to the normalization) or to take into account information in other spectral bands [FER 11].

Beyond the vegetation indices, other combinations of spectral bands are used to highlight other types of surfaces. Thus, there are indices for water, bare soil, built-up areas, etc.

3.3.2.2. *Local statistics*

The indices present in the preceding paragraph are calculated for each pixel of the image or by averaging within a region in the case of the object-based approach. Often, to distinguish certain types of land cover, individual pixel values or averages in a region are not enough. For example, to distinguish a homogeneous vegetation (lawn) from a heterogeneous vegetation (fallow, barren), information on the spatial variability is necessary.

A simple way to characterize this variability is to calculate the variance of the values around each pixel of the image. In general, up to four statistical moments (average, variance, skewness and kurtosis) may be helpful. For example, for quantity x (reflectance in a spectral band, NDVI, etc.) the moment of order p in a neighborhood (window) of $N \times M$ pixels can be calculated like this:

$$\mu_p = \frac{1}{N \times M} \sum_{i=1}^{N} \sum_{j=1}^{M} (x(i,j) - \mu_1)^p \qquad [3.8]$$

and μ_1 is simply the average in the window.

These statistical moments provide a first level of information about the spatial organization in the images. For a more detailed description of the spatial organization, texture coefficients can also be calculated (see pages 183–190 of [TUP 14]).

3.3.2.3. *Temporal features*

Some land cover classes can be distinguished only by their temporal dynamics. This is particularly the case for different types of vegetation: persistent forests in relation to deciduous forests, summer crops compared with winter crops, etc. In these simple cases, a judicious choice of observation dates (an image in the winter and an image in the summer, e.g.) may be sufficient to separate the classes. Often, we are interested in more subtle distinctions, such as corn and sunflower (two summer crops in France) or deciduous beech and oak. In these cases, a choice *a priori* between observation dates is difficult and it is best to work with a time series of high temporal resolution images. The description of a pixel is no longer given by the NDVI in summer and winter, but rather by the series of NDVI values for each acquisition available, once a month for instance.

In the same way as for radiometric indices, changing from a low level representation (reflectance vector) to the closest representation of the desired level of abstraction for interpretation (vegetation content, presence of water), the temporal information can also be summarized by the relevant features.

The first approach that we can think of is the use of statistics: for a pixel for which there is a time series, we can calculate the average of the series and the higher order statistical moments. We can also calculate other typical time series' descriptors in other signal processing areas (Fourier coefficients, wavelet transform, etc.) [MAL 08].

These techniques are of very general scope and can be useful, but in the case of remote sensing for Earth observation, we have *a priori* knowledge on the behavior of surfaces that allow us to develop specific indices. For example, we know that the surfaces that have a significant and steady dynamic correspond to vegetation. In this case, we can focus on the description of the time series of a vegetation index and describe the temporal profile to an appropriate representation level. It is thus customary to describe vegetation by some basic parameters that can be derived from a temporal NDVI profile: the cycle start date (emergence of crops, breaking in of the forest), the growth rate, maturity date, length of maturity, senescence date, the rate of senescence, and harvest date (for crops). A common way to model the annual vegetation cycle is to use a double sigmoid function:

$$\mathrm{f(x)} = \mathrm{A}\big(\mathrm{f}_1(\mathrm{x}) - \mathrm{f}_2(\mathrm{x})\big) + \mathrm{B} = \mathrm{A}\left(\frac{1}{1+\frac{e^{x_0-x}}{x_1}} - \frac{1}{1+\frac{e^{x_2-x}}{x_3}}\right) + B \qquad [3.9]$$

the graphic representation is given in Figure 3.3 for values $x_0 = 75$, $x_1 = 7$, $x_2 = 250$, $x_3 = 10$, $A = 0.95$, and $B = 0.1$. The interpretation of these parameters is as follows:

– x_1 and x_3 are the ascending and descending slopes respectively;

– x_0 and x_2 are the dates of maximum slopes;

– A is the amplitude of the profile;

– B is its minimum value;

– t_0 is the emergence date or the start of vegetation;

– t_1 is the maturity date of the vegetation;

– t_2 is the start of senescence;

– t_3 is the end of senescence;.

From the parameters of the double sigmoid, it is easy to estimate the relevant dates. For example, the starting date (t_0) can be obtained by calculating the upward slope, and its intersection with the minimum value of the profile:

$$t_0 = \frac{mx_0 - g(x_0)}{m} = x_0 - \frac{g(x_0)}{g'(x_0)} \qquad [3.10]$$

Similarly, we can get the maturity date:

$$t_1 = \frac{A + B - (g(x_0) - g'(x_0)x_0)}{g'(x_0)} \qquad [3.11]$$

Other key dates are similarly obtained. The only difficulty lies in estimating the parameters of the double sigmoid from the NDVI time series. This problem can be solved using numerical libraries to optimize the parameters of the function.

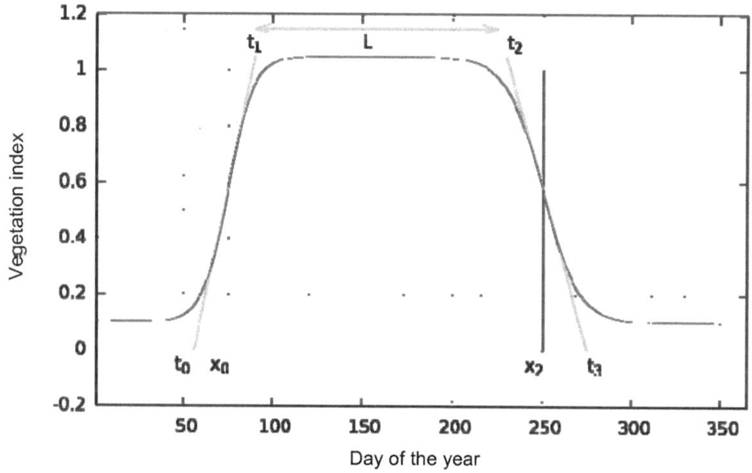

Figure 3.3. *Model of the growth cycle by a double sigmoid. Evolution of a generic vegetation index (NDVI for example) depending on the day of the year*

3.3.2.4. *Exogenous data*

We have seen in previous paragraphs how to transform the given image to provide information relevant to the identification of the characteristics of different observed surfaces. The objective of this transformation step was to simplify the task of algorithms producing land cover maps.

Also, for the same objective, we can use any available information outside of images that could be capable of contributing to the separation of land cover classes. We know, for example, that the altitude and sun exposure of a surface have an impact on the type of vegetation that can develop in a given area. These same parameters are correlated to the presence of buildings or roads. Soil type also plays an important role in terms of vegetation types (soil depth, useful water reserve for vegetation, rich in minerals, etc.). The local climate is also decisive in the presence of certain land cover classes.

If this information is spatially available, that is to say in map form, it can be used as features in the same way as the indices calculated from the images. It is therefore common to use digital terrain models (DTM) to calculate altitude, slope and exposure of each pixel. The DTM data are available almost anywhere in the world thanks to space missions such as SRTM and ASTER. Soil maps with information on the type of soil are less easily available and less often used. Climate information (seasonal or annual summaries of temperature, radiation, rainfall, derived from meteorological data) vary in availability according to the intended map scale.

Beyond this information on physical quantities, it is possible to use other information provided by geographic information systems. These can belong to categories such as distance to a road or a housing area (to, as appropriate, distinguish between natural vegetation and agricultural land), the distance to water surfaces (to assess the likelihood that an agricultural area is irrigated), population density, etc.

3.3.2.5. *Feature selection*

Among all features that it is possible to calculate, not all are useful and some are redundant. The use of many features to produce a land cover map can be problematic for two reasons. Firstly, the computation time can be important for the production of these features, but also subsequently in the procedure of map production where large volumes of data should be used.

The second problem may arise in the use of certain data classification algorithms whose quality may decrease significantly when the samples requiring classification are characterized by a large number of attributes. It is therefore useful to properly select the features that are required by the classification algorithm. The difficulty lies in obtaining the minimum set of features able to keep the maximum amount of information. Several approaches for feature selection are detailed below.

The first approach is to produce multiple land cover maps with different subsets of features, then compare validation metrics (section 3.3.1). The problem with this approach is that its cost is often prohibitive. Indeed, an exhaustive search of all combinations of features may be impossible to implement. We must therefore approach the problem differently.

Another way to determine the relevance of a feature is to look at the degree of correlation with the land cover classes. We can, for example, analyze the variance of a feature when computed for each class independently. For example, if the calculated NDVI variance on any set of pixels is larger than the variance calculated by class, the vegetation pixels may be used without being mixed with the non-vegetation pixels. More formally, we can use the correlation ratio:

$$\eta^2(I|J) = \frac{\mathrm{Var}(I)-\mathrm{Var}(E[I|J])}{\mathrm{Var}(I)} \qquad [3.12]$$

which shows how the variance of a Var(I) feature decreases when computed only on the samples of class J, Var(E [I|J]). By using this calculation for all classes, we can use the average correlation ratio to identify the features that provide the most information: those for which the correlation ratio is highest.

Among the most relevant features, some may be redundant. For example, vegetation indices are often very relevant, but two different vegetation indices provide very similar information. Redundancy between features can be measured by means of a correlation coefficient or any other statistical dependence measure.

One of the limits of univariate selection approaches (applied to each feature individually) is that they do not take into account the usefulness of several feature combinations. Indeed, two features in isolation may be somewhat helpful, but their combined use can be very powerful. In these cases, techniques based on principal component analysis [SRI 83] may be interesting.

Finally, very sophisticated techniques based on the stepwise addition or subtraction of features exist, but they are highly specialized and cannot be detailed in this book. For more details, the reader may consult [BIS 06], pages 55–67.

3.3.3. *Classification methods*

The heart of the land cover map production process is the classification stage. This step is used to assign each pixel or region of the image to the following land cover class. This allocation is a decision function *F* using the feature vectors to infer class *C*:

$$F:\vec{\Theta}\rightarrow C \qquad [3.13]$$

The difficulty of the procedure lies in the construction of the *F* function yielding small errors in the sense of the metrics presented in section 3.3.1.

While in some extremely simple cases, the classifiers can be constructed by experts, most problems require automatic methods using machine learning. The rare cases where the classification algorithm can be constructed manually are met when the number of classes is small and easily separable. In these cases, one can construct simple decision trees or small sets of rules operating on feature threshold. For example :

– if the average NDVI of April is less than 0.3, there is a summer crop;

– otherwise, if the average NDVI for the months of February and July is greater than 0.6, then it is a forest.

Understandably, this approach will be long and tedious to implement for more than a few classes. It will also be very difficult to reliably determine thresholds values. In practice, even for relatively simple cases, we turn to machine learning methods.

A detailed description of the machine learning theory for the classification is not possible in this chapter. The reader may consult specialist works such as [BIS 10] and [BIS 06] for further coverage of the subject. A synthetic treatment in the processing of remote sensing images is proposed in Chapter 5 of [TUP 14]. We will limit ourselves here to explaining the principles that guide the choice of approaches.

Classification methods are divided into two major groups, "supervised" and "unsupervised" methods. We will call the pixels or regions to be classified "individuals". These individuals are characterized by feature vectors.

Unsupervised methods, also called clustering methods, work on feature values to group individuals by similarity. The resulting groups form clusters. Once these clusters are obtained, the reference data can be used to classify them (assign semantics). These are the classes. In the case where there are more clusters than land cover classes, we group clusters in which there is no reference data with the closest classified clusters.

The difference between the various unsupervised classification methods lies in the type of similarity measure used to group the individuals and also in additional constraints added to avoid certain combinations. Thus, in the *k*-means algorithm the similarity between individuals is measured using the Euclidian distance on the features vector Θ with the additional constraint of minimizing the variance of the individuals within a cluster while maximizing the distance between the center of clusters. There are also statistical methods using probability densities, such as the EM algorithm (Expectation Maximization [BIS 10]).

This type of method is very easy to implement. However, there is no guarantee that the clusters obtained correspond to the desired thematic classes. Indeed, some clusters can contain multiple thematic classes, or a group can be represented by multiple clusters. While the latter does not pose a major problem, the first cannot be safely corrected. We can always apply the algorithm by finding an increasing number of clusters, but the separation of groups after the research is not guaranteed.

One way to ensure that all thematic groups of land cover are taken into account by the classification algorithm is to use the knowledge on the assignment of each individual of a given class during the learning process. This is what is done in supervised methods. In such methods, we attempt to directly construct the decision function that best separates individuals according to their affiliated class. They are therefore more efficient than unsupervised methods, but they require more reference data and require more thematic (knowledge and observation reference sites) and technical (georeferencing of sites, construction of learning algorithm) proficiency. Indeed, in the case of unsupervised methods, just a few reference samples are needed to label the clusters (a posteriori use of the reference data). In the supervised methods, all samples used for learning must be labeled.

Among the supervised classification algorithms commonly used for the production of land-cover maps, we can list the Random Forests (RF) [BRE 01], the Support Vector Machines (SVM) [CRI 00], [BIS 06] or multilayer perceptron neural networks [BIS 06], and maximum likelihood approaches [BIS 06].

The choice between a supervised or unsupervised approach is often determined by the amount and quality of available reference data. If these are few and of low quality (high probability that individuals are mislabeled), unsupervised methods will the chosen. If the reference data are of good quality but in small quantities, we will choose SVM type approaches. These methods are also very robust when the number of features used is high. If we have a large amount of reference data, but they contain a high noise level, RF type methods are often more efficient. Neural networks often require a very high amount of learning data, which limits their use in remote sensing. The maximum likelihood methods, widely used before the onset of SVM and RF, are still applied when the number of features is reduced.

In recent years, the use of semi-supervised methods that combine both families of approaches have been proposed for the cases where reference data are scarce [VAN 09].

Even if we can give general recommendations on the choice of methods depending on the type of data available, we prefere to compare different approaches for the same problem before making a final choice.

Finally, it is always possible to use several classification methods and to combine the results. The simplest approach is to implement a majority rules vote: the class assigned to each individual will be that which was produced by the majority of classifiers used. Other more sophisticated fusion approaches taking into account the individual performance of each classifier can be used [HAL 97, BIS 06].

Another post-processing method often used after classification is spatial regularization. This consists of applying spatial coherence rules to reduce the number of misclassified pixels. A simple version of this processing is to replace the class of isolated pixels (belonging to a class not shown in their close vicinity) by the majority class in their neighborhood. More complex techniques taking into account the spatial correlation between classes or shapes of objects can also be used.

3.3.4. *Change detection*

One of the main advantages of remote sensing compared with other mapping approaches is the ability to frequently repeat observations with the same perspective. This allows the implementation of land cover maps at regular intervals, providing a useful temporal depth for many uses.

In the case of land cover maps, comparisons between maps produced with data acquired at different periods allow for analysis of land cover changes. Thus, in the case of the study of urban sprawl, we can compare the built-up areas or the roads between two maps produced 2 or 3 years apart. In the case of scrub encroachment on former farmland, maps separated by 5 or 10 years will be useful. In some jurisdictions or for certain land cover classes, higher frequencies may be required (natural disasters, countries with high population growth). In most cases, in addition to the time difference between two maps, the reference date with respect to which changes are compared is also very important. It seems, therefore, that producing land cover maps with an annual frequency can be useful in most uses.

The usual approach for monitoring land cover changes is that of post-classification comparison. This approach consists of mapping the class changes. Thus, the result is a map with a nomenclature that represents the transitions between classes at time t and those at time $t + 1$: agricultural to urban, forest to agriculture, etc.

3.4. Use examples

In this section we illustrate examples of land cover maps produced by satellite imagery classification.

3.4.1. *Generic land cover mapping*

By generic mapping, we mean one that uses a nomenclature that is not specific to a particular application. This type of maps is suitable for many uses where the user wishes to know the extent and location of large groups of land cover. The different types of generic land cover maps are distinguished primarily by the spatial resolution (where the choice determines the size of the objects it is possible to detect) and the extent of the mapped area. Often there is a tradeoff between these two characteristics. We will give two examples of these compromises.

3.4.1.1. *Mono-date images at high spatial resolution*

This example illustrates the use of a Pleiades HR multi-spectral image (2.80 m resolution sampled at 2 m and 4 visible and near infrared spectral bands). The high resolution imaging is necessary to distinguish small objects such as roads and hedges (trees or shrubs). However, as explained in section 3.2.1, it is difficult and expensive to obtain several images of the same area at different times to highlight the temporal behavior of surfaces. In this case, it is difficult to distinguish deciduous tree forests from evergreen trees. Similarly, it is not possible to characterize the agricultural land precisely at the level of crop classes, as they are characterized by a succession of vegetation cover and bare soil.

To obtain different vegetation classes, the texture features (similar to local statistics presented above) are very relevant for high spatial resolution imaging. In addition to spectral features, they make it possible to distinguish herbaceous, shrub and tree vegetations.

Figure 3.4 illustrates the type of results that can be obtained with this type of approach. Note the accuracy of the detection of fine landscape elements. The average FScore for this map is 0.77.

3.4.1.2. *Multi-date images at medium spatial resolution*

Some vegetation classes can be distinguished by their temporal behavior. For this, it is necessary to use a series of images covering at least 12 months in order to observe a complete phenological cycle. The satellites that provide systematic temporal revisits have decametric spatial resolutions.

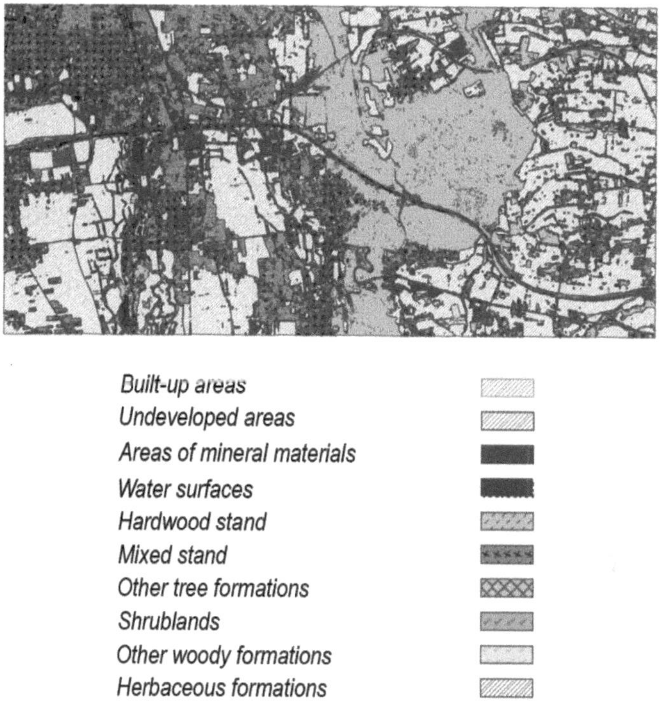

Figure 3.4. *A generic land cover map from mono-date high spatial resolution imaging. Illustration derived from [MAL 08] on the Tarbes region (France), latitude 43.2218°N, longitude 0.1197°E. For a color version of this figure, see www.iste.co.uk/baghdadi/3.zip*

This approach is illustrated here using Landsat images acquired from January to December 2010 with a revisit of about 16 days. The systematic revisit also provides images over large areas. The example chosen here covers all the Pyrenees (France) and its foothills on an area of 500 km × 200 km.

The nomenclature chosen here (Figure 3.5) reflects the temporal dynamic because it distinguishes summer crops, winter and woody crops (permanent), grasslands, and two forest types. These latter are classified according to their phenology and not according to the leaf type as was the case in the previous mono-date example. However, because of the low spatial resolution (30 m), the fine landscape elements (roads, hedgerows, etc.) are missing from the map. The overall accuracy of this example is 0.75.

The arrival of new satellites, for example, the family of Sentinels of the European Space Agency, such as Sentinel-2, with a spatial resolution of 10 m and temporal revisit every 5 days, is expected to improve the quality and the detail of produced maps.

Finally, note that it is possible to combine the mono-date high spatial resolution approach with the multi-temporal approach to produce more detailed maps. This, however, requires delicate work to ensure data coherence (especially the geometric superposition).

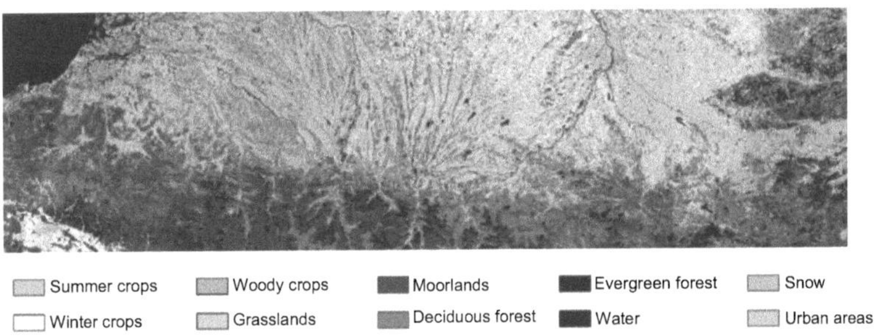

Figure 3.5. *Land cover map produced from Landsat images acquired during 1 year over the Pyrenees area (France). latitude 43.2732°N, longitude 0.5246°E. For a color version of this figure, see www.iste.co.uk/baghdadi/3.zip*

3.4.2. *Detailed mapping of agricultural crops*

The two examples above illustrate generic land cover mapping. For some applications, maps with a very detailed nomenclature for certain land cover classes families are required, while many other classes of the generic nomenclature can be ignored.

To illustrate this use case, we have chosen the mapping of agricultural land. In this case, all non-agricultural land can be grouped, but within agricultural areas, we want to know the exact type of crop for each plot. Specifically, the objective is to produce a map of the main annual crops in the area (wheat, barley, rape seed, corn, and sunflower). To achieve this degree of fineness (in the previous example, these five classes were regrouped into two classes), we need:

– a high spatial resolution to distinguish each agricultural plot;

– a high temporal resolution to adequately characterize periods of growth, maturity, and senescence of each crop.

However, this is often not enough, because crops such as winter wheat and winter barley are very similar at all levels. In this case, we must introduce *a priori* information into the classification process, ensuring that the considered varieties are morphologically separable in the field. A crop implanted 1 year on a plot is normally chosen based on established crops in previous years (crop rotation). If this knowledge is available, it can be used as a primary input of the classifier.

Figure 3.6 shows a map of crops obtained from Formosat 2 [POU 11] images. The map was produced using 1 year of Formosat-2 acquisitions (2012, 8 m spatial resolution and four spectral bands in the visible and the near infrared). The features used for the classification are spectral indices for each date, as well as information on the present crops in each plot over the previous 3 years. The overall accuracy of the map is 85%.

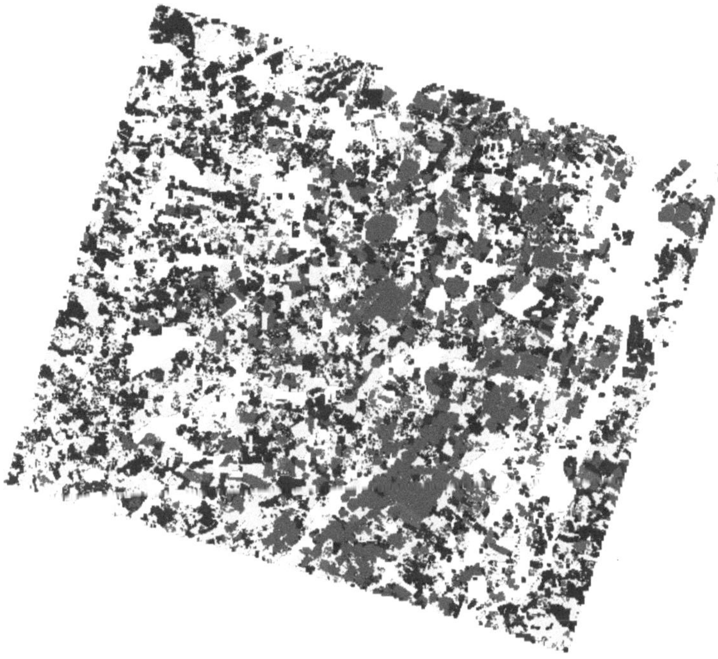

Figure 3.6. *Map of annual crops produced in 2012 from 1 year of Formosat-2 acquisitions on a site covering 400 km² in the southwest of Toulouse (France), latitude 43.4509° N, longitude 1.1501° N. Wheat is in yellow, corn is in red, green is barley, rape seed is purple, and sunflower is brown. The white areas correspond to non-agricultural land. For a color version of this figure, see www.iste.co.uk/baghdadi/3.zip*

3.5. Key points

Land cover maps are needed for various purposes ranging from land cover planning to their use in models that describe environmental processes (water cycle, climate).

The legend of the land cover maps (choice of classes) must be adapted to the intended thematic application. To achieve good quality, the land cover maps are produced from appropriate imaging data in terms of spatial, spectral, and temporal resolutions. Various classification methods allow us to produce these maps. They require reference data to be calibrated and validated. Therefore, the availability, quality, and the thematic and spatial representativeness of reference data should not be neglected.

The new missions with high spatial, temporal, and spectral resolution such as Sentinel-2 are expected to achieve new levels of quality to date. The appropriate combination of these data with high spatial resolution imagery may help to refine certain themes of the nomenclature (dense urban areas, hedges, small streams).

3.6. Bibliography

[BEN 12] BENHADJ I., DUCHEMIN B., MAISONGRANDE P, *et al.*, "Automatic unmixing of MODIS multi-temporal data for inter-annual monitoring of land use at a regional scale (Tensift, Morocco)", *International Journal of Remote Sensing*, vol. 33, no. 5, pp, 1325–1348, 2012.

[BIS 06] BISHOP C.M., *Pattern Recognition and Machine Learning*, Springer-Verlag, New York, 2006.

[BRE 01] BREIMAN L., "Random forests", *Machine Learning*, vol. 45, no. 1, pp, 5–32, 2001.

[CRI 00] CRISTIANINI N., SHAWE-TAYLOR J., *An Introduction to Support Vector Machines and Other Kernel-based Learning Methods*, Cambridge University Press, 2000.

[FER 11] FÉRET J.B., FRANÇOIS C., GITELSON A. *et al.*, "Optimizing spectral indices and chemometric analysis of leaf chemical properties using radiative transfer modeling", *Remote Sensing of Environment*, vol. 115, no. 10, pp. 2742–2750, 2011.

[GIR 10] GIRARD M.C., GIRARD C.M., *Traitement des données de télédétection*, Dunod, Paris, 2010.

[HAL 97] HALL D.L., LLINAS J., "An introduction to multisensor data fusion", *Proceedings of the IEEE*, vol. 85, no. 1, pp. 6–23, 1997.

[LAP 03] LAPORTERIE F., FLouzat G., "The morphological pyramid concept as a tool for multi-resolution data fusion in remote sensing", *Integrated Computer-Aided Engineering*, vol. 10, no. 1, pp, 63–79, 2003.

[LEG 96] LEGROS J.P., *Cartographie des sols : de l'analyse spatiale à la gestion des territories,* Presses polytechniques et universitaires romandes, 1996.

[MAL 08] MALLAT S., *A Wavelet Tour of Signal Processing*, Academic Press, 2008.

[POU 11] POULAIN V., INGLADA J., SPIGAI M. *et al.,* "High-resolution optical and SAR image fusion for building database updating", *IEEE Transactions on Geoscience and Remote Sensing*, vol. 49, no. 8, pp, 2900–2910, 2011.

[ROB 10] ROBIN A., MOISAN L., LE HEGARAT-MAscle S., "An a-contrario approach for subpixel change detection in satellite imagery", *IEEE Transactions on Pattern Analysis and Machine Intelligence*, vol. 32, no. 11, pp, 1977–1993, 2010.

[SRI 83] SRIVASTAVA M.S., CARTER E.M., *Introduction to Applied Multivariate Statistics*, Elsevier, 1983.

[TUP 14] TUPIN F., INGLADA J., NICOLAS J.M. (eds), *Remote Sensing Imagery*, ISTE John Wiley & Sons, New York, 2014

[VAN 09] VANDEWALLE V., Estimation et sélection en nomenclature semi-supervisée, PhD thesis, Lille University of Science and Tehcnology, Lille I, 2009.

[VAU 14a] VAUDOUR E., BAGHDADI N., GILLIOT J.M., "Mapping tillage operations over a peri-urban region using combined SPOT4 and ASAR/ENVISAT images", *International Journal of Applied Earth Observation and Geoinformation*, vol. 28, no. 1, pp. 43–59, 2014.

[VAU 14b] VAUDOUR E., NOIROT-COSSON P.E., MEMBRIVE O., "Apport des images satellitaires de très haute resolution spatial Pléia des à la caractérisation des cultures et des operations culturales en début de saison", *Revue française de photogrammétrie et de télédétection*, vol. 208, pp.97–103, 2014.

4

Contribution of Remote Sensing for Crop and Water Monitoring

4.1. Introduction

Agriculture brings with it a number of issues: agricultural production and food security, water and soil resource conservation, limiting the impact of farming on the quality of our environment (water, air, soil). In the context of global changes and the challenges they pose for agriculture sustainability, our ability to characterize how croplands function in terms of water, carbon and particle fluxes is crucial. Developments of agro-ecosystems modeling, including their interactions with the atmosphere and the anthropogenic factors are valuable tools for progress in this direction. Remote sensing, with the high variety of spectral ranges and the fine spatial and temporal resolution currently available, is a tool of great value for various applications in agriculture. The availability of robust inverse methods that allow surface biophysical variables to be assessed, combined with modeling approaches, makes it a high performance tool. The major contributions of remote sensing include:

– its ability to cover large stretches of land and to provide information on the various land uses and practices generated by agriculture. These uses and practices are important to know, both for census purposes

---

Chapter written by Dominique COURAULT, Valérie DEMAREZ, Martine GUÉRIF, Michel LE PAGE, Vincent SIMONNEAUX, Sylvain FERRANT and Amanda VELOSO.

(agricultural statistics, agri-environmental monitoring, etc.) and for modeling the behavior of agro-hydrosystems;

– providing frequent variables characterizing soil and vegetation properties that allows us to monitor the status of crops, their production potential, their irrigation requirements. This monitoring is a highly strategic issue, both for forecasting purposes, food security and good resource management;

– the possibility, from the same information, of assessing the contribution of agricultural lands to net emissions of CO_2 and other greenhouse gases (GHGs); this assessment is essential for proposing alternative agricultural scenarios for mitigating the contribution of croplands to climate change;

– thanks to the fine spatial and temporal resolution of information, the possibility of providing a decision support for farming activities according to the intra-field variability (precision farming). This dimension represents an important lever for enabling agricultural systems to achieve better efficiency and economical use of inputs for an agriculture that respects the environment.

Before presenting the structure of this chapter, it is important to remember the definition of a few key terms that will be used often throughout the chapter.

Land uses include both land cover, that is to say, the type of vegetation or crop types (see Chapter 3) and associated agricultural practices. Agricultural practices include a set of operations carried out by the farmer from plowing and soil preparation, sowing the main or intermediate crops, various applications of fertilizers, various phytosanitary treatments (herbicides, fungicides, etc.), irrigation with determination of dates and doses, the harvest and crop residue management. These practices vary according to region, climate, soils and production systems (cereals, meat, milk, vegetables, etc.). They therefore have a high spatial variability that is impossible to know with precision and to regularly monitor over a long period because of the many factors involved. Satellite images, particularly from high spatial and temporal resolution sensors, allow us to detect some practices at the plot level. Among the sensors with spectral bands in the optical domain, we can cite FORMOSAT-2, which can provide daily images from the same angle of view with a spatial resolution of 8 m for certain areas, the ESA Sentinel-2 tandem, which has a spatial resolution of 10 m and a repeatability of 10 days with Sentinel-2A (launched in June 2015) and 5 days after the launch of Sentinel-2B (in 2016), the constellation of five Rapid-Eye satellites which can observe any area of the globe within 24 hours, and cover the entire globe in just 8 days with a resolution of 5 m (https://earth.esa.int/web/guest/missions/3rd-party-missions/current-missions/rapideye) and among the sensors operating in the field of microwaves, TerraSAR-X and COSMO-SkyMed, which are able to obtain images with a very fine spatial resolution [DED 10, ELH 14].

Two types of approaches exist to describe cultural practices based on:

– direct methods relying on the analysis of spectral signatures of the fields;

– indirect methods that rely on the inversion of functioning models through data assimilation.

These two approaches are illustrated in the different sections of this chapter.

Crop monitoring often refers to its phenological stage. A succession of events defines the plant's "internal clock" and allows us to track its development from bud-break to senescence, through various levels of "greenness" characterizing the status of vegetation and the accumulation of biomass in different organs. These stages are highly variable in space, according to the crop management and interactions with the climate. Many studies have shown an interest in using radiometric indexes such as the normalized vegetation index (NDVI) [TUC 79], the normalized difference water index (NDWI), the global vegetation index (GVI), or the enhanced vegetation index EVI [ROC 09]. The NDVI has many advantages: it is a robust and reliable indicator [PRI 13] and the spectral bands used for its computation are available on all optical satellites (see Chapter 2). It is therefore widely used in the scientific community and its simple formulation makes it easily accessible to non-specialist remote sensing users. Figure 4.1 shows NDVI maps derived from Duchemin *et al.* [DUC 09] and Chirouze [CHI 13], obtained over a small 4 km × 4 km agricultural sector north of Mexico in the Yaqui Valley (27.263°N, 109.892°W), observed every 3 days by the FORMOSAT-2 satellite.

On these maps, each field can be identified thanks to the fine spatial resolution of the images (8 m). Winter wheat is predominant. Its development is limited at the first stages in November–December (the blue color represents low NDVI values). Leaf growth starts in January–February, increasing the percentage of green color detected by the satellite in each pixel of the images, the maximum growth is represented in red (the highest NDVI values).

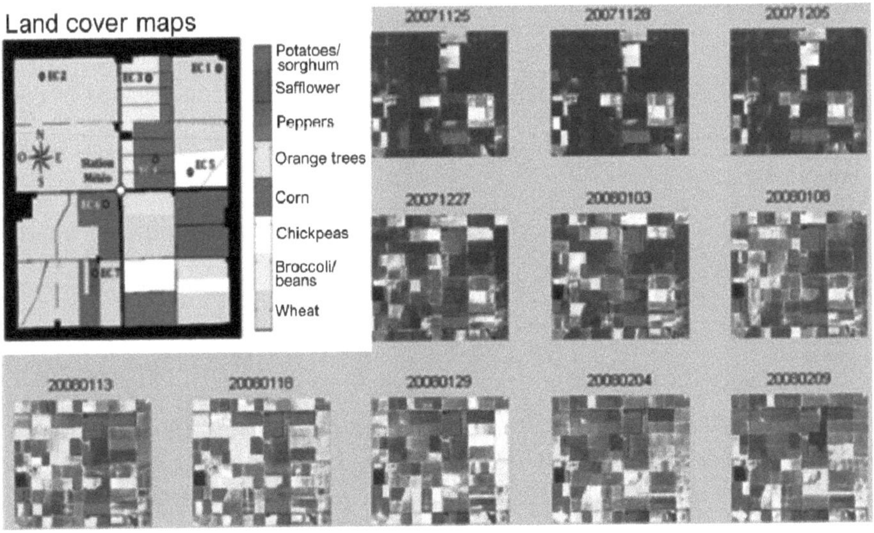

Figure 4.1. *Land use map of a small 4 x 4 km agricultural area located in the northwest of Mexico [CHI 13], and NDVI maps obtained from FORMOSAT-2 images acquired in 2007 and 2008 [DUC 09]. The more the crops are green and well developed, the brighter the colors and they tend to be yellow-red (high NDVI values). Inversely, bare soils or surface with low vegetation appear dark and blue. For a color version of this figure, see www.iste.co.uk/baghdadi/3.zip*

This first example already shows the potential of remote sensing data to provide first level spatial and temporal information for crop monitoring.

This chapter aims to detail the different methods of using this data for various applications. Since the range of uses is very wide, we have restricted ourselves to examples obtained with optical domain data, which are the most commonly used. Other examples can be found in chapters on radar (see Chapter 6) and the thermal infrared domains [LAG 16]. We have chosen to illustrate the most common themes:

– presenting in the first part simple indicators which allow crop development monitoring;

– then introducing some agricultural practice detection examples, with a particular focus on irrigation practices;

– the estimation of the surfaces and water needs of crops has been largely investigated, because it should allow to optimze water resource;

– different agricultural production estimation methods based on the combination of models and satellite data are presented in section 4.5;

– the use of high spatial and temporal resolution sensors to support crop management at plot scale (precision agriculture) is the subject of section 4.7;

– finally, the chapter concludes with some examples of estimation of the environment impact of agriculture.

4.2. Indicators for crop monitoring

Remote sensing is a very valuable tool for tracking development stages and crop growth. The analysis of the dynamics of these states allows us:

– to characterize and compare situations between different crop management choices and weather conditions and to assess their impact on growth;

– to better characterize some specific agricultural practices that may lead to more or less progressive (such as an annual crop sowing) or sudden (such as grazing or cutting grassland) changes in growth (in section 4.3).

Some indicators can be calculated from NDVI temporal profiles (Figure 4.2) such as the emergence date, the rate of crop development, the date of the maximum NDVI value, the beginning of the senescence stage and the cycle's duration [LAR 14]. To determine these key stages and these parameters, NDVI profiles must first be interpolated to obtain daily NDVI values [BAD 84, GUE 96a, LAR 14]. Gaussian or logistics functions are most commonly used to perform this temporal interpolation.

Figure 4.2 shows an example of an NDVI interpolated profile, calculated from FORMOSAT-2 images acquired every five days on a corn plot in the southwest of Toulouse (France). The temporal interpolation model shown here is a double logistic function that allowed us to reproduce the peak of growth of the main crop and to derive the following key parameters: emergence period, cycle duration, maximum growth and senescence. Changes in the NDVI in the early stages corresponding to winter re-growth are not considered. The sowing period can be estimated when the season is over, or earlier if the number of available images in the early cycle is large enough.

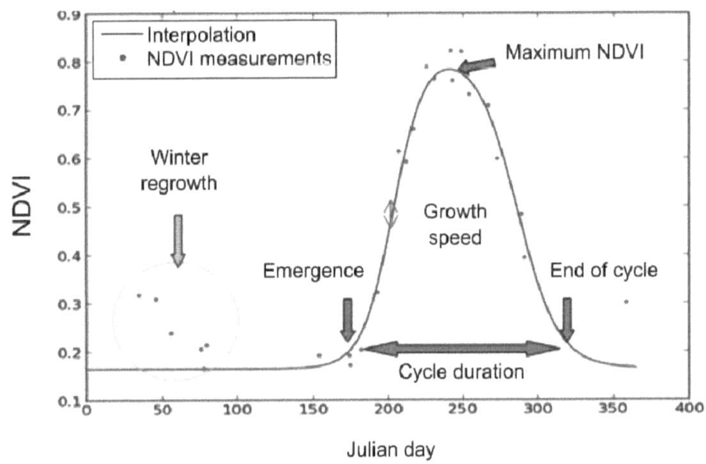

Figure 4.2. *Detection of the main development phases of a corn plot obtained from the analysis of a temporal NDVI profile derived from FORMOSAT-2 images. The interpolation was done from a double logistic [BOI 15]*

Figure 4.3 shows examples of indicators (cycle duration, maximum NDVI) obtained from the NDVI profiles, extracted from agricultural plots in the southwest of Toulouse (France). These indicators differ depending on the crop type (corn, soybean, sunflower, sorghum) and allow us to detect some practices such as irrigation. We can see, for example, that the maximum NDVI value differs significantly between irrigated (silage and irrigated grain corn, soybean) and rainfed (sunflower, sorghum, unirrigated corn) crops the cycle duration helps to descriminate between silage and grain corn.

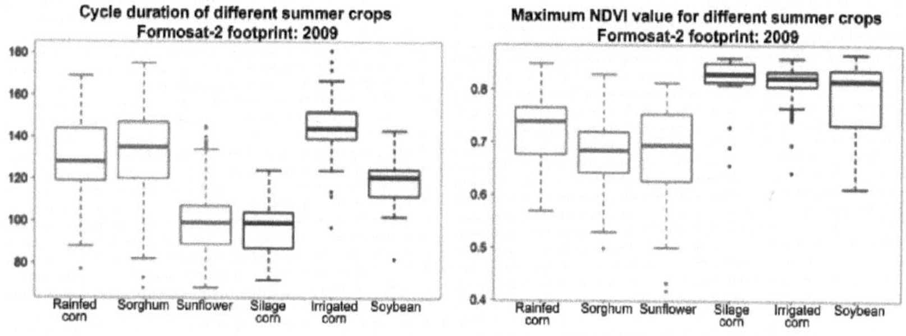

a) b)

Figure 4.3. *Indicators derived from the analysis of an NDVI profile calculated for different fields from FORMOSAT-2 images acquired in 2009 on an agriculture area in the southwest of Toulouse (France). a) Cycle duration (in days); b) maximum NDVI on the main summer crops [LAU 13]. For a color version of this figure, see www.iste.co.uk/baghdadi/3.zip*

These indicators characterize the crop behavior that differs depending on the species, the climate (dry or wet year) and the agricultural practices (e.g. silage or not, irrigated or rainfed).

4.3. Indicators of agricultural practices at the territory level

Analysis of farming systems, that is to say, the crop types, rotation and farming practices in sustainable, organic or conventional agriculture, are all factors to be considered when interested in estimating yields. When comparing the performance of different farming systems to optimize the available resources, it is important to identify the different crops and the associated practices. The first step is crop type identification.

4.3.1. *Crop type identification and succession*

There are different methods to provide the land use classification. They are summarized in Chapter 3. Monitoring the crop succession on agricultural lands allows us to establish typologies based on land cover dynamics, themselves linked to the cropping system. This can be done through statistical data mining methods leading to 2D segmentation (space and time) or spatial-temporal segmentations (Carottage method [MAR 13]). Figure 4.4 illustrates the application of this method on a watershed located in Brittany (France). Classifications with segmentation were performed on past multi-year high spatial resolution images (SPOT-HRV, IRS-Liss, etc.) throughout the watershed leading to a typology of temporal dynamics and a mapping of of cropping systems.

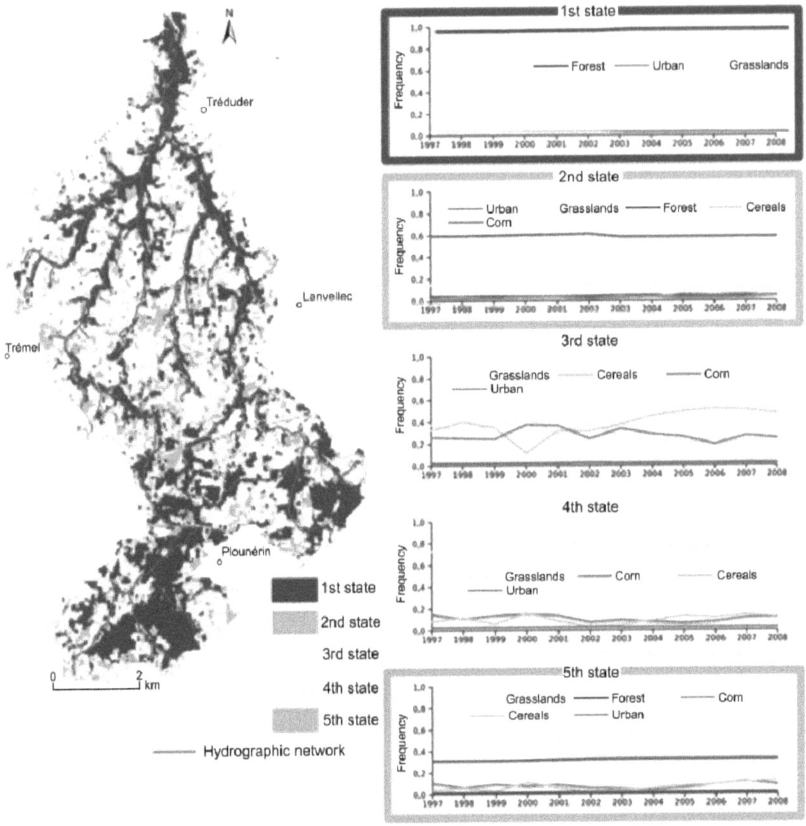

Figure 4.4. *Map of types of agricultural land temporal dynamics in connection with the crop systems. Watershed case of Yar in Brittany. The territory is segmented into five types ("States") of land use (LU) dynamics and corresponds to the five colors of the map. The main LU groups are the grassland, corn, wheat grain type, urban and forest. Each state is described by a diagram of the LU succession evolution observed six times from 1997 to 2008. States 1 (shown in dark green on the map) and 2 (gray) correspond to stable units in time (mainly covered by forests and urban). The other three states (3–4–5) have a greater LU diversity and a more pronounced temporal variation. State 3 (yellow on the map), the grassland, corn and cereals evolve the same way until the 3rd period, then the meadow and corn are gradually replaced by cereals, denoting a change in the crop system. States 4 and 5 represent stable areas predominantly covered with grasslands and forests (source: [MAR 13]). For a color version of this figure, see www.iste.co.uk/baghdadi/3.zip*

4.3.2. *Characterization of agricultural practices*

4.3.2.1. *Soil tillage*

Detecting plots that have just been plowed, sown or irrigated is useful information for crop models used to estimate yield, water or carbon fluxes, which will be described below (section 4.5) or for the agro-environmental management of irrigated areas. Figure 4.5 shows a map that distinguishes these different operations on a Moroccan site. It was obtained from the combination of FORMOSAT-2 and Envisat radar data/ASAR images. It results from the analysis of the color composite made from the reflectance measured in the red spectral band on three different dates: December 4, 8 and 12, and two radar images. The spectral signatures of different surfaces differ significantly, especially deep plowing, showing reflectance values much lower than those of soils that have been sown. Deep plowing tends to increase the surface roughness, and surface moisture increases compared with an undisturbed soil. These two factors have the effect of decreasing the reflectance. Inversely, when a soil is prepared for sowing, generally, the surface is characterized by a lower roughness, it tends to dry up quickly and the reflectance is slightly higher than the same freshly plowed soil [HAD 09]. Adding information from radar images allows us to detect the irrigated land from unirrigated land. A classification can then be proposed from these image combinations.

Initial state	Detection period	Type	Number of plots	Legend
Unsown and untilled	10 – 12	Tillage	4	
	13 – 16		1	
	Unplowed plots		38	
Tilled	8 – 12	Tillage	6	
	12 – 16		9	
	Untilled plots		16	
Sown	12 – 13	Irrigation	4	
	13 – 16		2	
	Non-irrigated plots		44	

Figure 4.5. *Maps of surface states and agricultural operations (e.g. deep plowing, sowing, irrigation) on an agricultural region in Morocco obtained from three FORMOSAT-2 images and two ENVISAT/ASAR images acquired at the beginning of the season, December 4–12, 2005. The detection periods noted in the table correspond to the periods in which the cultural practices noted in column 3 were made [HAD 09]. For a color version of this figure, see www.iste.co.uk/baghdadi/3.zip*

4.3.2.2. *Emergence period*

The emergence period during which the plantlet's first leaves arise from the germinated seedling up to the soil surface can be estimated from the analysis of the temporal NDVI profiles as shown in Figure 4.2. Figure 4.6 shows a map of the estimated emergence dates for summer crops (corn, soy, sunflower) from FORMOSAT-2 images (2009). Emergence dates are spread mainly between late March (Julian Day "JD" 90) and the end of July (JD 200). The spatial resolution of this map is 8 m, which allows us to access intra-field variability. Significant spatial heterogeneity is sometimes visible within the same plot. Such disparity is especially observed in sunflower plots (zoom). This variability may be because of the heterogeneity of the soil, or sowing and/or irrigation. These maps allow us to obtain dynamic and spatial information on the crop development schedule.

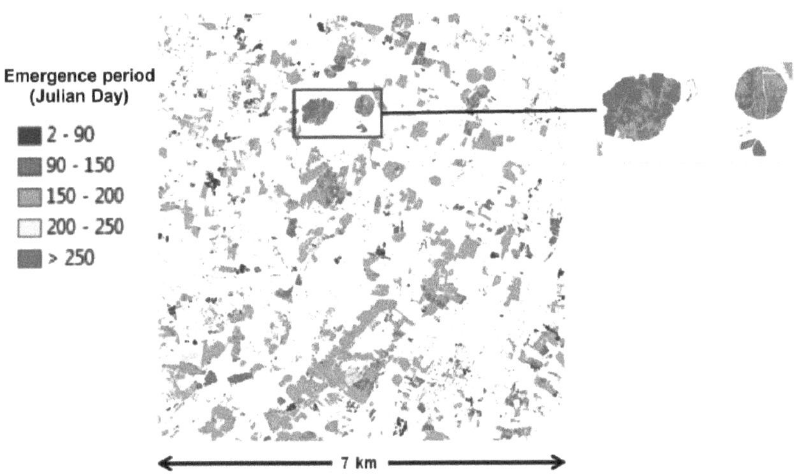

Figure 4.6. *Map of the emergence dates expressed in Julian Day Number for summer crops (corn, soy, sunflower) obtained from FORMOSAT-2 images acquired in 2009. The white areas correspond to winter crops, bare soils, wood, urban surfaces and water [BOI 15] (images @ Demarez). For a color version of this figure, see www.iste.co.uk/baghdadi/3.zip*

4.3.2.3. *Harvest dates*

From the analysis of the temporal NDVI profiles derived from SPOT images, Lebourgeois *et al.* [LEB 07] proposed an operational prototype product for monitoring the cutting and plowing of sugar cane on the island of La Reunion (Figure 4.7). The product is intended for sugar factory managers to ensure they have strategic information for organizing of the harvest campaign. The analysis is based on modeling the temporal NDVI profile and the detection of sudden changes due to harvesting. This example shows that a simple approach makes it possible to respond to sugar factories managers questions through an expert analysis based on NDVI profiles.

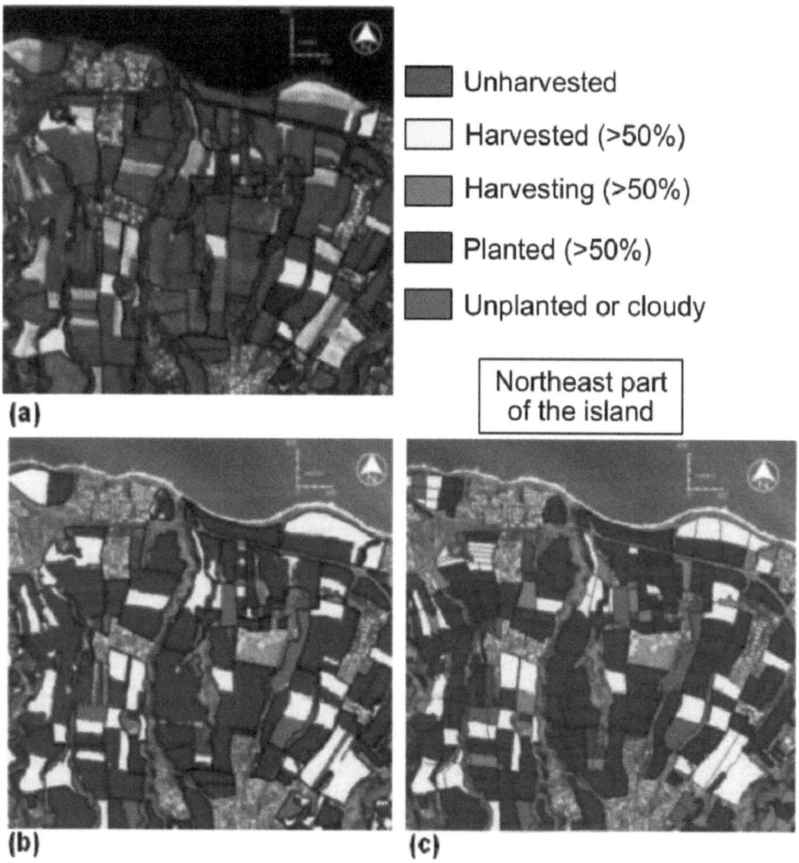

Figure 4.7. *Example showing the mapping of harvested, plowed or unharvested sugar cane areas on a SPOT image of 19 August 2004, 10 m resolution (source: SUCRETTE/CIRAD – Réunion [LEB 07]): a) colored composition (NIR/Red/Green) where photosynthetically active vegetation appears red, cane residues in light blue, bare soil in dark blue; b) classified blocks of cane plots; c) classified cane plots. For a color version of this figure, see www.iste.co.uk/baghdadi/3.zip*

4.3.2.4. *Grassland management: mowing date identification*

Figure 4.8 shows another example of practice detection. Here, the identification of the mowing date of irrigated grasslands located in the Crau region (in south-eastern France). The monitoring of these agricultural systems is particularly important in this area because the grasslands play a crucial role in the hydrological functioning of the groundwater and are a major economic capital because of their controlled origin designation (COD) label thanks to their high-quality hay. Three harvests per year are done. The grasslands are irrigated by flooding from the channels from the Durance, every 10 days on average, from March to September. The hay produced at each harvest is certified provided that different criteria defined by strict decree are respected. In particular the harvesting period must be during a specified period and the amount of irrigation and fertilization cannot exceed defined thresholds. Remote sensing can help detect the harvest period.

A method based on the analysis of the NDVI profiles calculated for each grassland plot was developed to estimate the mowing dates from FORMOSAT-2 images acquired every three days in this area. All plots are not harvested at the same time. The bottom right section of Figure 4.8 shows the NDVI profiles obtained for grasslands of the entire region. We observe that the first harvest is spread over a 16 day period the second over a 14 day period. The determination of the harvest date is based on the detection of the date corresponding to the minimum NDVI value followed by an increase of NDVI for the following three dates. The grassland harvest is made up of several steps: mowing, then forming windrows to dry the hay and finally putting it into bundles or bales depending on the farming equipment. All these operations can take from 3 to 6 days. This period was taken into account in the automatic cut detection algorithm applied to the FORMOSAT-2 images. The maps obtained (Figure 4.8 top right) have been validated from surveys conducted on 120 plots. The results showed a good accuracy of the model (<2 days [COU 10]). A linear relationship was observed between the date of the first grassland harvest and the date of the first irrigation after mowing. The first plots mowed are indeed the first irrigated. Knowing that irrigation is then performed every 10 days on average in the sector, following the water distribution schedule defined by the channel managers, it is possible to establish irrigation maps on this region. This spatial information can then be used in different models (crop or hydrological models) to monitor the hydrological behavior of these agricultural systems.

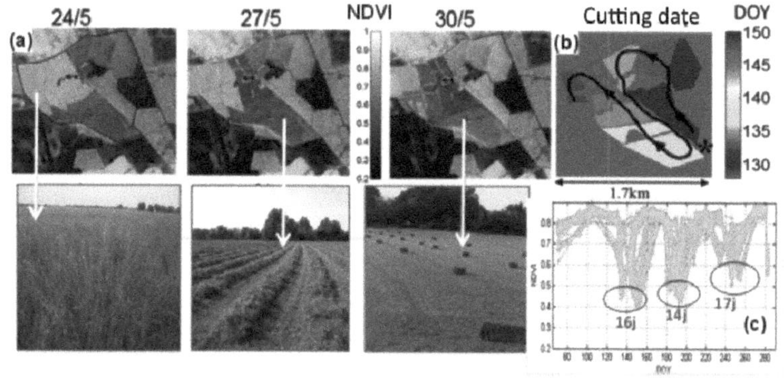

Figure 4.8. *a) NDVI calculated for three consecutive days (May 24, 27 and 30) from FORMOSAT-2 images acquired on the Crau region in 2006 (zoom on a small area of irrigated grasslands of the same farmer); b) NDVI changes can track the path of the farmer and identify plot mowing dates (expressed here in calendar days "DOY"). The automatic detection algorithm described in Courault et al. [COU 10] establishes maps of harvest dates. (c) NDVI profiles are calculated for all plots (the green curves clearly distinguish the three harvests: the first cut is spread over 16 days, the second over 14 days and the last 17 days) (images @ Courault). For a color version of this figure, see www.iste.co.uk/baghdadi/3.zip*

4.3.2.5. *Grass land management: management modes*

Dusseux *et al.* [DUS 14] differentiate the grassland management modes over a watershed in Brittany (France) by analyzing the time profile of the leaf area index (LAI) obtained after inverting five SPOT and Quick Bird images (see Chapter 2). Grasslands represent 60% of the cultivated land in the area. The objective was to identify the three modes: grazing, mowing and mixed mode, which have all a different impact on the environment. Figure 4.9 shows the variability of the profiles obtained following the three management modes. Different classification methods (including original methods, that take into account the possible time shifts between plots within the same mode) described in Dusseux *et al.* [DUS 14] have been used to differentiate the various profiles as well as possible.

4.3.2.6. *Irrigation detection*

Knowledge of irrigated areas is essential for water managers who are responsible for the distribution of this resource over large areas. For example, in France, the irrigated area has increased from 500,000 ha in 1970 to 1.576 million ha in 2000 (Common Agricultural Policy) and is stabilized, representing 5.8% of the Useful Agricultural Area in 2010. About 60% of irrigated lands are located in the southwestern part of France. Currently, managers do not have operational tools able to deliver near real time information about the irrigated areas in their territory. Having such maps could enable them to better manage the crop water needs and supplies, which is a critical issue, especially during episodes of severe droughts, which happen, for example, one year out of five in the Midi-Pyrenees region (France). Irrigation management is also crucial in semi-arid countries.

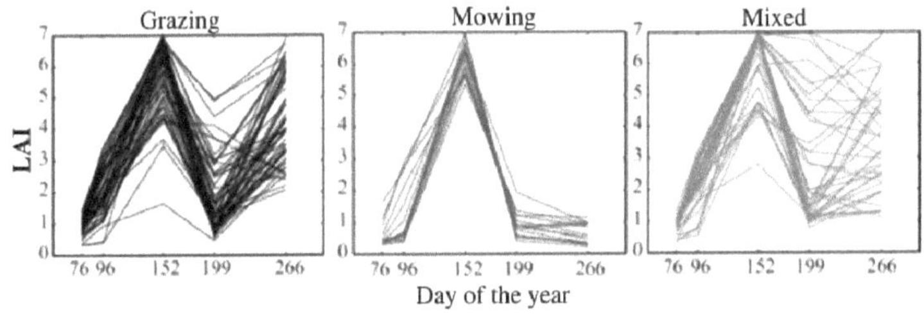

Figure 4.9. *Examples of real time LAI profiles obtained for the three grassland management modes: a) grazing; b) mowing; and c) mixed; (source: [DUS 14]). A LAI around six indicates a grassland where the canopy is high and well developed. When grasslands are mowed or grazed, LAI falls quickly. There is a regrowth after grazing or mixed operations after mowing*

The difficulties related to the detection of irrigated areas are numerous. Firstly irrigation is a quick process. Irrigation may last a short time (generally a few hours per hectare), and when remote sensing images are acquired, the soil surface moisture might not be detectable over the longterm. Then, there are different irrigation techniques inducing different effects on reflectances. There are three major categories of irrigation methods: drip irrigation, gravity flood irrigation and aspersion (sprinklers). Moreover, the effect of irrigation on the development of vegetation is not always visible. Finally, water supplies depend on the availability of water resources but also on human decisions, which explains the disparities observed between departments and countries. Thus, the distinction between irrigated and rainfed crops remains a challenge.

The challenge to better managing water resources is to develop tools to map irrigated areas which must be robust enough, operational and valid for different territories. Several approaches have been proposed to obtain this information. Ozdogan *et al.* [OZD 10] present an overview of the methods used to map irrigated areas using remote sensing. Most operational methods use vegetation indexes such as NDVI or the Green Index (GI) and classification algorithms. They are distinguished by the spectral range of the images used (optical, thermal, radar) and their spatial and temporal resolution. Maps of irrigated areas are available at a 10 km spatial resolution. This is the Global Irrigated Area Map (GIAM) obtained by Thenkabail *et al.* [THE 06] from the combination of images from different sensors (AVHRR, VGT, JERS-1 and MODIS). The spatial resolution of such a map is low and thus difficult to validate and necessarily contains a high degree of uncertainty, as discussed by Brown *et al.* [BRO 09]. These authors conclude on the need to process high and temporal spatial resolution images to access more reliable results.

The examples in the following sections illustrate some products that are currently available or under development.

4.3.2.6.1. Optic data methods – global scale

A "land use" product has recently been developed from MERIS images: CCI Land cover. It is freely available to everyone all over the world with a spatial resolution of 300 m (http://maps.elie.ucl.ac.be/CCI/viewer/).

The processing method is described in http://maps.elie.ucl.ac.be/CCI/viewer/download/ESACCI-LC-PUG-v2.5.pdf.

The major irrigated regions in the world, such as the Nile Delta or the Indus Basin can be identified on this map. However, many irrigated areas such as the Lerma Basin in Mexico, Tensift in Morocco, or south-west France (shown in Figure 4.10) are not detected. Moreover, the last map dates to 2010 with an update every five years.

4.3.2.6.2. Optic data methods – regional scale

Le Page *et al.* [LEP 12] proposed a map of irrigated areas derived from MODIS products (MOD13Q1, NDVI synthesis to 16 days) over a semi-arid region of southern Morocco. The authors propose to define irrigated areas analyzing the dynamics of the growing cycles using a method based on decision trees based on the frequency of crops for a period of 9 years. The presence of annual cycles indicates irrigated crops which differ from more erratic rainfed crops. A similar technique is to perform a multi-annual analysis directly on the NDVI signal, and calculate the relevant metrics (number of seasons, overall trend, integration of NDVI above a certain threshold, etc.). These data are combined with auxiliairy data, such as slopes, to be used as inputs to a classification algorithm. In the example in Figure 4.11, a non-supervised classification was performed on the plain of Merguellil in Tunisia. Except for an untapped area located in the north of the city of Kairouan, actual irrigated areas cover all irrigated areas that are managed by public authorities (vertical lines). However, areas that were identified as private irrigation areas by public authorities (horizontal lines) are very different. This is partly because of incorrect mapping of these private areas, and secondly, due to the relatively recent expansion of private irrigated areas fed by deep wells. Graphs b, c and d in Figure 4.11 show three mean NDVI time profiles, the first being irrigated areas indicated on the map.

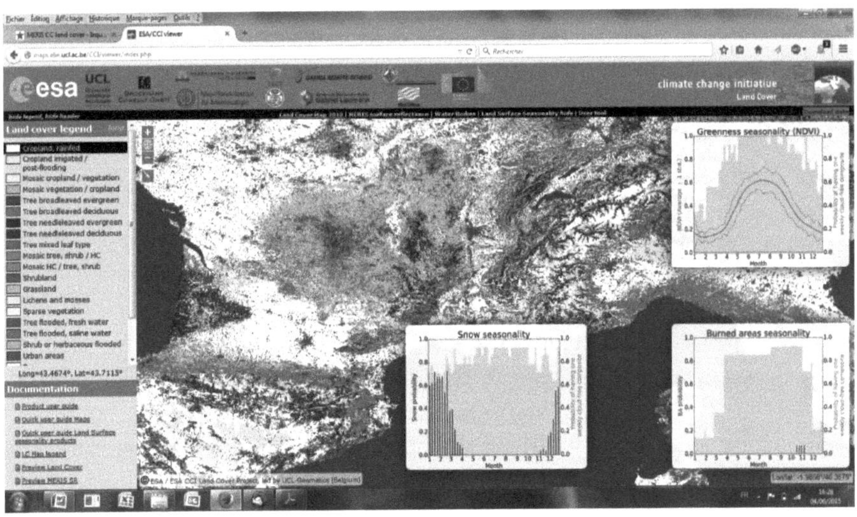

Figure 4.10. *Land use map data obtained from the MERIS sensor on the southwest region (France). The agricultural plain is occupied by rainfed crops (rainfed crops in yellow), when in fact much of the area is covered by irrigated crops (http://maps.elie.ucl.ac.be/CCI/ viewer/). For a color version of this figure, see www.iste.co.uk/baghdadi/3.zip*

There are also other less operational approaches that are the subject of long term research. These include, for example, the work of Peña-Arancibia *et al.* [PEN 14] using the difference between evapotranspiration and rain to identify pixels with a notable excess of evapotranspiration. However, this approach does not take into account any contributions by surface or subsurface water fluxes. It is also less relevant when there is an overlap of rainfed and irrigated crops.

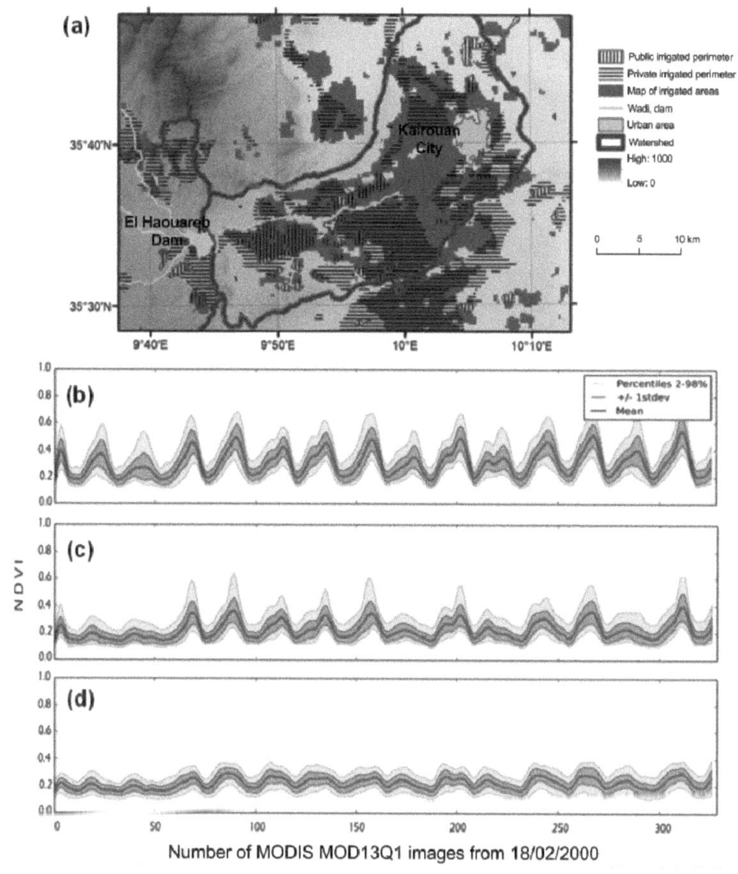

Figure 4.11. *Map of irrigated areas (in blue, map obtained from past 2000–2014 MODIS (MOD13Q1) images, the Kairouan plain in Tunisia). The shaded areas are from the Kairouan GIS. The graphs (b, c and d) illustrate the NDVI profiles corresponding to three of the classes obtained by the model, b) frequent recurrence of contrasting annual cycles: this is probably irrigated crops (blue on the map), c) episodic return of annual cycles, this is probably rainfed, d) lack of annual cycle, this is bare soil (images @ Le Page). For a color version of this figure, see www.iste.co.uk/ baghdadi/3.zip*

4.3.2.6.3. Optic data methods – local scale

a) Using data acquired in the visible and near infrared spectral range (VNIR)

Several studies have been conducted on the early detection of irrigation using FORMOSAT-2 and ENVISAT/ASAR [DUKE 08, DUKE 09]. For example, Boizet [BOI 15] processed maps of irrigated areas (Figure 4.12(a)) using a time series of NDVI extracted from FORMOSAT-2 images. They used the differences on the temporal NDVI profiles of irrigated and rainfed crops during the maximum growth stage (Figure 4.12(b)).

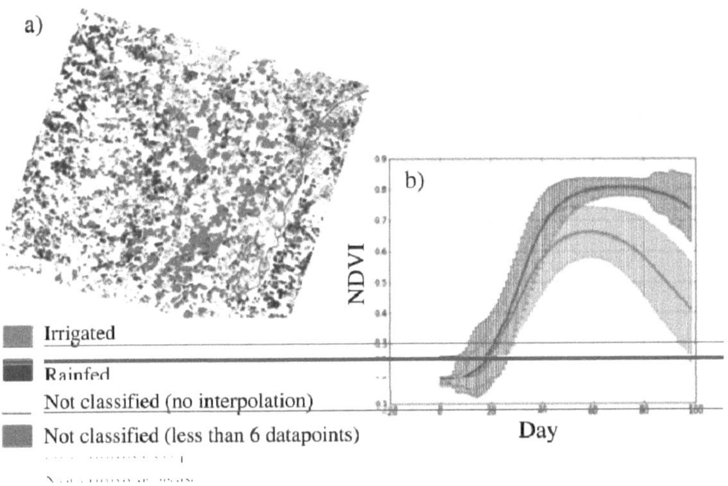

Figure 4.12. *a) Map of irrigated areas located in the southwest of France computed from FORMOSAT-2 images in 2009; (b) average NDVI profiles calculated on different plots of irrigated (in blue) and rainfed (in green) crops. The standard deviations are represented by the green and blue areas. Irrigated crops are distinguished from rainfed ones thanks to the maximum NDVI value. The x-axis is the number of days after emergence [BOI 15]. For a color version of this figure, see www.iste.co.uk/baghdadi/3.zip*

The distinction between the NDVI of irrigated and rainfed corn is not always possible, especially during wet years or when the presence of clouds on optical data make the signal unusable. Radar images can acquire images in cloudy conditions. Current research is therefore investigated towards the use of optical imaging (such as Sentinel-2) combined with thermal and radar images, to improve irrigated area detection capabilities. Different thermal and radar data are already freely available covering the globe: such as

ASTER, Landsat-8 (Optical + thermal) and Sentinel-1 (radar) images. Figure 4.13 shows the interest in and complementarity of thermal images compared with optical images.

Figure 4.13. *On the left, a Landsat color composite (3, 4, 7 TM bands) on the of Haouz plain (Morocco), NDVI in the center and the corresponding thermal image on the right. Compared with the colored composite and the NDVI, the temperature accentuates the contrast between the irrigated areas (dark thus cold) and rainfed areas (light thus warmer) (images @ Simonneaux). For a color version of this figure, see www.iste.co.uk/baghdadi/3.zip*

b) Using thermal data

The data acquired in the thermal infrared range (TIR) provides the surface temperature which is a key variable for monitoring the energy and water balance. When fine spatial resolution images are acquired in this spectral range (e.g. from an airborne camera, as shown in the center of Figure 4.14), we can clearly distinguish the different forms of irrigation. Methods based on threshold surface temperatures can then be implemented to classify irrigated and rainfed areas.

One of the current problems is the inadequacy between spatial and temporal resolutions for thermal sensors currently in orbit, which are not suitable for water management at plot or farm scales. Sensors such as Landsat (LDCM) or ASTER have thermal bands at relatively fine spatial resolutions (60 m and 90 m, respectively), but their revisit time for the same site (16 days) is insufficient to monitor irrigation in many sectors. On the other hand, the MODIS sensor offers thermal images twice a day across the globe, but the spatial resolution of 1 km does not allow water management at plot scale. Various development projects of high spatial and temporal resolution thermal sensors have been proposed to solve this problem in recent years: among the most recent are: MISTRIGRI, TIREX, THIRSTY [LAG 13], HYs PIRI http://hyspiri.jpl.nasa.gov/ [LAG 16].

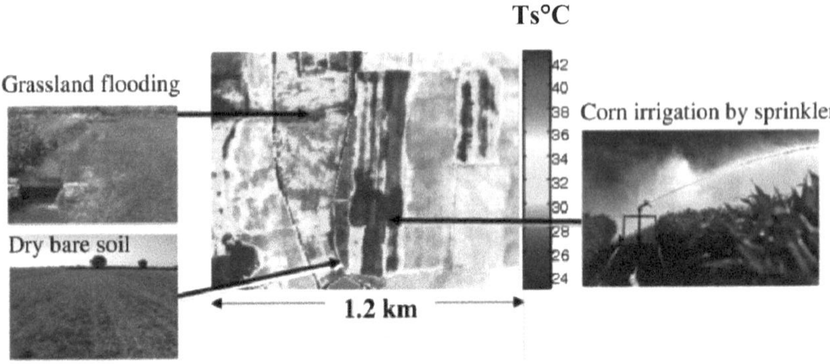

Figure 4.14. *Thermal image (resolution 3.5 m) acquired on 27 July 2006, from an airborne FLIR thermal imaging camera (8–11 μm) at 1500 m above the ground surface. The various Irrigation, gravity, aspersion practices are clearly distinguishable by lower surface temperatures across the plot (images@Courault). For a color version of this figure, see www.iste.co.uk/baghdadi/3.zip*

c) Using radar data

Radar data also offers a strong potential for detecting changes in surface humidity [BAG 16]. An example based on the use of SAR (X-band) data, shown in Figure 4.15, shows that it is possible to identify the progress of the blade of water on grasslands irrigated by flooding [ELH 14].

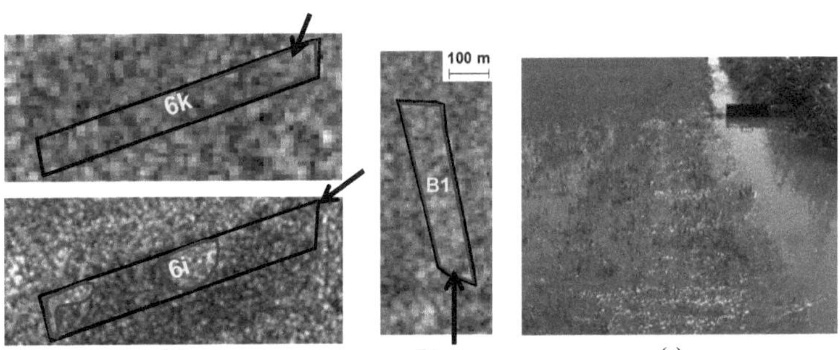

Figure 4.15. *Irrigation monitoring from radar images on three plots a) 6 k, 6i ((b) B1) of the Domaine du Merle near Salon-de-Provence (France). Radar imagery is used to identify flooding areas (areas delineated by a red outline on figures a and b); plot 6 k represents the blade of water in progression as shown on photo c, plot B1 is fully submerged and in plot 6i, irrigated grassland after a few days are still visible [ELH 14]*

In the same way as thermal data, automatic methods based on threshold detection can be developed to detect certain irrigation practices. The advantage here is that radar sensors can operate day and night, regardless of weather conditions. Satellites at fine spatial resolution, such as TerraSAR-X and COSMO-Sky Med deliver images with 1 m pixels with daily revisits. They are complemented by Sentinel-1, covering of Europe at 10m every five days. The radar signal can also be used in models that monitor or manage irrigation, such as the PILOT model [ELH 16] via assimilation methods which will be presented in section 4.4.

4.4. Estimating water status and the water needs of crops using models

Biophysical variables estimation is described in Chapter). Even if certain variables such as the Green Area Index (GAI) may saturate for very developed and dense fields, we can estimate that globally the different methods currently proposed (Chapter 2) allow us to follow crop development with good precision for a lot of situations. On the other hand, the estimate of the water status of surfaces still remains problematic at regional scale, on the one hand because of the high spatial variability of soil and cropping practices, and on the other hand because of a lack of specific local measurements. Monitoring the crop water status is essential because it enables us to check if plants are sufficiently supplied with water for their optimal development. It also allows us to rapidly detect water stress in order to better manage the irrigation and to properly assess the production and exchange of water and CO_2 within the atmosphere, and N_2O emissions from the ground.

Actual evapotranspiration (AET), which represents the actual crop water consumption, is one of the key variables for the monitoring of the surface's water balance. Different methods have been proposed to estimate AET using remote sensing data [COU 05, KAL 08]. Evapotranspiration maps produced from MODIS data are currently freely available, but provide only low resolution and/or temporal integrations at best (product MOD16, http://www.ntsg.umt.edu/project/mod16, at a resolution of 1 km every eight days). More or less operational tools have been developed. They are based on the use of available high resolution satellite data via various platforms (e.g. for Landsat https:// landsat.usgs.gov//Landsat_Search_and_Download.php, or ASTER https:// asterweb.jpl.nasa.gov/data.asp). These tools include the EVASPA chain process developed at the INRA in

Avignon ([GAL 13], Boulet, Chapter 10, Volume 4). EVApotranspiration assessment from SPAce requires input data acquired in optical and thermal spectral ranges and considers different algorithms to estimate the main parameters involved in the models (albedo, or fCover, soil heat flux and net radiation). The processing chain allows the assessment of AET uncertainty. Maps generated by EVASPA (Figure 4.16) allow us to distinguish the dry lands or the areas affected by water stress, and the well irrigated areas showing high evapotranspiration (in Figure 4.16 corresponding to the Camargue region in southeastern France, rice fields are well developed in July and have high AET values, the fields appear in red on the images while the dry and stony grassland areas located in the center of the image (blue zone) have low AET values).

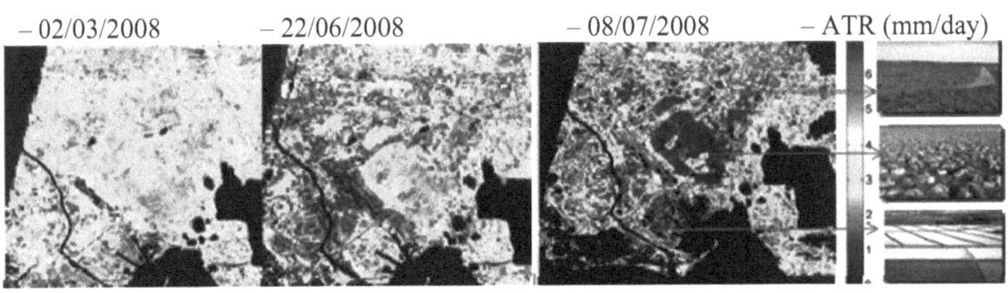

Figure 4.16. *Evapotranspiration map (expressed in mm/day) obtained from the EVASPA chain applied to Landsat-7 images acquired on the Crau-Camargue region in 2008 (images@Courault). For a color version of this figure, see www.iste.co.uk/baghdadi/3.zip*

4.4.1. *Evapotranspiration estimation using a simplified model of the water balance*

The methods of AET assessment based on thermal imaging have the advantage of taking into account the real state of the vegetation and the water content of the soil at the time of image acquisition, but they are difficult to extrapolate over time. The surface temperature varies very quickly depending on many different factors (surface humidity, wind, etc.) and there is currently no thermal satellite data in orbit with high spatial and temporal resolution. This constraint is a real issue when applying these methods at farm scale for crop and water monitoring.

Another way to produce a spatial AET map is using transfer models simulating the exchanges between the soil, vegetation and atmosphere (commonly called SVAT models). They require climate forcing, information on the current ground vegetation and the hydrodynamic characteristics of the soil. Among the different modeling approaches, a family of widely used methods for agricultural applications to estimate the water balance for crops is the FAO approach proposed by Allen [ALL 98]. One of these methods (called the double coefficient method) calculates AET by distinguishing transpiration and evaporation from the following equation:

$$AET = (Kcb\ Ks + Ke)ET0 \qquad [4.1]$$

where ET0 is the reference evapotranspiration of a standard surface of well watered grass (ET0 is calculated using temperature and humidity, net radiation and wind speed using the Penman–Monteith formula); Kcb is the basic crop coefficient reflecting the potential transpiration of the plant under unstressed water conditions (usually ranges from a minimum of 0 in the absence of vegetation to a maximum of 0.7 to 1.15 depending on the crops); Ks is a stress factor which modulates transpiration according to the water availability in the root zone (Ks varies between 1 in the absence of stress and 0 in case of maximum water stress); Ke represents the evaporation coefficient which controls the evaporation of the bare soil fractions depending on the surface moisture (Ke equals 0 in the absence of evaporation and has a physically constrained maximum value of 1.1 to 1.3). A simplification of the dual coefficient method, called the single coefficient method, includes transpiration and evaporation for unlimited water conditions into a single coefficient Kc (Kc = Kcb + Ke). In Allen *et al's* publication [ALL 98], various tables of Kc and Kcb values are given for many crops, which vary depending on their phenological stage (Figure 4.17).

This method presents a remarkable complementarity with satellite information in the visible range. It has been shown that the Kcb parameter can be estimated from a vegetation index [BAU 87, NEA 89, CHO 94]. The major limitation of this method is the lack of knowledge of soil moisture because this factor varies considerably in space and is for the time not easily obtained from satellite. It is possible to measure it, more or less precisely only at few punctual locations. Additionally, rainfall data can be obtained from meteorological stations but irrigation amounts cannot be known plot by plot over large areas. Another limitation is the absence of satellite data during the frequent cloudy periods [SAA 15]. It is also important to note that soil evaporation simulations do not take into account variations in surface conditions such as the presence or absence of residue, which could have an effect on both the energy budget and on the resistance to water transfer within the soil. Regarding the problem of irrigation, one solution is to simulate the irrigation water supply using a model simulating the water balance on a daily scale, as proposed in the SAMIR tool [SIM 09]. Assumptions about the farmer's irrigation practices must be defined to apply this approach (threshold to start irrigation, amounts of water brought to each field, end date of irrigation, minimum time between two irrigation events, etc.).

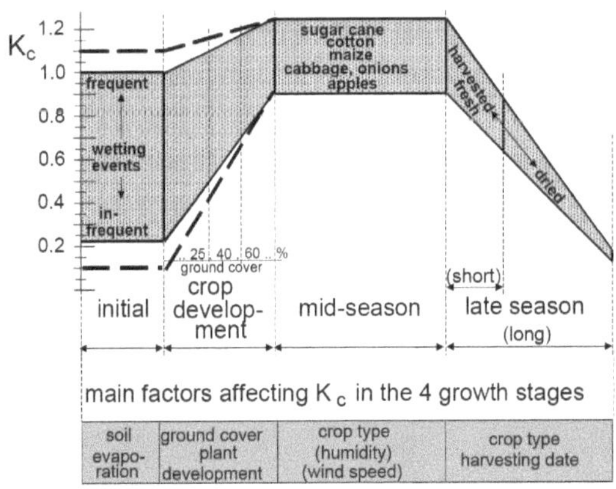

Figure 4.17. *Kc variation for the four main crop stages (from Allen et al. [ALL 98])*

An experiment in the Tensift basin (Morocco) showed that the use of the SAMIR model with a NDVI time series computed from Landsat TM images (seven dates from January to June 2003) could correctly reproduce the total amount of water used when growing wheat through irrigation. Figure 4.18 shows the simulation results between January 2002 and June 2002 for three wheat plots on which measurements were performed to validate the model. Simulated and observed irrigation water supplies were respectively 216, 207, 223 mm and 251, 220, 180 mm, indicating a correct agreement between observations and simulations. However, the simulated irrigation dates were often shifted compared with the observations, because the farmer has many constraints that are not included in this simplified approach. The first plot (Figure 4.18(a)) shows a good response of the model (shown in red) compared with observations (shown in black). The second plot (Figure 4.18(b)) shows a strong discrepancy between simulated and observed evapotranspiration in early February because the observed irrigation water input was not simulated by the model. For low vegetation cover, evaporation of bare soil is the dominant process and evapotranspiration is therefore strongly influenced by the dates of water input. For dense vegetation, the time shift for irrigation dates has less impact, because transpiration is the main process and it remains stable as long as there is no stress. For the third plot (Figure 4.18(c)), a sharp drop was observed between March and April, which corresponds to a 10-day stress period that has not been reproduced by the model which uses an automatic irrigation mode (explaining the significant overestimation of the simulated water amount on this plot).

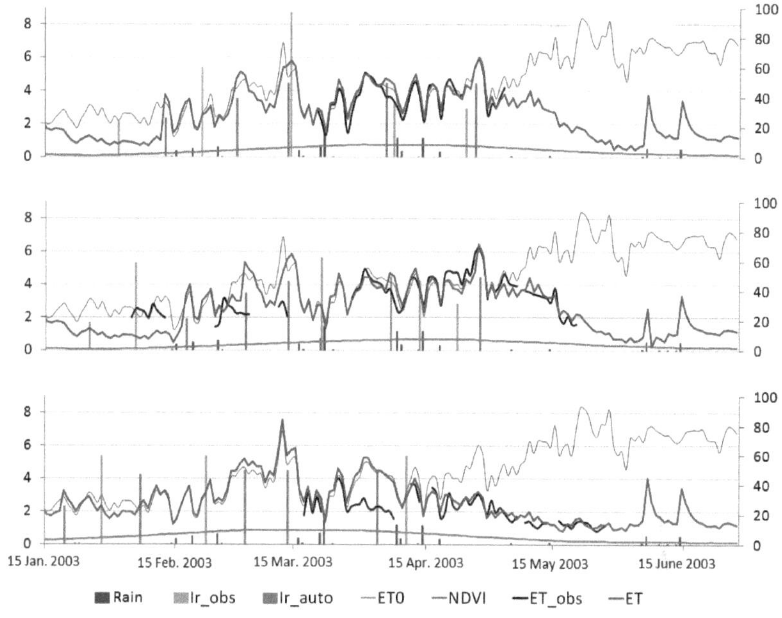

Figure 4.18. *Simulation of water balance of three irrigated wheat plots in the Haouz plain (Morocco) by the SAMIR model used in automatic irrigation mode. On the left (in mm): ET0 – reference evapotranspiration, ETobs (black line) – evapotranspiration measured using the turbulent method, ET (red line) – the simulated evapotranspiration. NDVI interpolated from images is represented by a green line. On the right (mm): green vertical lines represent actual irrigation (Ir_obs) and blue vertical lines represent simulated irrigation (Ir_auto). Rainfall is shown by the brown vertical lines. For a color version of this figure, see www.iste.co.uk/baghdadi/3.zip*

These results show that if the model provides overall satisfactory information on evapotranspiration and the amount of water supplied by irrigation after aggregation across several plots or a season, the application of this model for piloting irrigation at field scale requires additional information, in particular to take into account the actual irrigations applied by the farmer.

4.4.2. *Irrigation management in operational mode in near real time at plot level*

Following the successful launch of Landsat-8 and Sentinel-2, a series of high spatial resolution optical images are now-available for free to a wide community. A model such as that described above can thus be fed by satellite data in near real time, enabling quick decision-making concerning irrigation needs. The SPOT4-Take5 experiment [HAG 13] acquired high spatial and temporal resolution images on the Tensift basin (Morocco) during the Winter–Spring season of 2013. This is the season in which the winter wheat is produced in the Marrakech region in Morocco [LEP 14]. An experiment in near real time (less than a week after acquiring the image) was conducted on an irrigated 4 hectare plot of winter wheat. This plot was irrigated depending on the water balance calculated using satellite data acquired throughout the season. The crop development was forced by using the images acquired every five days at 20 m spatial resolution. Surface reflectance and the average NDVI were calculated at each date for each plot. A weather station set up close to the studied fields provided information on rainfall and on reference evapotranspiration. A comparison was made between this plot and a reference plot, irrigated in the traditional way by the farmer. The results in terms of the performance observed on the two plots were very similar (Figure 4.19). Two water turns representing a total of $800m^3$/ha (12%) were saved using remote sensing data, meaning a water saving of 1.34 m^3/kg against 1.52 m^3/kg (Figure 4.19).

This study led to the development of an operational tool to help irrigation management: SAT-IRR (Figure 4.20). After drawing up the outline of the studied plot on a satellite map via an Open Layers interface, the user must enter the crop type, the sowing date, define the soil type, and give information on irrigation. The tool then selects the corresponding empirical models. Various relationships can be chosen for the cultural coefficient and the NDVI. The tool also selects an optimal trajectory of the evolution of these two variables over time from the FAO library. The user can refine these parameters according to local conditions. The server loads climatic daily measurements from the nearest weather station or from weather forecast models (in the example shown here, the forecast model is HIRLAM). The tool processes satellite images as soon as they are acquired (Landsat-8 or Sentinel-2). The user must input information on irrigation (date and dose brought) necessary to correctly compute the water balance. The main difficulty is managing the impact of cloud images in the processing chain. SAT-IRR extrapolates the plant growth using standard profiles proposed by the FAO. On this last point, note that more specific plant models could also be used to carry out this work. Finally, weather forecasting allows to choose an irrigation date in advance.

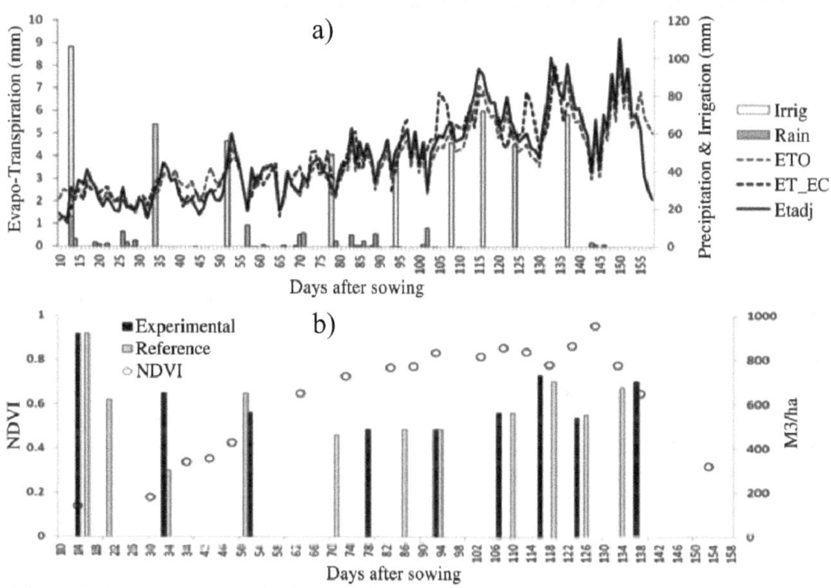

Figure 4.19. *Results of experiment conducted on a wheat plot in Morocco to predict irrigation using the SAMIR tool: a) simulated evapotranspiration of the test plot (ET adj top) is consistent with measurements made from an eddy-covariance station (method of turbulent flow, ET_EC), ET0 is the potential evapotranspiration, b) comparison of irrigations performed on the test plot compared with the reference plot shows that two water turns were saved*

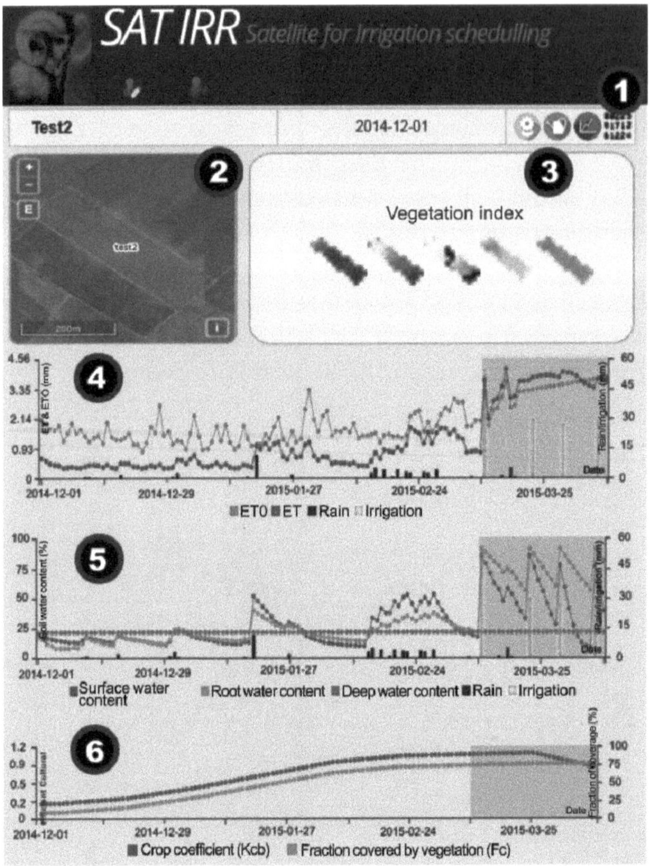

Figure 4.20. *Screenshot of the SAT-IRR tool (web interface). (1) The four icons allow to change parameters and to define the plot contours, enter the parameters related to irrigation and view the results (graphs or tables). (2) An Open layers window to navigate the plot. (3) The NDVI time series is displayed. Three graphs provide information about this. (4) The atmospheric part (precipitation, reference evapotranspiration and crop evapotranspiration). (5) Changes in the water content of the soil separated into three layers. (6) The evolution of the basic crop coefficient (Kcb) and fraction cover (Fc). The shaded areas on the right of the graphs are forecasts for the coming months, including irrigation recommendations, which are indicated by green bars. For a color version of this figure, see www.iste.co.uk/ baghdadi/3.zip*

4.5. Agricultural production quantification

Estimation of agricultural production is currently based on the use of crop models. The implementation of crop models at local scale is relatively easy, as information on the soil and agricultural practices is easily accessible. This does not hold true for their broader implementation scales (catchment, production basin, country) is more difficult as the soil and crop characteristics vary a lot in time and space [BOO 96, MOU 98, ISP 04]. Different types of models have thus been developed to solve this problem. They are characterized by their level of complexity and their ability to exploit different levels of information associated to variable level of precision and therefore, to be spatialized [CON 15]. Remote sensing is a valuable tool to "feed" these models and compensate for the lack of local information and for the simplifications introduced into some models.

4.5.1. *Reminder on crop models*

In the category of complex models specific to crops we can find the so-called "ecophysiological" models that simulate the major plant processes such as photosynthesis, respiration, decomposition, energy and water balance in a mechanistic way (SUCROS, WOFOST, see Bouman *et al.* [BOU 96]). In the category of complex models we can also find "agronomical" models (ARCWHEAT, CERES, STICS [JAM 98, BRI 03]) that describe the effect of agricultural practices (fertilization, irrigation, intercropping, tillage etc.) on a large number of crops. In these models, some physiological processes, including those related to photosynthesis, have been simplified. They mostly use the Monteith equation [MON 72] that simulates the daily increase in biomass from the global incident radiation (RG) and three efficiencies (climatic, light absorption and radiation use).

$$BM_t = \varepsilon_{brt} \cdot \varepsilon_{it} \cdot \varepsilon_{ct} \cdot RG_t \qquad [4.2]$$

with, t, the time step (usualy daily), BM_t , the biomass production (g/m^2), ε_{brt}, radiation use efficiency in g/MJ/j [SIN86]), ε_{it}, light absorption (or Fraction of Absorbed Photosynthetically Active Radiation (FAPAR)), ε_{ct}, climatic efficiency, the ratio of photosynthetically Active Radiation (PAR) to RG (global radiation expressed in MJ/m^2/j).

These models are called "complex", given the number of input parameters and simulated processes. The STICS model, for example, needs in order to simulate a wheat crop, more than 227 parameters divided into three groups: 129 parameters relating to the characteristics of the plant, 23 parameters for agricultural practices and 75 parameters describing soil properties. This information may be available at a local scale, but not over large areas.

Apart from complex models, there are "simple" models that calculate the biomass from the Monteith equation [MON 72]. In such models, the ε_{it} efficiency is determined using a vegetation index such as the NDVI, computed from satellite data (see the early work of Kumar and Monteith [KUM 81] and Tucker and Sellers [TUC 86]). Baret and Olioso [BAR 89] and Steinmetz *et al.* [STE 90] have shown that in the case of annual crops, the proportion of radiation intercepted by the canopy is more or less linearly related to NDVI or other vegetation indexes. These models are easy to spatialize and are well suited for estimating the net primary production of natural ecosystems such as forests [DON 03] or grasslands [PRI 91].

The third category of crop models (classified in "semi-empirical approaches") combines the Monteith equation with some major processes considering the main stages of plant development (emergence, growth and senescence) and the water availability in the soil, thereby overcoming in large part the limitations of previous models, having a limited number of parameters. These agro meteorological models (Aqua Crop, AqYield, SAFY, SAFYE-CO_2) have a restricted number of formalisms and restricted parameters. As simple models, they are suitable for the study of large areas. Many studies have combined these models with remote sensing data, and demonstrated the relevance of this approach to analyze the crop functioning [BOO 96, MOU 98, LOB 03, ISP 04, DUK 08, DUC 08, LIU 10, CLA 12, CHA 14]. Most of these studies conclude on the need for remote sensing data acquired at high spatial and temporal resolution (HRST) to improve the predictions of these models. One of the main limitations of these approaches is the ability to evaluate the soil water content over large areas, which is not information that is easily accessible everywhere.

4.5.2. *Review on the use of remote sensing data combined with crop models*

The various methods combining remote sensing data with crop models have been described in detail in different publications [GUE 96b, DOR 07]. Below is a brief reminder of the key term definitions:

– *forcing*: biophysical variables derived from remote sensing data are used directly as inputs for the crop model (case shown in sections 4.5.3 and 4.5.5);

– *recalibration*: biophysical variables derived from remote sensing data are used as control variables in order to retrieve some parameters which are difficult to estimate (case shown in section 4.5.5);

– *sequential assimilation*: biophysical variables are injected sequentially when they are available and through filtering techniques such as variational or ensemble technics allow us to correct the states predicted by the model. Only a few uses of this type of methods, widely applied in the fields of hydrology and more recently in SVAT [OTT 16], have been implemented in crop models. We can include the work of Naudet al. [NAU 07] and Chen and Cournède [CHE 14] who do not use remote sensing data.

4.5.3. *Biomass and yield estimate by forcing simple models*

The possibility to force the ε_i efficiency of the simple Monteith model using reflectance measurements has been implemented across instrumented rice field plots [LEB 91] and has demonstrated the relevance of this approach. A calibration of the ε_i– NDVI relationship was proposed in the form of:

$$\varepsilon_i = 1 - \left(\frac{0.9 - NDVI}{0.9 - NDVI_{back}}\right)^a \qquad [4.3]$$

where $NDVI_{back}$ is the NDVI value of the surface under the crop corresponding to soil plus water. The coefficient a varies between 1.48 and 1.7 depending on the rice varieties [LEB 91]. This relationship was used to estimate the biomass produced on a validation set of field plots from a model similar to that described in equation [4.2] with a correlation coefficient r^2 of 0.7 (see [LEB 91] for more details). The grain yield was also predicted by adjusting a harvest index (ratio of grain yield and total biomass), but its predictive ability proved less robust, so far as the limiting factors of grain weight, involved in the reproductive period, are not always linked to the evolution of the Green Leaf Area Index (GLAI).

The same approach has been applied and extended to take into account the water stress in this simple model. The objective was to estimate cereal production in semi-arid conditions over large areas in Algeria, using data from NOAA-AVHRR in the optical (NDVI) and thermal (surface temperature) domain [GUE 93a].

The model is the following:

$$BM_t = \varepsilon_{bt} \cdot \varepsilon_{it} \cdot \varepsilon_{ct} \cdot RG_t$$
$$\varepsilon_{bt} = \varepsilon_{bpt} \cdot ISH_t \qquad [4.4]$$

where t is the time considered (usually daily), BM_t is the aerial biomass production (g/m^2), ε_{bt} is the actual efficiency biomass conversion at time t (g/MJ/d), ε_{bpt} is a potential efficiency biomass conversion at time t, in the absence of water stress.

ISH is a water stress index that applies to biomass conversion efficiency by reducing ε_b from its potential value. This index is calculated as the ratio of actual evapotranspiration (AET) and maximum evapotranspiration (ETM), itself estimated from the climatic potential evapotranspiration (ETP) by a crop coefficient (Kcb) (see part 4.4.4 of [GUE 93a] for details). The conceptual diagram of the model is shown in Figure 4.21. The Kcb coefficient, which depends on the stage of crop development, is expressed here as a linear function of LAI, itself estimated from the NDVI. The AET is calculated using the difference between the surface temperature (*Ts*) estimated from the thermal spectral bands of AVHRR and the air temperature (*Ta*) according to the formula proposed by Seguin *et al.* [SEG 93]:

$$AET_p = Rn_p - 0.3 \sum_p (Ts - Ta)_p \qquad [4.5]$$

in which *Rn* is the net radiation and *p* represents time periods in the order of a decade.

This model was applied to the Wilaya, of Sidi Bel Abbes, Algeria (administrative area of approximately 3200 km^2) in a semi-arid environment (annual rainfall of 350 mm), where the cultivation of cereals (barley and durum wheat) is dominant. NOAA-AVHRR images acquired for two years (27 images in 1989, 33 in 1990) have provided NDVI and Ts input variables. These data have been filtered and interpolated to obtain daily values. This simple model was used to explain a significant portion of the variability in biomass and yield affected by water stress (Figure 4.22). It is a useful tool to dynamically estimate these quantities in semi-arid environments from optical and thermal satellite data acquired at medium resolution.

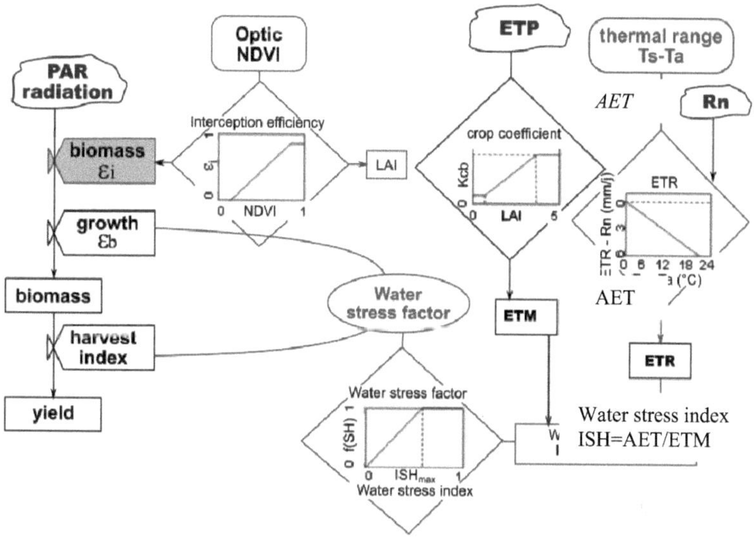

Figure 4.21. *Diagram of the biomass and yield estimation model for cereals forced by AVHRR sensor data. For a color version of this figure, see www.iste.co.uk/baghdadi/3.zip*

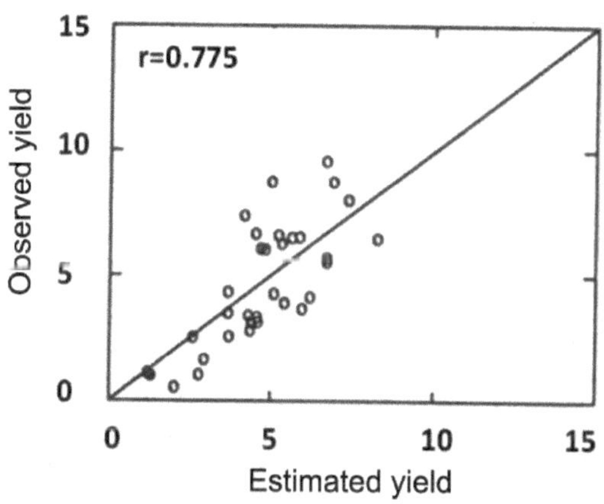

Figure 4.22. *Estimated and observed yield values (quintal/ha) of the 44 communes of the Wilaya area of Sidi Bel Abbes (Algeria) in 1990 [GUE 93b]. For a color version of this figure, see www.iste.co.uk/baghdadi/3.zip*

4.5.4. *Biomass and yield estimate by forcing semi-empirical models*

Among the third category of models classified in semi-empirical approaches, the SAFY model (simple algorithm for yield estimate [DUC 08]) was designed to be used with high spatial and temporal resolution remote sensing data. Recent studies have shown that this model provides reliable predictions of biomass and correct yields on irrigated wheat crops in a semi-arid region and on rainfed wheat [VEL 14], corn and sunflower in temperate areas [DUK 15, DUK 08, HAD 09, CLA 12, HCA 14]. SAFY simulates the daily dynamics of the GLAI and above ground biomass (Dry Aerial Matter: DAM). The calculation of the daily biomass is based on the use of the Monteith equation, supplemented by a thermal stress function (FT). The FT function, as in the STICS model [BRI 03], depends on four parameters: air temperature (*Ta*), optimal temperature, and two extreme temperatures, below and above which plant growth stops. In SAFY, the radiation use efficiency shown in the Monteith equation (ε_{brt}) is replaced by an effective light use efficiency (ELUE), which reflects the ability of the plant to convert the absorbed photosynthetically active radiation into dry aerial matter (DAM) under various agro-environmental constraints such as the availability of water and nitrogen. The GLAI is simulated from the dry biomass of the day (*d*) using a foliar partition function (Pl) which varies from 0 to 1, and a specific leaf area index (SLA in g/cm^2):

$$DAM(d) = ELUE.\varepsilon_{it}\varepsilon_{ct} \cdot RG_t.FT\ (Ta) \quad [4.6]$$

$$GLAI(d+1) = \Delta DAM\ .\ Pl\ .SLA \quad [4.7]$$

The yield (RDT) is estimated from the total dry aerial matter (DAM tot) multiplied by a harvest index (HI) specific to the crop studied:

$$RDT = DAM_{tot}\ .HI \quad [4.8]$$

HI is estimated from measurements or bibliographic references.

Studies were conducted by Claverie *et al.* [CLA 12] and Veloso [VEL 14] in Haute-Garonne (France) to assess the ability of the SAFY model for production assessment of summer crops (corn and sunflower) and wheat. The model was spatialized using FORMOSAT-2 imagery covering an area of 20 km^2. The yields estimated by the model were compared with departmental statistics provided by the French Ministry of Agriculture (Agreste database) from 2006 to 2009. The results obtained from corn and sunflower are shown in Figure 4.23. Good consistency was found between model outputs and agricultural statistics with a maximum deviation of 0.5 t/ha (same difference for wheat, not shown here) except for corn in 2008 and 2009 where the difference was greater.

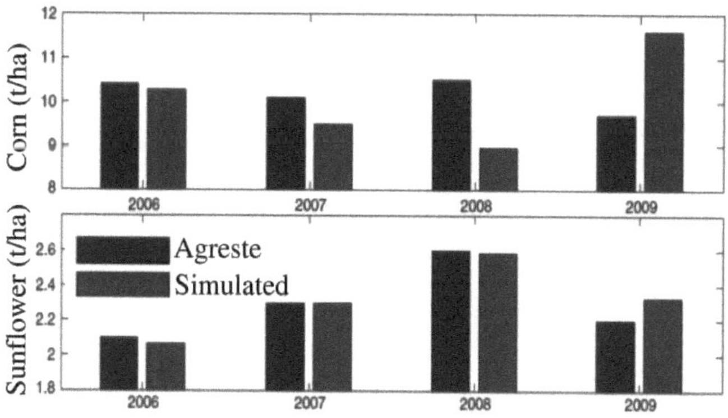

Figure 4.23. *Comparison between yields (ton/ha) simulated by SAFY and those provided by Agreste statistics (department values). For a color version of this figure, see www.iste.co.uk/baghdadi/3.zip*

4.5.5. *Biomass and yield estimates from complex models*

Two examples of remote sensing data assimilation in different crop models are presented.

4.5.5.1. *Example of forcing STICS crop model for the assessment of grassland production from FORMOSAT-2 data*

Various thematic maps were produced in the Crau region (Southern France, 43.55N latitude, 4.876E longitude) by combining crop models, meteorological data, satellite images, the land register, soil and irrigation maps. The crop model STICS has been applied to each plot of irrigated grassland. STICS was developed at INRA [BRI 03] to simulate the crop production taking into account climate, soil and agricultural practices. It can reproduce the functioning of many crops such as wheat, corn, tomatoes, sugarcane, rice, or grasslands. It simulates, among other things, the vegetation development, the production and the water and nitrogen balances.

In the case of irrigated grasslands, the main inputs concern the irrigation dates and water doses, fertilization (dose and dates) and mowing dates. These practices present a high spatial variability throughout the region. Mowing and irrigation dates were estimated from FORMOSAT-2 images following the methods described in section 4.2.1 [COU 10]. This information was then used as forcing inputs in the eSTICS model to estimate production on each plot (Figure 4.24). To assess the contribution of remote sensing data compared with a standard use of the model without inputs from remote sensing, different simulations were performed considering either the model without remote sensing data forcing, or the model forced with various spatial data (LAI, mowing dates, irrigation dates). The results described in Courault *et al.* [COU 10] showed that the best simulations were when most of the spatial information is derived from remote sensing and used as inputs in the crop model.

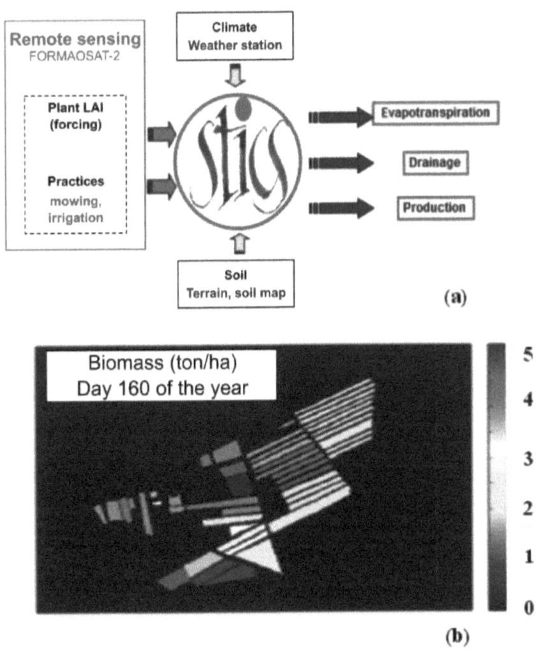

Figure 4.24. *a) Schematic diagram showing the remote sensing data forcing used in the STICS crop model for estimating yields, drainage and evapotranspiration over the Crau region; b) simulated biomass map showing biomass produced on June 10 for each plot. For a color version of this figure, see www.iste.co.uk/baghdadi/3.zip*

4.5.5.2. *Recalibration of the SUCROS model to predict sugar beet yields in northern France, in Picardy*

Monitoring crop growth and forescating yields are important issues in production basins. Thus, the monitoring of the sugar beet development for the sugar industries may advantageously be based on a tool to monitor the production of beetroots. The agronomic services of sugar factories could learn much from the real-time monitoring or the a posteriori analysis of the growth conditions of different plots and better understand the yield variation and the related causes.

A tool consisting of a crop model (SUCROS [SPI 89], version including water stress) and high spatial resolution remote sensing data in the optical domain (provided by SPOT-HRV images and an airborne Pushbroom sensor, equiped with the same spectral bands as SPOT-HRV operated by the French National Space Studies Centre (CNES)) was proposed for two sugar beet production areas in Picardy in France [GUE 03, LAU 05]. The analysis of sugar beet production in these production areas has shown the impact of two major factors in the variability of yields:

– the quality of the success of the crop establishment phase in spring (negative effect of increased seedling mortality in case of slaking crust formation at soil surface);

– the summer water supply (negative effect of an insufficient water storage of soil in case of high water stress).

A diagram has been proposed (Figure 4.25) where the SUCROS model, in which input parameters are fed from various sources (meteorological network, soil map of the Aisne 1:10,000 and associated pedo transfer functions, and sowing dates provided by the farmers) was recalibrated for each unit of land for each plot. A series of images in red and near infrared reflectance acquired from the airborne radiometer and SPOT images were used and the coupling of different models was necessary. After correcting atmospheric effects with the radiative transfer model (SMAC, [RAH 94]), the surface reflectances obtained were transformed into the TSAVI vegetation index [BAR 89] to minimize the impact of residual errors. These data could then be compared with the TSAVI values simulated from LAI (by SUCROS) through the use of the radiative transfer model SAIL [VER 84] applied to simulate the spectral reflectance from the field biophysical variables.

The target parameters in this recalibration were those involved in the two types of limiting process for the production:

– three parameters that represent the quality of crop establishment: duration (in °C day) between the planting and emergence stage, emergence rate and the value of LAI at 500°C day (scale in degree days calculated from planting);

– the maximum depth reached by the roots and the speed of root growth (equivalent parameters to represent the evolution of the available water capacity of soil and to correctly simulate the water supply of the crop and the possible water stress).

Figure 4.26 shows, for a given plot, how the assimilation of four images allows to re-estimate the implantation parameters into more favorable values than the default parameter set the results provide a better simulation of LAI (which largely determines the root biomass and sugar production).

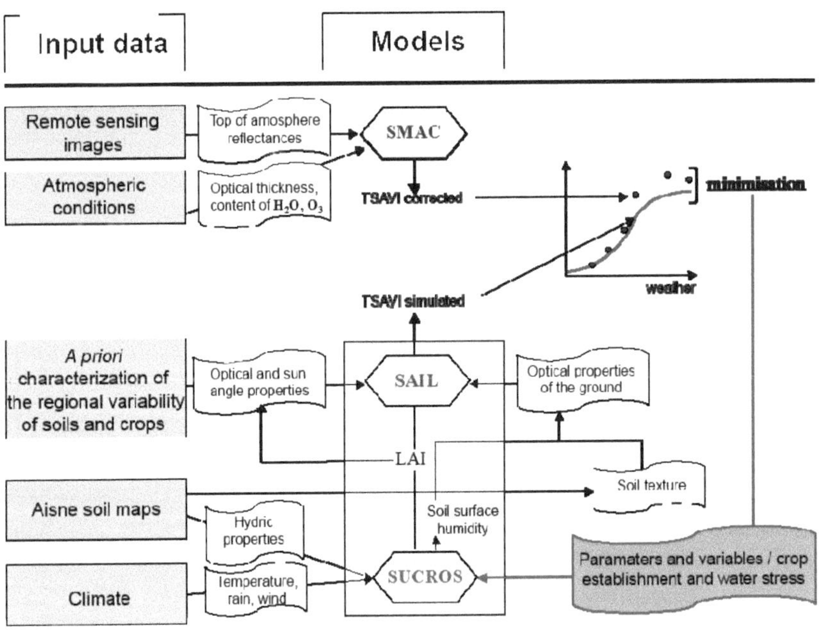

Figure 4.25. *Models and data used in the assimilation procedure [LAU 05]. For a color version of this figure, see www.iste.co.uk/baghdadi/3.zip*

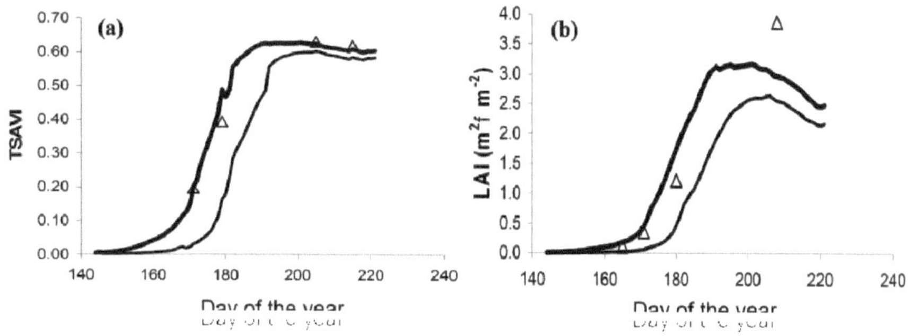

Figure 4.26. *Example of implantation parameter recalibration; thin lines simulation before recalibration, bold lines simulation after recalibration observations are represented by triangles). The result of the re-estimation on a) TSAVI index adjustment, b) the LAI. In this case, the success of crop establishment was underestimated (simulated start slower than observation); the recalibration of crop establishment parameters (sowing-emergence duration and number of emerged plants) allowed to simulate a LAI dynamics in agreement with observations*

The implementation of this procedure, with only four remote sensing images, has significantly improved the estimation of sugar beet production: the average relative error of prediction is 18% with only the model SUCROS, 12% with remote sensing data assimilation into the model. This improved accuracy represents an appreciable gain, even if it is not considered sufficient by the sugar factory managers who still prefer forecasting methods based on heavy campaigns of field measurements. The availability of many more images with the SENTINEL program should achieve higher precision and thus give a higher advantage to these methods.

4.6. Some cases studies of environmental impacts of agriculture: spatial modeling of water, nitrogen and CO_2 fluxes

This section presents various examples of environmental impact studies obtained from a studies conducted at different spatial scales involving crop models coupled with other models (hydrological, geochemical, etc.) and remote sensing data using assimilation methods as described above. The objectives of these studies were either to help farmers and water managers with decision-making, to better estimate the permanent soil properties, or to make diagnoses on the agro-hydro system state in order to predict future evolutions.

4.6.1. *Analysis of the impact of irrigation on the recharge of groundwater (integrated modeling approach)*

In many regions, irrigation affects the groundwater that is used for various purposes.

Global changes (change of land use and climate change) raise many questions about the future of water resources, the potential and future trends of natural ecosystems. The identification of irrigated areas, knowledge on the evolution of water consumption by surfaces, and the variation of water stocks in ground tables are key parameters of an integrated water resources management. The use of models combined with remote sensing data that can spatialize the main variables of interest helps to better understand the functioning of agro-ecosystems and analyze the impact of the global changes on water resources. Remote sensing products such as landuse, LAI or AET maps (shown above) can be used as input data in integrated models, coupling, for example, a hydrological model and crop model as shown in Figure 4.27(a). Such a tool makes it possible to diagnose the current state of the water level of the groundtable. Once calibrated from current or past situations, these models can then be used to predict the scenarios to answer questions such as:

– if we reduce the surface percentage of irrigated grasslands by decreasing the irrigated surfaces by 12% (which is the mean value observed in Crau the last 10 years, because of urban sprawl), then what will be the impact on the groundwater?

– if we reduce irrigation by 30% because of the increased drought recurrent in the region, what will be the grassland productions?

Figure 4.27 shows some outputs of this integrated approach. The evapotranspiration maps (Figure 4.27(b)) obtained by the EVASPA station (presented in section 4.4) are used in the hydrological model which simulates the level of the groundwater (Figure 4.27(b) bottom). The historical MODIS series have been used since 2001 to continuously monitor the AET over the whole region (Figure 4.27(c) red curves). AET variations are analyzed considering cumulative rainfall during the same periods and the variations of the water level of the groundtable (Figure 4.27(c) black line). The comparison of these three variables allows to see the relationship between the water transfer processes at the surface and the groundtable. The calibrated and validated models can be used to simulate different scenarios.

The map at the bottom of Figure 4.27(c) shows yield difference simulated with the STICS model on the grasslands between 2030 and 2010 according to a scenario with a decrease in irrigation of 30%. Production of grasslands generally decreases over the entire area (7% compared with the current situation).

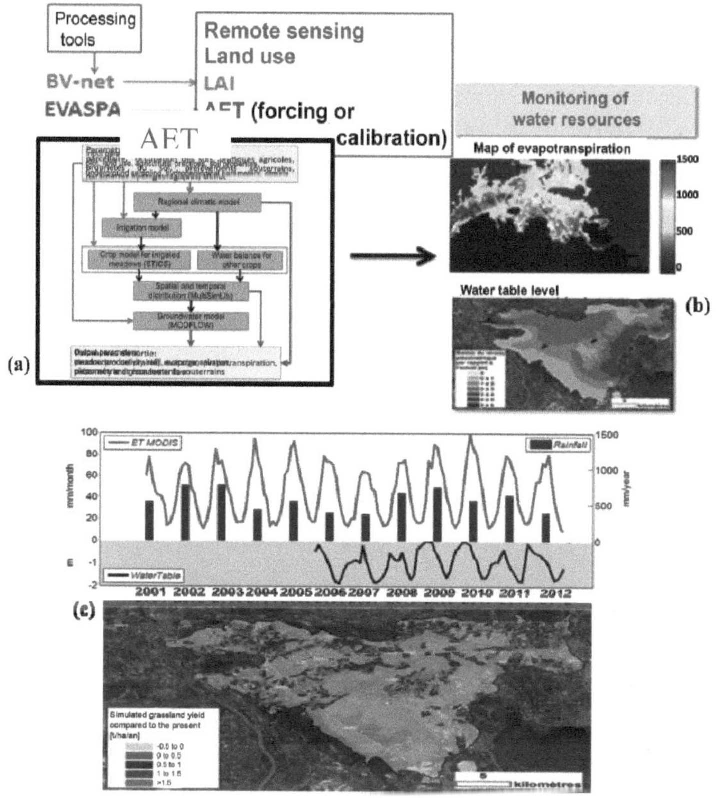

Figure 4.27. *a) Diagram showing how remote sensing data can inform various models simulating the evolution of water resources and production at a regional scale; b) some outputs of the integrated model: top, cumulative evapotranspiration map for one year (in mm) over the Camargue region (South of France) obtained from EVASPA tool and MODIS data and bottom, the map corresponding to the level of the groundwater in the area of Crau; c) bottom: graph showing the relationships between the groundwater level (black line) and the AET (red line) estimated from MODIS and the corresponding rainfall (blue bars) map below simulated grassland yield differences between 2030 and 2010 according to a scenario with a 30% irrigation reduction [OLI 13]. For a color version of this figure, see www.iste.co.uk/baghdadi/3.zip*

In irrigated semi-arid areas, the use of water for agriculture can reach 90% of the total water demand and groundwater often provides a large part of this. In the model developed on the plain of Haouz in Morocco (6000 km^2), satellite imagery was used to estimate water needs according to the simplified method described in section 4.3, and to determine the irrigated areas during the period of 2001–2008. These water requirements for agriculture were compared with recorded irrigated surfaces to indirectly deduce the groundwater uptake through a conceptual model (Figure 4.28). The surface model was coupled with a groundwater flow model. The results provided were satisfactory in regards to the surface runoff and the level of the estimated groundwater compared with observations.

4.6.2. *The spatial modeling of water and nitrogen flows*

The water and nitrogen cycles are closely related. Modern agricultural systems may receive an excess of nitrogen in the form of agricultural inputs that secure crop production at economically viable levels. These excesses lead, in some contexts, to an increasing pollution of water resources, followed by ecological impacts on aquatic ecosystems and socio-economic issues regarding the protection of drinking water, that both question conventional farming practices [MOR 13].

Couplings between crop and hydrological models have been developed to assess the various types of pollution due to agricultural practices, especially nitrogen pollution. The aim of these simulations is twofold: to estimate the excess of nitrogen in agricultural watersheds and its transfer into the hydrological system. To do so, interactions between hydrology, crop growth and multiple nitrogen transformation processes in the soil must be taken into account. The TNT2 model (topography based nitrogen transfer and transformation [BEA 02, FER 11]) is an example of a fully spatially distributed coupling modeling approach. It can simulate the water transfer within the soil from the surface up to the ground table and dissolved nitrogen (in the form of nitrate NO_3^-) between the cells of the grid of a digital elevation model (DEM) describing the topography. The model models simulates the water and nitrogen flow in the river. A crop model (using the formalism of the STICS model) simulates the water and nitrogen consumption associated with crop growth. Nitrogen in its mineral form is available in the soil and is produced by the mineralization of soil organic matter simulated by STICS and application of fertilizers. The spatially integrated model uses spatial information derived from soil maps (depth, texture, organic matter, etc.) and agricultural practice information (crop succession for each plot, fertilization practices, sowing dates and crop type, etc.), derived from field surveys and satellite observation, to spatially represent the process of water and nitrogen transfers through vertical and lateral flows towards the groundwater and the river.

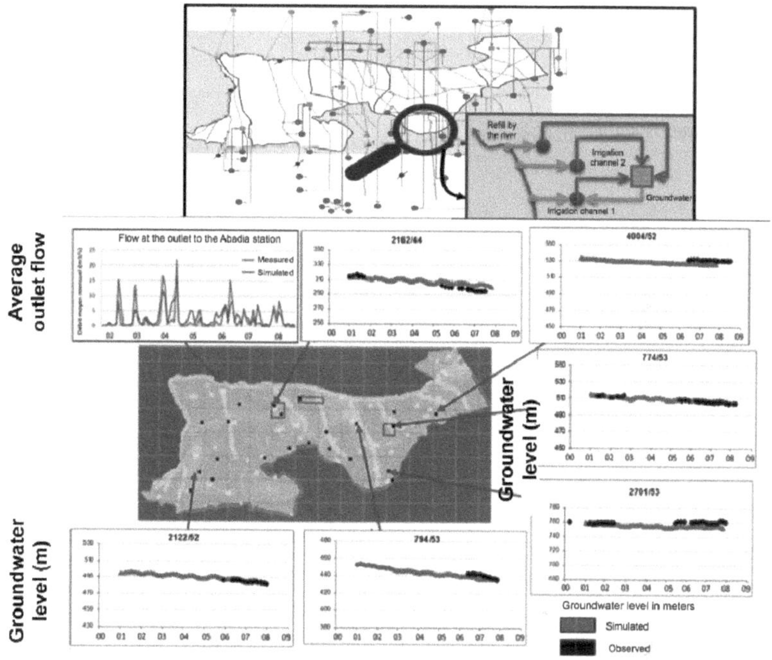

Figure 4.28. *The conceptual model (top) of the Haouz plain in Morocco was developed under WEAP-21 [YAT 05]. It aims to reproduce, in a simplified way, the fluxes between supply and demand. Besides the water needs (for municipalities and industry), the model includes a set of sub-models simulating agricultural areas similar to that of the zoom shown on the upper right: the red dots mean water demand for sites (here two irrigated sectors are fed by a river, then groundwater recharged by the river), the green squares represent water resources (here from groundwater). Fluxes between resources and demand are symbolized by green arrows, returns (excess irrigation, various ground table recharge) are symbolized by red arrows. Water requirements for agricultural surfaces (AET-rain-stock of water in the soil) were obtained using the analysis of MODIS remote sensing images (NDVI in the Kc calculation [4.1]). Below, the map shows the variable water levels of the groundwater (green = permanent, yellow = pronounced drop) at the end of the 2001–2008 simulation. The validation showed values close to the observations (verification of results in some areas such as the flow at the outlet and groundwater level measurement) [LEP 12]. For a color version of this figure, see www.iste.co.uk/baghdadi/3.zip*

The Sentinel-2 and FORMOSAT-2 time series provide estimates of the leaf area index (LAI derived from the BV-NET tool, (see Chapter 2) at appropriate spatial scales for intra-plot observation. LAI is also simulated by the crop model at the DEM cell grid level. This information can be used at the field crop level to retrieve the sowing dates using a recalibration method (see section 4.5.2 [FER 14]) and improves the simulation of the dynamics of evapotranspiration and water content in the soil [MAR 16] or to estimate soil parameters controlling growth, following methods similar to those presented in the next section [FER 16]. The high resolution of the temporal observations allows to retrieve through inversion methods the heterogeneities of soil characteristics (such as the water depth and field porosity) which play a crucial role in the water transfer in the plots (Figure 4.29 after Ferrant *et al.* [FER 16]).

Figure 4.29(a) shows the intra-plot heterogeneities of crop development, visible on an aerial photograph (a) in the small experimental watershed in the village of Auradé, located in Gers (France). A LAI series derived from FORMOSAT 2 images was interpolated for each pixel in order to calculate maximum LAI (LAX) reached during a sunflower season in 2006. The the spatial variability of the maximum LAI (LAX) (Figure 4.29(b)) is representative of the heterogeneity of the growth shown by the high resolution aerial photograph (a). These maximum LAI values are then used to recalibrate some model parameters, including soil properties (among them the available soil water content, (AWC)) (Figure 4.29(c)). Using this AWC map, the model reproduces a satisfactory spatial heterogeneity of the maximum LAI values (Figure 4.29(b)). Therefore, the model more accurately simulates the consumption of nitrogen during crop growth. The map (Figure 4.29(d)) shows that the nitrogen stored in the crop before harvest varies according to:

– the soil water content correctly estimated (Figure 4.29(c));

– the agricultural practices on each plot (intra-plot heterogeneity shown by rectangles);

– hydrological transfers (black ellipses for example).

The evaluation of the amount of nitrogen consumed by crops and exported during the harvest is essential to assess the excess nitrogen (i.e. say the amount of nitrogen unused by the crops) which may contaminate the environment. This contamination is also highly dependent on the physical characteristics of watersheds (hydrology, saturated areas, soil types). This example illustrates the importance of satellite images in the evaluation of nitrogen surpluses across agricultural land.

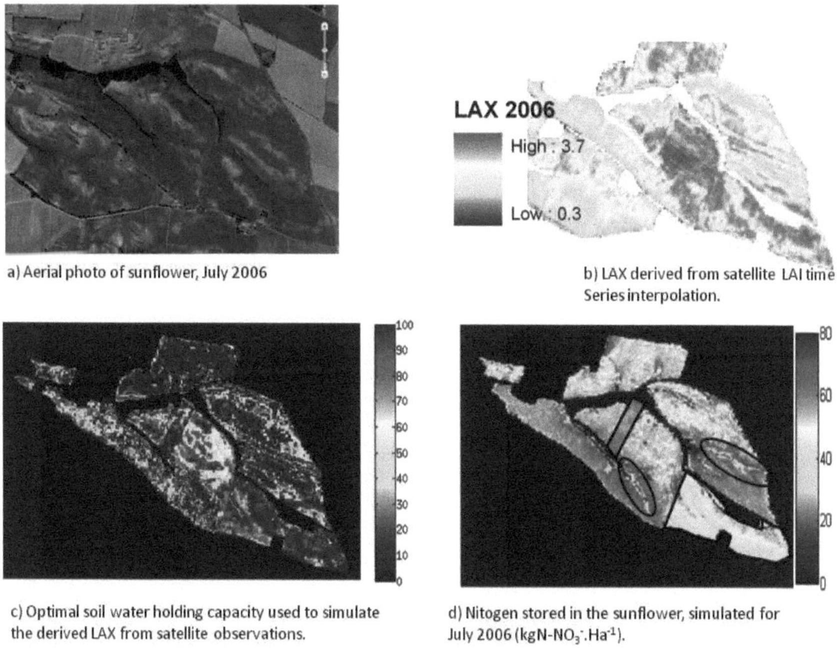

Figure 4.29. *a) Aerial view of a farming area in Gers (2 km wide) consisting of large plots with steep slopes. The white areas visible on the aerial photograph correspond to delays in plant emergence for sunflowers and hydromorphic areas. We find these emergence delay of the sunflower on ridges and in hydromorphic areas. Remote sensing of LAI reveal these heterogeneities (LAI max map, b). Maximum Water Holding Capacity (FER 16], c); The resulting spatial heterogeneity of nitrogen content in crops simulated by the model TNT2 [FER 11] is shown in d).*

The estimation of the LAI maximum (LAX) through remote sensing constrains the nitrogen losses to the environment estimated by an agro-hydrological model by improving the estimated sowing dates (plot scale) and soil parameters (pixel scale) controlling the available soil water content.

4.6.3. *The spatial modeling of CO_2 fluxes and carbon budget*

Agricultural land occupies more than a third of the land surface and is at the heart of sustainable development issues because it contributes to global change by emitting GHG and using water resources.

In this context, the major challenges are:

– maintaining an adequate level of agricultural production, by taking into account the specific pedoclimatic and socio-economic characteristics of each territory;

– reducing the environmental impacts of agro-ecosystems.

To face these challenges, it is important not only to quantify and spatially analyze the dynamics of crop production and environmental indicators, such as the carbon budget (or GHG budget), but also to understand the influence of environmental conditions and management practices on those processes. It will allow us to find solutions for a better management of the agro-ecosystem resources and to identify levers for reducing the environmental impacts of agriculture at different scales (local, regional, global, etc.).

Most models used for the study of net CO_2 fluxes, the carbon (C) budget and its components are either: (i) general biogeochemical models (SPA [WIL 96], DNDC [ZHA 02], ISBA-Ags [CAL 98], ORCHIDEE [KRI 05], ORCHIDEE-STICS [GER 08]) that are generally not able to take into account the complexity and characteristics of agricultural systems, in particular the effects of farm management, or (ii) agro-meteorological models (CERES [GAB 98], STICS [BRI 03], SAFY-WB [DUC 15]) that were originally designed to simulate crop production (NPP, yield) but that have been adapted to simulate the net CO_2 fluxes.

The joint use of some of these models with optical and/or radar satellite data at high spatial and temporal resolution has already helped to provide an accurate spatial description of crop growth and phenology, including the assessment of the components of net CO_2 fluxes (photosynthesis, soil and plant respiration) and the carbon budget (net fluxes, C exports at harvest) (SPA model [REV 13, SUS 10], SAFYE-CO_2 model [VEL 14]). A study was carried out on winter wheat crops in the southwest of France using the

SAFYE-CO_2 model. The GAI time series derived from FORMOSAT-2 and SPOT data were used to calibrate the model parameters related to crop phenology (such as the emergence date, beginning of senescence, leaf partition, see section 4.1). The model was evaluated in terms of biomass and yield, Gross Primary Production (GPP), ecosystem respiration (Reco) and net CO_2 fluxes (Net Ecosystem Exchange, NEE) estimates. Finally, the carbon budgets were calculated. Annual crop carbon budgets take into account the carbon inputs through organic fertilization (C_{inputs}), carbon export at harvest ($C_{exports}$) and the ecosystem annual net CO_2 fluxes (Net Ecosystem Production, NEP), which is the sum of the daily NEE [4.9]. The NEP term is negative if the plot is an annual CO_2 sink (if not, it is positive) and the C_{inputs} term has a negative sign. The carbon budget will therefore be negative if the ecosystem acts as a carbon sink, and it will be positive if the ecosystem is a carbon source.

$$C\ budget = NEP + C_{inputs} + C_{exports} \qquad [4.9]$$

Figure 4.30(a) compares the components of the annual carbon budget observed (left bar) and that simulated by the SAFYE-CO_2 model (hatched bars to the right in each figure).

The differences (in percentage) between observations and simulations of these components are displayed and indicate good model performances. However, this is not the case for the sum of these components in the final carbon budget (84% difference, Figure 4.30(a)). This example illustrates the importance of having a highly accurate estimation of each term that makes up the carbon budget (especially for the NEP and $C_{exports}$ terms that largely compensate for each other) to obtain accurate estimates. This study also showed that the performance of crop models could be further improved by systematically taking into account the development of spontaneous re-growth and weeds and the presence of cover crops (Figure 4.30(b)). The inclusion of inter-crops in cropping systems can limit carbon losses and thus increase soil carbon storage [CES 10].

This type of modeling approach combined with remote sensing data is faced with two important limitations. The first concerns straw management. At present, from remote sensing images, it is very difficult to differentiate the cultivated plots on which straw is exported from those on which straw is buried. As a result, carbon export estimates may have a large uncertainty (which affects the carbon budget). In the future, studies based on a combined use of multi-sensor data (optical, hyper spectral and radar) are expected to overcome this limitation and provide information at the plot scale concerning straw management. The second issue concerns the estimates of carbon inputs through organic fertilization. Because, it is not possible to obtain an estimate of this term for each plot at the regional scale from remote sensing data, the current solution is to have open access to regional and national databases and inventories regarding organic fertilization practices.

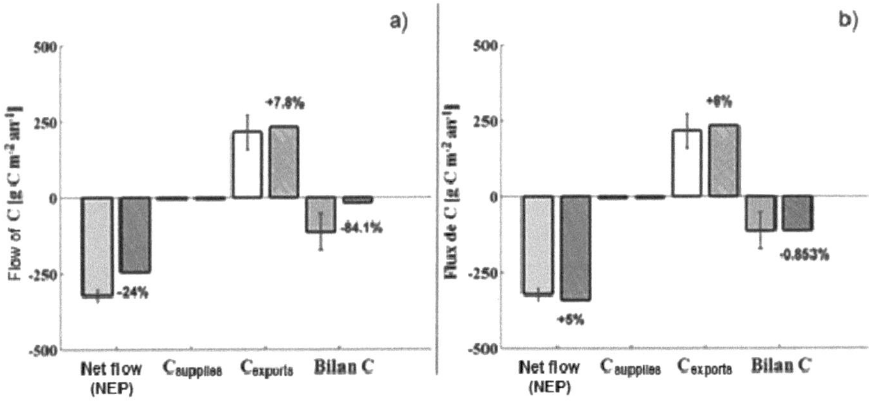

Figure 4.30. *The components of the annual carbon budget, by considering for the NEP estimates a) only the main crop, and b) by taking into account the main crop and spontenous re-growth. For net ecosystem for net ecosystem production (NEP, gray bars) and the amount of carbon exported at harvest ($C_{exports}$, white bars), the first bar of each pair represents the measurements and the information provided by the farmer, and the second bar, which is hatched, shows the results obtained with the model. The black bars represent carbon imported through seeds and organic fertilization (C_{inputs}). The values of C inputs on the right and left bars are identical since this variable cannot be estimated by the model. The green bars represent the carbon budget calculated from observations (left bar) and model outputs (right bar). Error bars represent standard deviations associated with each variable. For a color version of this figure, see www.iste.co.uk/baghdadi/3.zip*

4.7. Precision agriculture

The research field that takes into consideration plot heterogeneity and its temporal evolution in technical operations management, with the aim to optimize production and reduce environmental pollution, is in the domain of "precision agriculture". For 20 years, the development of satellite positioning systems, methods of geophysical measurement, remote sensing, yield mapping, the option for a growing number of techniques (soil tillage sowing, fertilizer and pesticide applications, etc.) have all been advances that make it possible to measure plot heterogeneity and to take this into account in the management of technical operations [GUE 01]. The challenge of the research focused on formalizing and automating the process that leads from the collection of relevant information on the crop state (soil/plant) to a decision tool for the spatial application of a technique up to control of the farming equipment. Remote sensing plays an important role in the collection of information upstream of the decision process. Given the spatial resolution necessary to access the within-field variability, for now it concerns essentially data in the optical domain, which provides information on leaf development and chlorophyll content of vegetation. Operational services have been created by several operators in France and internationally. Some use embedded measurement systems for tractors such as N-Sensor (Yara http:// www.yara.fr/fertilisation/outils-et-services/n-sensor/, a sensor that estimates the state of nitrogen nutrition in crops) while others use satellite or airborne sensors such as Farmstar (Airbus, Arvalis, Cetiom, https//www.farmstar-conseil.fr/). Accuracy varies depending on the sensor used and can descend to the level of the plant when the sensors are embedded on a tractor.

To assist nitrogen fertilization monitoring for example, the estimation of biophysical variables (LAI, chlorophyll content) using remote sensing is used to diagnose the crop state in comparison to a reference situation (N-Sensor) or through the implementation of proven agronomic models (Farmstar) and specific rules are accurately applied. Figure 4.31 shows a Farmstar product, a plot map of recommended nitrogen rates for the 3rd nitrogen application on wheat. The method can be applied to various types of high-resolution data acquired by satellites, airplanes or drones.

4.7.1. *Indicator-based advice for crop management*

In Figure 4.31, the calculation of the nitrogen rate is based on the calculation of an indicator, which reflects the nitrogen nutrition status of the crop.

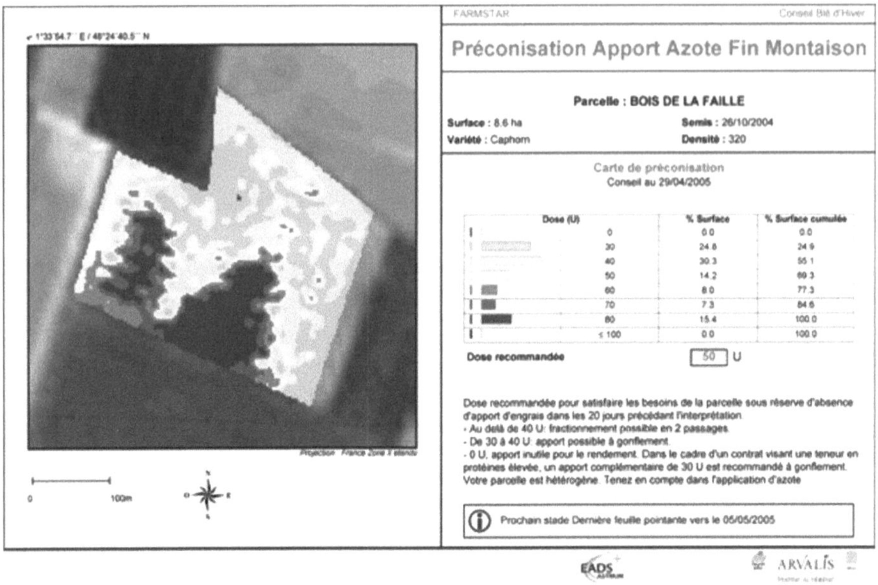

Figure 4.31. *Example of nitrogen variable rate application advice for wheat by Farmstar (Airbus, Arvalis, Cetiom). The different colors indicate the different recommended rates and an average recommendation is given to farmers who do not have the opportunity to make a variable rate application. For a color version of this figure, see www.iste.co.uk/baghdadi/3.zip*

The nitrogen nutrition status of the crop is conventionally assessed by the nitrogen nutrition index, or NNI, the ratio between the nitrogen content of the aerial parts of the crop to the critical content defined as the value above which the crop does not suffer from nitrogen deficiency. The critical curve expressing this optimal state is also called the nitrogen dilution curve (Figures 4.32(a) and (b)). The dilution of the nitrogen content of the aerial parts of the crop over time (concomitantly with the increase in biomass) is linked to self-shadowing effects that appear in the field with plant development and the evolution of the leaf/stem ratio (for details, see [JUS 94]).

Attempts have been made to directly connect the chlorophyll content of leaves, estimated by remote sensing, to the NNI [BLO 05], and then apply appropriate nitrogen supply rules. Established relationships vary greatly over time and require many calibrations.

Houlès *et al.* [HOU 07] proposed to directly estimate the fertilization deficit through the absorbed nitrogen ΔQN deficit. This is defined (Figure 4.32(c)) as the difference between the amount of nitrogen absorbed by the field (leaves + stems), here denoted as QN, and the amount of critically absorbed nitrogen (QN_c). The critical absorption curve is derived from the dilution curve: the QN quantities (QN_c, respectively) are calculated as the product of the actual nitrogen N_R (respectively critical nitrogen content N_c) and the plant biomass W_R (we consider here amounts of nitrogen in kg per hectare).

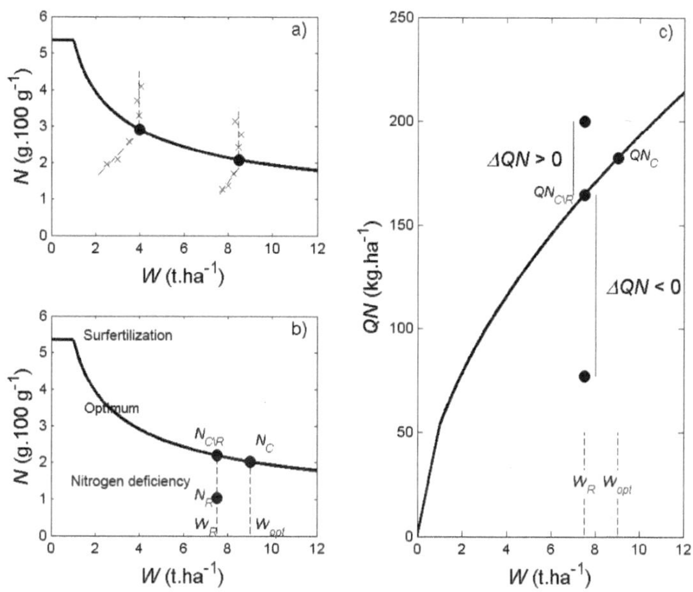

Figure 4.32. *Dilution curve and absorption deficit: a) construction of points of the dilution curve (solid circles) from experimental points (crosses): each family of experimental points corresponds to an observation date of couples (nitrogen dose provided; crop biomass measured); the curve is defined by the set of points for which an increase in the nitrogen rate no longer provides an increased biomass production; b) the three areas defined by the dilution curve: definition of critical nitrogen content from an actual biomass measurement (W_R) and the actual nitrogen (N_R). $N_{C\backslash R}$ is the W_R projection on the dilution curve, while N_C is determined based on the biomass that would have reached the field if it had never suffered from nitrogen deficiency, W_{opt} (solid circles); c) critical absorption curve and absorption deficit: the deficit or excess of absorption will be calculated based on the amount of nitrogen that would have absorbed the crop at the observed biomass rather than the optimal biomass W_{opt} (source [HOU 07])*

This indicator built on ecophysiological considerations can be approached by remote sensing. Indeed, the reflectances in the optical domain give access to the LAI, the leaf chlorophyll content (Cab) and the integrated content over the canopy (QCab) (see Chapter 2). From these variables, and an empirical allometric relationship set calibrated on large experimental data sets, the biomass of aerial parts (W_R), the nitrogen content of the aerial parts (N_R), and the critical values (N_C) can be estimated. Therefore, estimate the nitrogen absorption deficit (ΔQN) which is the standard for determining the nitrogen rate to be applied. Figure 4.33 shows different ways to estimate the nitrogen absorption deficit.

Houlès *et al.* [HOU 07] compared the different ways of calculating ΔQN in Figure 4.33. This is the method that directly connects the chlorophyll content (QC_{ab}) to the nitrogen content of the field (Q_{NR}) and the LAI to the critically absorbed nitrogen (Q_{NC}, method 3) which gave the best results in capacity prediction (the prediction error is 18 kg N ha^{-1} for a range variation of ΔQN from 10 to 190 kg N ha^{-1}). The method is described in Houlès *et al.* [HOU 07]. The authors also showed that the method is more robust when used with other wheat varieties than on those in which empirical relationships have been established.

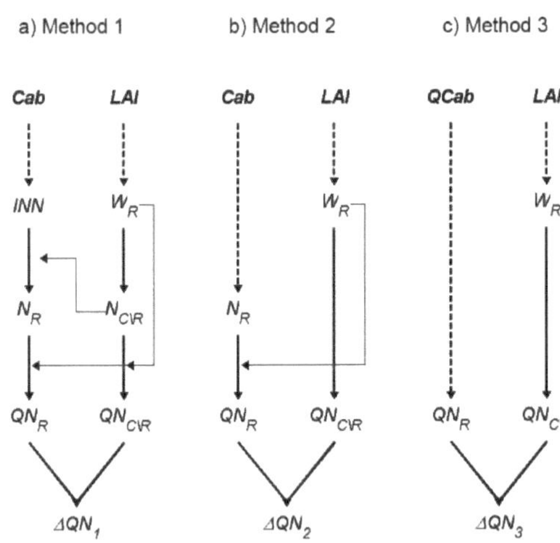

Figure 4.33. *Three methods for estimating the nitrogen absorption deficit for wheat from the chlorophyll content and leaf area indexes accessible through remote sensing (NNI is a nitrogen nutrition index, calculated as N_R/N_C) [HOU 07]*

Subject to the availability of appropriate remote sensing images (acquired at fine resolution and at key periods when fertilization advice must be given), and calibration relationship sets, the ΔQN indicator is easily integrable into a decision support system and delivers spatial information on the nitrogen supply to be provided. This method has been integrated into the Farmstar "Nitrogen supply at end of stem elongation" product for wheat, relying on airborne campaigns. The recent availability of Sentinel-2 images, equipped with optical spectral bands in the visible to near infrared giving access to the chlorophyll content (following the methods described in Comar *et al.* [COM 10]) allows us to consider an easier (and less expensive) implementation of this method on a wider scale.

Drones, which are being used increasingly, can be a very good alternative to satellite sensors. Indeed, even though the theoretical temporal frequency of their acquisitions has increased significantly with the arrival of Sentinel-2, satellite images will still remain subject to the hazard of clouds for local solutions. Drones use various types of sensors, work at a fine scale, and operate at selected periods (see special issue of technical specifications from the INRA: https://www6.inra.fr/cahier_des_techniques/Les-numeros-thematiques/GPS-and-SIG).

4.7.2. *Advice based on the use of crop models*

One of the limits of the previous approach is the need to acquire images at key moments of the crop calendar, which is not always feasible given the weather conditions. Another limit is that it does not include, in the diagnosis, the state of the soil–plant system as a whole (only the nitrogen content of leaves is considered). So, one may have to advise a nitrogen input to correct a nitrogen deficiency observed while a large amount of mineral nitrogen exists in the soil but is unavailable because of water stress. Using a crop model, however, allows us to update all the system state variables (including the mineral nitrogen content of the soil, mineralization potential, etc.) and give a daily prediction (even in the absence of remote sensing data). Still in the case of the nitrogen fertilization of wheat, the concept of a diagnosis and a recommendation based on the predictions of a crop model was tested at INRA in Avignon (France). The difficulty of the implementation of a crop model to a sub-plot scale lies in the availability of information on spatial inputs, mainly the parameters describing the soil properties, responsible for plot heterogeneity. The assimilation of remote sensing data into the model allows for recalibration, either at the pixel level, or for homogeneous areas of soil properties, thus obtaining a spatialization of growth variables prediction in the plot to formulate fertilization recommendations. In this case, remote sensing data may not be acquired at key dates; it is enough to have a sufficient number of images allocated to the growing season.

For this exploratory work, four hyperspectral images acquired with an airborne CASI (Compact Airborne Spectrographic Imager 3, Itres Inc.) sensor (made available by Astrium) and one SPOT image acquired during the wheat growing season were used [GUE 07]. The LAI and leaf chlorophyll content values were derived from these images (Figure 4.34). The observations of these two types of variables were used to achieve a spatial recalibration (with a spatial resolution of 20 m) of some of the model input parameters, mainly those describing soil properties. The use of Bayesian methods (such as the Importance Sampling method) using prior information on the parameter values can solve this ill-posed problem of estimation where multiple equivalent solutions are possible.

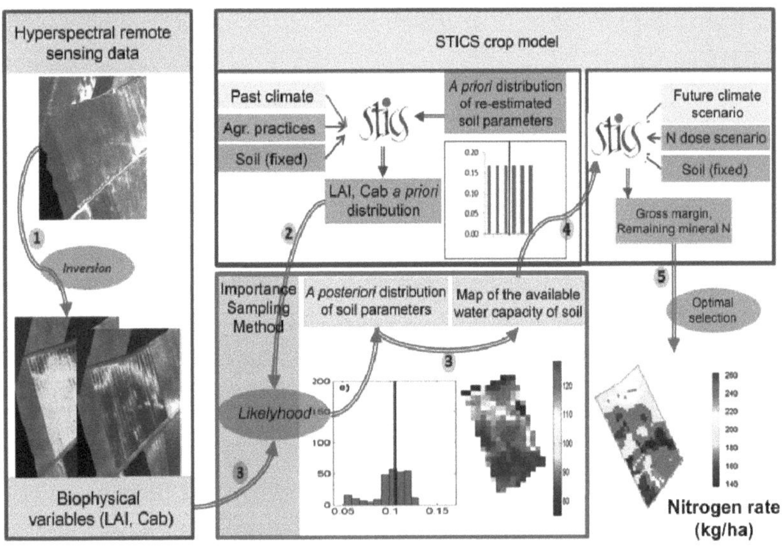

Figure 4.34. *The five steps of the development of the spatialized nitrogen rate recommendation advice using the STICS crop model and hyperspectral remote sensing data [GUE 07]. For a color version of this figure, see www.iste.co.uk/baghdadi/3.zip*

Once the model is spatially recalibrated (Figure 4.34, steps 2 and 3), it is used to assess nitrogen rate application scenarios crossed with climate scenarios (Figure 4.34, step 4). The selected rate is the one optimizing an agri-environmental criterion based on the maximizing gross margin (related to the crop yield and grain protein content) constrained to a threshold of remaining mineral nitrogen in the soil at harvest (Figure 4.34, step 5) [HOU 04]. Note that this constraint, which simulates the implementation of more environmentally friendly practices, can take different forms, but is not, to date, regulatory in nature. At the end of this process, a map of the recommended nitrogen rates is therefore produced.

4.8. Results and prospects

Many spatial observations are already available over a wide spatial and temporal domain. The launch of Sentinel-2 satellites, and soon Venμs, will add a multi-temporal dimension to the high spatial resolution data that has existed until now to cover the globe (SPOT, Ikonos, etc.). This clearly opens the way for new development for operational applications over lands such as:

– land use mapping, characterization of crop systems, change detection;

– estimation of biophysical products (LAI, fAPAR, chlorophyll content, biomass);

– crop and water monitoring taking into account the variability of agricultural practices (irrigation, harvest, etc.)

– estimation of agricultural production, water and carbon fluxes and budgets at different scales from the plot to the territory.

More and more services and platforms are being developed to deliver free satellite data with pre-processing (atmospheric corrections, georeferencing, biophysical products) let us note the Theia initiative (thematic pole continental surfaces: https://www.theia-land.fr/) in France and the Copernicus services for Europe (http://www.copernicus.eu/main/services). The aim of the latter program is to develop ways of collecting and disseminating data and integrating them into systems for monitoring and forecasting the state of the environment.

4.9. Key points

It is possible to monitor crop development from optical data using the analysis of temporal profiles of vegetation indices or from the LAI with a relatively good accuracy.

The detection of certain agricultural practices (emergence date, date of harvest, irrigation) is now accessible thanks to the arrival of new sensors with high spatial and temporal resolution such as FORMOSAT-2 and Sentinel-2.

Radar and thermal data can significantly improve the performances for crop and water management (particularly the detection of irrigation practices).

Various remote sensing data assimilation methods have been developed in crop models to study a wide range of agro-ecosystem situations for purposes such as estimating permanent soil characteristics, yield estimation, water and carbon balances, monitoring the groundwater level, or the adaptation of agricultural practices (nitrogen fertilization, irrigation).

Some applications can run in nearly real time mode and can assist in the management of these essential resources.

4.10. Bibliography

[ALL 98] ALLEN R.G., PEREIRA L.S., RAES D. *et al.*, "Crop evapotranspiration: guidelines for computing crop requirements", *Irrigation and Drainage Paper No. 56*, FAO, Rome, Italy, p. 300, 1998.

[BAD 84] BADHWARG D., "Use of LANDSAT-derived profile features for spring small-grains classification", *International Journal of Remote Sensing*, vol. 5, no. 5, pp. 783–797, 1984.

[BAG 16] BAGHDADI N., ZRIBI M., "Characterization of soil surface properties using radar remote systems", in BAGHDADI N., ZRIBI M. (eds), *Surface Remote Sensing in Continental Hydrology*, ISTE Press, London and Elsevier, Oxford, 2016.

[BAR 89] BARET F., OLIOSO A., "Photosynthetically absorbed radiation by a wheat canopy estimated from spectral reflectance", *Agronomie*, vol. 9, no. 9, pp. 885–895, 1989.

[BAU 87] BAUSCH W.C., NEALE C.M.U., "Crop coefficients derived from reflected canopy radiation: a concept", *Transactions ASAE*, vol. 30, pp.703–709, 1987.

[BEA 02] BEAUJOUAN V., DURAND P., RUIZ L. *et al.*, "A hydrological model dedicated to topography-based simulation of nitrogen transfer and transformation: rationale and application to the geomorphology-denitrification relationship", *Hydrological Processes*, vol. 16, pp. 493–507, 2002.

[BLO 05] BLONDLOT A., GATE P., POILVIE H., "Providing operational nitrogen recommandations to farmers using satellite imagery", *Proceedings of the 5th European Conference on Precision Agriculture*, Uppsala, Sweden, pp. 345–352, 2005.

[BOI 15] BOIZET A., Cartographie des dates de levée et des surfaces irriguées dans le Sud-Ouest de la France, à partir d'images SPOT4 et FORMOSAT 2, Master report CESBIO, 2015.

[BOO 96] BOOTE K.J., JONES J.W., PICKERING N.B., "Potential uses and limitations of crop models", *Agronomic Journal*, vol. 88, no. 5, pp.704–716, 1996.

[BOU 96] BOUMAN B.A.M., VANKEULEN H., VANLAAR H.H. *et al.*, "The 'School of de Wit' crop growth simulation models: apedigree and historical overview", *Agricultural Systems*, vol. 52, nos. 2–3, pp. 171–198, 1996.

[BRI 03] BRISSON N., GARY C., JUSTES F. *et al.*, "An overview of the crop model STICS", *European Journal of Agronomy*, vol. 18, nos. 3–4, pp. 309–332, 2003.

[BRO 09] BROWN J.F., MAXWELL S., PERVEZ S., "Mapping Irrigated Lands across the United States using MODIS satellite imagery", in THENKABAIL P., LYON J., TURRAL H.M.C. (ed.), *Remote Sensing of Global Croplands for Food Security*, CRC Press, 2009.

[CAL 98] CALVET J.C., NOILHAN J.L., ROUJEAN J.L. *et al.*, "An interactive vegetation SVAT model tested against data from six contrasting sites", *Agricultural and Forest Meteorology*, vol. 92, no. 2, pp. 73–95, 1998.

[CES 10] CESCHIA E., BÉZIAT P., DEJOUX J.F. *et al.*, "Management effects on net ecosystem carbon and GHG budgets at European crop sites", *Agriculture, Ecosystems & Environment*, vol. 139, no. 3, pp. 363–383, 2010.

[CHA 14] CHAHBI A., ZRIBI M., LILI-CHABAANE Z. *et al.*, "Analysis Of Optical Remote Sensing Potential For Yields Of Cereals Estimation", *1st International Conference on Advanced Technologies for Signal and Image Processing (ATSIP)*, pp. 367–372, 17–19 March, 2014.

[CHE 14] CHEN Y., COURNÈDE P.H., "Data assimilation to reduce uncertainty of crop model prediction with convolution particle filtering", *Ecological Modeling*, vol. 290, pp.165–177, 2014.

[CHI 13] CHIROUZE J., Estimation spatialisée de l'évapotranspiration et suivi de l'irrigation en milieu semi-aride à l'aide des données de l'infrarouge thermique, PhD thesis, University Paul Sabatier Toulouse, France, Cesbio, 2013

[CHO 94] CHOUDHURY B.J., AHMED N.U., IDSO S.B. *et al.*, "Relations between evaporation coefficients and vegetation indices studied by model simulations", *Remote Sensing Environment*, vol. 50, pp. 1–17, 1994.

[CLA 12] CLAVERIE M., DEMAREZ V., DUCHEMIN B. *et al.*, "Maize and sunflower biomass estimation in southwest France using high spatial and temporal resolution remote sensing data", *Remote Sensing Environment*, vol. 124, pp. 844–857, 2012.

[COM 10] COMAR A., BARET F., SOLAN B.D. *et al.*, "Estimation of green area index and chlorophyll content based on 3D canopy architecture models and the combination of gap fraction and hyperspectral reflectance measurements. Application to high throughput wheat phenotyping", *International Conference on Agricultural Engineering – AgEng: Towards Environmental Technologies*, Clermont-Ferrand, France, pp. 464–464, 6-8 September, 2010.

[CON 15] CONSTANTIN J., WILLAUME M., MURGUE C. *et al.*, "The soil-crop models STICS and AqYield predict yield and soil water content for irrigated crops equally well with limited data", *Agricultural and Forest Meteorology*, vol. 206, pp. 55–68, 2015

[COU 05] COURAULT D., SEGUIN B., OLIOSO A., "Review on estimation of evapotranspiration from remote sensing data: from empirical to numerical modeling approaches", *Irrigation and Drainage Systems*, vol. 19, no. 3–4, pp. 223–249, 2005.

[COU 10] COURAULT D., HADRIA R., RUGET F. *et al.*, "Combined use of FORMOSAT-2 images with a crop model for biomass and water monitoring of permanent grassland in Mediterranean region", *Hydrology and Earth System Sciences*, vol. 14, no. 9, pp. 1731–1744, 2010.

[DED 10] DEDIEU G., HAGOLLE O., HUC M. *et al.* "The VENμS mission and the benefits of earth observation with high spatial and temporal resolutions", *3rd International Symposium Recent Advances in Quantitative Remote Sensing*, available at http://ipl.uv.es/raqrs/, Valencia, Spain, 27 September 1 October 2010.

[DON 03] DONG J.R., KAUFMANN R.K., MYNENI R.B. *et al.*, "Remote sensing estimates of boreal and temperate forest woody biomass: carbon pools, sources, and sinks", *Remote Sensing Environment*, vol. 84, no. 3, pp.393–410, 2003.

[DOR 07] DORIGO W.A., ZURITA-MILLA R., DE WITT A.J.W. *et al.*, " A review on reflective remote sensing and data assimilation techniques for enhanced agroecosystem modeling ", *International Journal of Applied Earth Observation and Geoinformation*, vol. 9, no. 2, pp. 165–193, 2007.

[DUC 06] DUCHEMIN B. *et al.*, "Monitoring wheat phenology and irrigation in Central Morocco: On the use of relationships between evapotranspiration, crops coefficients, leaf area index and remotely-sensed vegetation indices", *Agricultural Water Management*, vol. 79, no. 1, pp. 1–27, 2006.

[DUC 08] DUCHEMIN B. *et al.*, "Agrometerological study of semi-arid areas: an experiment for analysing the potential of time series of FORMOSAT-2 images (Tensift-Marrakech plain) ", *International Journal of Remote Sensing*, vol. 29, nos. 17–18, pp. 5291–5300, 2008.

[DUC 09] DUCHEMIN B., BENHADJ I., HADRIA R. *et al.*, "Evaluation of irrigation water amount in semi-arid croplands using time series of FORMOSAT-2 images", *IEEE International Geoscience and Remote Sensing Symposium*, pp. III-474–III-477, 12-17 July, 2009.

[DUC 15] DUCHEMIN B., FIEUZAL R., AUGUSTIN RIVERA M. *et al.*, "Impact of sowing date on yield and water use efficiency of wheat analyzed through spatial modeling and FORMOSAT-2 images", *Remote Sensing*, vol. 7, no. 5, pp. 5951–5979, 2015.

[DUS 14] DUSSEUX P., GONG X., HUBERT-MOY L. *et al.*, "Identification of grassland management practices from leaf area index time series", *Journal of Applied Remote Sensing*, vol. 8, no. 1, p. 083559, 2014.

[ELH 14] EL HAJJ M., BAGHDADI N., BELAUD G. *et al.*, "Irrigated grassland monitoring using a time series of TerraSAR-X and COSMO-SkyMed X-Band SAR data", *Remote Sensing*, vol. 6, no. 10, pp. 10002–10032, 2014.

[ELH 16] EL HAJJ M., BAGHDADI N., CHEVIRON B. *et al.,* "Integration of remote sensing derived parameters in crop models: application to the PILOTE model for hay production", *Agricultural Water Management,* vol. 176, pp. 67–79, 2016.

[FAI 04] FAIVRE R., LEENHARDT D., VOLTZ M. *et al.*, "Spatialising crop models", *Agronomie*, vol. 24, no. 4, pp.205-217, 2004.

[FER 11] FERRANT S., OEHLER F., DURAND P. *et al.*, "Understanding nitrogen transfer dynamics in a small agricultural catchment: comparison of a distributed (TNT2) and a semi distributed (SWAT) modelling approaches", *Journal of Hydrology*, vol. 406, pp. 1–15, 2011.

[FER 14] FERRANT S., GASCOIN S., VELOSO A. *et al.*, "Agro-hydrology and multi-temporal high-resolution remote sensing: toward an explicit spatial processes calibration", *Hydrology and Earth System Sciences*, vol. 18, no. 12, pp. 5219–5237, 2014.

[FER 16] FERRANT S., BUSTILLO V., BUREL E. *et al.*, "Extracting soil water holding capacity parameters of a distributed agro-hydrological model from high resolution optical satellite observations series", *Remote Sensing, Special issue Remote Sensing in Precision Agriculture*, vol. 8, no. 2, p. 154, 2016.

[GAB 98] GABRIELLE B., DENOROY P., GOSSE G. *et al.*, "Development and evaluation of a CERES-type model for winter oilseed rape", *Field Crops Research*, vol. 57, no. 1, pp. 95–111, 1998.

[GAL 13] GALLEGO-ELVIRA B., OLIOSO A., MIRA M. *et al.*, "EVASPA (EVApotranspiration Assessment from SPAce) tool: An overview", *Procedia Environmental Sciences*, vol. 19, pp. 303–310, 2013.

[GER 08] GERVOIS S.B., CIAIS P., DE NOBLET-DUCOUDRÉ N. *et al.*, "Carbon and water balance of European croplands throughout the 20th century", *Global Biogeochemical Cycles*, vol. 22, no. GB2022, p. 13, 2008.

[GUE 01] GUÉRIF M., BARET F., MOULIN S. *et al.*, "Prise en compte de l'hétérogénéité parcellaire et de son evolution temporelle dans la gestion des interventions techniques: potentiel de la télédétection", in MALÉZIEUX E. (ed), *Modélisation des agro-écosystèmes et aide à la décision*, CIRAD, Montpellier 2001.

[GUE 07] GUÉRIF M., HOULÈS V., MARY B. *et al.*, "Modulation intra-parcellaire de la fertilisation azotée du blé fondée sur le modèle de culture Stics. Intérêt de la démarche et méthodes de spatialisation", *Agriculture de précision*, GUÉRIF M., KING D. (eds), éditions Quae, 2007.

[GUE 03] GUERIF M., LAUNAY M., "Assimilating remote sensing data in a sugar beet model to describe crop variability: an application to sugar beet yield prediction and precision agriculture", *Colloquium on Sugar Beet Growth and Modeling*, p. 135–142, Lille, France, 12 September, 2003.

[GUE 93a] GUÉRIF M., DE BRISIS S., SEGUIN B., "Complementarity of SPOT-HRV and NOAA-AVHRR data for crop yield assessment in semiarid environments", *EARSeL Advances in Remote Sensing*, vol. 2, no.2-VI, pp. 110–123, 1993.

[GUE 93b] GUÉRIF M., DELÉCOLLE R., "Introducing remotely sensed estimates of canopy structure into plant models", in VARLET-GRANCHER C., BONHOMME R., SINOQUET H. (eds), *Canopy Structure and Light Microclimate, Methods of Characterization and Applications*, INRA, 1993.

[GUE 96a] GUÉRIF M. *et al.*, "Identification de parcellesagricoles à partir de la forme de leurévolutionradiométrique au cours de la saison de culture", *Photo-interprétation*, vol. 34, pp.12–22, 1996.

[GUE 96b] GUÉRIF M., COURAULT D., BRISSON N., "Assimilation des données de télédétectiondans les modèles de fonctionnement des cultures", *Actes de l'Ecole-Chercheurs INRA en Bioclimatologie, Le Croisic*, INRA Editions, Du pp.169–191, 25–29 March 1996.

[HAD 09] HADRIA R., DUCHEMIN B., BAUP F. *et al.*, "Combined use of optical and radar satellite data for the detection of tillage and irrigation operations: case study in Central Morocco", *Agricultural Water Management*, vol. 96, pp. 1120–1127, 2009.

[HAG 13] HAGOLLE O., SYLVANDER S., HUC M. *et al.*, "SPOT4 (Take5): Simulation of SENTINEL-2 Time Series on 45 Large Sites", *Proceedings of ESA's Living Planet Symposium*, Edinburgn, 2013.

[HOU 04] HOULÈS V., MARY B., GUÉRIF M. *et al.*, "Evaluation of the crop model STICS to recommend nitrogen fertilisation rates according to agro-environmental criteria", *Agronomie*, vol. 24, pp. 339–349, 2004.

[HOU 07] HOULÈS V., GUÉRIF M., MARY B., "Elaboration of a nitrogen nutrition indicator for winter wheat based on leaf area index and chlorophyll content for making nitrogen recommendations", *European Journal of Agronomy*, vol. 27 no.1, pp. 1–11, 2007.

[JAM 98] JAMIESON P.D., PORTER J.R., GOUDRIAAN J. *et al.*, "A comparison of the models AFRCWHEAT2, CERES-wheat, Sirius, SUCROS2 and SWHEAT with measurements from wheat grown under drought", *Field Crops Research*, vol. 55, nos. 1–2, pp. 23–44, 1998.

[JUS 94] JUSTES E., MARY B., MEYNARD J.M. *et al.*, "Determination of a critical nitrogen dilution curve for winter-wheat crops", *Annals of Botany*, vol. 74, no.4, pp. 397–407, 1994.

[KAL 08] KALMA J.D., MCVICAR T.R., MCCABE M.F., "Estimating land surface evaporation: areview of methods using remotely sensed surface temperature data", *Surveys in Geophysics*, vol. 29, nos. 4–5, pp. 421–469, 2008.

[KRI 05] KRINNER G., VIOVY, N., DE NOBLET-DUCOUDRE N. *et al.*, "A dynamic global vegetation model for studies of the coupled atmosphere-biosphere system", *Global Biogeochemical Cycles*, vol. 19, pp.1–33, 2005.

[KUM 81] KUMAR M., MONTEITH J.L., SMITH H. (ed.), *Remote Sensing of Crop Growth, Plants and the Daylight Spectrum*, Academic Press, 1981.

[LAG 13] LAGOUARDE J.P. *et al.*, "The MISTIGRI thermal infrared project: scientific objectives and mission specifications", *International Journal of Remote Sensing*, vol. 34, nos. 9–10, pp. 3437–3466, 2013.

[LAG 16] LAGOUARDE J.-P., BOULET G., "Energy balance of continental surfaces and the use of surface temperatures", in BAGHDADI N., ZRIBI M. (eds), *Surface Remote Sensing in Continental Hydrology*, ISTE Press, London and Elsevier, Oxford, 2016.

[LAR 14] LARUE F., Apport de la télédétection à hautes résolutionsspatiale et temporelle pour l'estimation des besoins en eau des cultures irriguées sur le bassin Versant de la Neste, Master Thesis, Cesbio, Toulouse, France, 2014.

[LAU 05] LAUNAY M., GUÉRIF M., "Assimilating remote sensing data into a crop model to improve predictive performance for spatial applications", *Agriculture, Ecosystems and Environment*, vol. 111, pp. 321–339, 2005.

[LAU 13] LAURENS S., Apport de l'imagerie optique multi-temporelle pour la cartographie des surfaces irriguées, Sud-ouest de Toulouse (Midi Pyrénées, France), Master Thesis, Cesbio, Toulouse, France, 2013.

[LEB 91] LEBLON B., GUERIF M., BARET F., "The use of remotely sensed data in estimation of PAR use efficiency and biomass production of flooded rice", *Remote Sensing of Environment*, vol. 38, no.2, pp. 147–158, 1991.

[LEB 07] LEBOURGEOIS V., BÉGUÉ A., DEGENNE P. *et al.*, "Improving harvest and planting monitoring for smallholders with geospatial technology: the Reunion Island experience", *International Sugar Journal,* vol. 109, no. 1298, pp. 109–119, 2007.

[LEP 12] LE PAGE M., BERJAMY B., FAKIR Y. *et al.*, "An integrated DSS for groundwater management based on remote sensing.The case of a semi-arid aquifer in Morocco", *Water Resources Management*, vol. 26, pp. 3209–3230, 2012.

[LEP 14] LE PAGE M., TOUMI J., KHABBA S. *et al.*, "A Life-size and near real-time test of irrigation scheduling with a SENTINEL-2 like time series (SPOT4-Take5) in Morocco", *Remote Sensing*, vol. 6, pp. 11182–11203, 2014.

[LOB 03] LOBELL D.B., ASNER G.P., ORTIZ-MONASTERIO J.I. *et al.*, "Remote sensing of regional crop production in the Yaqui Valley, Mexico: estimates and uncertainties", *Agriculture Ecosystems & Environment*, vol. 94, no. 2, pp. 205–220, 2003.

[LIU 10] LIU J., PATTEY E., MILLER J.R. *et al.*, "Estimating crop stresses, aboveground dry biomass and yield of corn using multi-temporal optical data combined with a radiation use efficiency model", *Remote Sensing of Environment*, vol. 114, no. 6, pp. 1167–1177, 2010.

[MAR 16] MARTIN E., GASCOIN S., GRUSSON Y. *et al.*, " On the use of hydrological models and satellite data to study the water budget of river basins affected by human activities: examples from the Garonne basin of France", *Surveys in Geophysics*, vol. 37, no. 2, pp. 223–247, 2016.

[MAR 13] MARI J.F., LAZRAK E.G., BENOÎT M., "Time stochastic modeling of agricultural landscapes for environmental issues", *Environmental Modeling & Software*, vol. 46, pp. 219–227, 2013.

[MON 72] MONTEITH J.L., "Solar radiation and productivity in tropical ecosystems", *Journal of Applied Ecology*, vol. 9, pp. 747–766, 1972.

[MOR 13] MOREAU P. *et al.*, " CASIMOD'N: an agro-hydrological distributed model of catchment-scale nitrogen dynamics integrating farming system decisions", *Agricultural Systems*, vol. 118, pp. 41–51, 2013.

[MOU 98] MOULIN S., BONDEAU A., DELECOLLE R., "Combining agricultural crop models and satellite observations: from field to regional scales", *International Journal of Remote Sensing*, vol. 19, no. 6, pp. 1021–1036, 1998.

[NAU 07] NAUD C., MAKOWSKI D., JEUFFROY M.H., "An interacting particle filter to improve model-based predictions of nitrogen nutrition index for winter wheat", *Ecological Modeling*, vol. 207, pp. 251–263, 2007.

[NEA 89] NEALE C.M.U., BAUSCH W., HEERMANN D., "Development of reflectance-based crop coefficients for corn", *Transaction ASAE*, vol. 32, pp. 1891–1899, 1989.

[OLI 13] OLIOSO A., LECERF R., BAILLIEUX A. *et al.*, "Modeling of drainage and hay production over the Crau aquifer for analysing impact of global change on aquifer recharge", *Procedia Environmental Sciences*, vol. 19, pp. 691–700, 2013.

[OLI 99] OLIOSO A., CHAUKI H., COURAULT D. *et al.*, "Estimation of evapotranspiration and photosynthesis by assimilation of remote sensing data into SVAT models", *Remote Sensing Environment*, vol. 68, no. 3, 341–356, 1999.

[OTT 16] OTTLÉ C., MAHFOUF J.-F., "Data assimilation of observations from space", in BAGHDADI N., ZRIBI M. (eds), *Microwave Remote Sensing of Land Surfaces*, ISTE Press, London and Elsevier, Oxford, 2016.

[OZG 10] OZGODAN M., YANG Y., ALLEZ G. *et al.*, "Remote sensing of irrigated agriculture: opportunities and challenges", *Remote Sensing*, vol. 2, pp. 2274–2304, 2010.

[PEN 14] PEÑA-ARANCIBIA J.L., MCVICAR T.R., PAYDAR Z. *et al.*, "Remote Sensing of Environment Dynamic identi fi cation of summer cropping irrigated areas in a large basin experiencing extreme climatic variability", *Remote Sensing of Environment*, vol. 154, pp. 139–152, 2014.

[PRI 91] PRINCE S.D., " A model of regional primary production for use with coarse resolution satellite data", *International Journal of Remote Sensing*, vol. 12, no. 6, pp. 1313–1330, 1991.

[PRI 13] PRINGLE M.J., "Robust prediction of time-integrated NDVI", *International Journal of Remote Sensing*, vol. 34, no.13, pp. 4791–4811, 2013.

[RAH 94] RAHMAN H., DEDIEU G., "SMAC: a simplified method for the atmospheric correction of satellite measurements in the solar spectrum", *Remote Sensing of Environment*, vol. 15, pp. 123–143, 1994.

[REV 13] REVILL A., SUS O., BARRETT B. *et al.*, "Carbon cycling of European croplands: aframework for the assimilation of optical and microwave Earth observation data", *Remote Sensing of Environment*, vol. 137, pp. 84–93, 2013.

[ROC 09] ROCHA A.V., SHAVER G.R., "Advantages of a two band EVI calculated from solar and photo synthetically active radiation fluxes", *Agricultural and Forest Meteorology*, vol. 149, no. 9, pp. 1560–1563.

[SAA 15] SAADI S., SIMONNEAUX V., BOULET G. *et al.*, "Monitoring irrigation consumption using high resolution NDVI image time series. Calibration and validation in the Kairouan plain (Tunisia)", *Remote Sensing*, vol. 7, no. 10, pp. 13005–13028, 2015.

[SEG 93] SEGUIN B., COURAULT D., GUÉRIF M., "Surface temperature and evapotranspiration from local to regional scales", *Remote Sensing of Environment*, vol. 49, pp. 287–295, 1993.

[SIM 09] SIMONNEAUX V., LEPAGE M., HELSON D. *et al.*, "Spatialized estimates of evapotranspiration of irrigated crops using remote sensing: Application to irrigation management in the Haouz plain (Marrakech, Morocco)", *Revue Sécheresse*, vol. 20, no. 1, pp. 123–130, 2009.

[SIN 86] SINCLAIR T.R., "Water and nitrogen limitations in soybean grain production.1. Model development", *Field Crops Research*, vol. 15, no. 2, pp. 125–141, 1986.

[SPI 89] SPITTERS C.J.T., VAN KEULEN H., VAN KRAALINGEN D.W.G., "A simple and universal crop growth simulator : SUCROS87", in RABBINGE R., WARD S.A., VAN LAAR H.H. (eds), *Simulation and System Management in Crop Protection, Simulation Monographs*, Pudoc, Wageningen (Netherlands), 1989.

[STE 90] STEINMETZ S., GUERIF M., DELECOLLE R. *et al.*, "Spectral estimates of the absorbed photosynthetically active radiation and light-use efficiency of a winter-wheat crop subjected to nitrogen and water deficiencies", *International Journal of Remote Sensing*, vol. 11, no.10, pp.1797–1808, 1990.

[SUS 10] SUS O., WILLIAMS, M., BERNHOFER C. *et al.*, "A linked carbon cycle and crop developmental model: Description and evaluation against measurements of carbon fluxes and carbon stocks at several European agricultural sites", *Agriculture, Ecosystems & Environment*, vol. 139, no. 3, pp. 402–418, 2010.

[THE 06] THENKABAIL P.S., BIRADAR C.M., TURRAL H. *et al.*, An irrigated area map of the world (1999) derived from remote sensing, Research Report 105, International Water Management Institute, Colombo, Sri Lanka, p. 65, 2006.

[TUC 79] TUCKER C.J., "Red and photographic infrared linear combinations for monitoring vegetation", *Remote Sensing Environment*, vol. 8, pp.127–150, 1979.

[TUC 86] TUCKER C.J., SELLERS P.J., "Satellite remote-sensing of primary production", *International Journal of Remote Sensing*, vol. 7, no. 11, pp.1395–1416, 1986.

[VEL 14] VELOSO A., Regional estimates of the production, fluxes and budgets of carbon and water for winter wheat by using high resolution remote sensing data combined with a crop model: application to southwest France, PhD thesis, University of Toulouse III- Paul Sabatier, Toulouse, France, 2014.

[VER 84] VERHOEF W., "Light-scattering by leaf layers with application to canopy reflectance modeling –the sail model", *Remote Sensing of Environment*, vol. 16, pp. 125–141, 1984.

[WIL 96] WILLIAMS M., RASTETTER E.B., FERNANDES D.N. *et al.*, "Modeling the soil-plant-atmosphere continuum in a Quercus-Acer stand at Harvard Forest: the regulation of stomatal conductance by light, nitrogen and soil/plant hydraulic properties", *Plant, Cell & Environment*, vol. 19, no. 8, pp. 911–927, 1996.

[YAT 05] YATES D., SIEBER J., PURKEY D. *et al.*, "WEAP21-A demand-, priority-, and preference-driven water planning model: Part 1, model characteristics", *Water International*, vol. 30, pp. 487–500, 2005.

[ZHA 02] ZHANG Y., LI C., ZHOU X.*et al.*, "A simulation model linking crop growth and soil biogeochemistry for sustainable agriculture", *Ecological Modelling*, vol. 151, no. 1, pp. 75–108, 2002.

5

Contribution of Remote Sensing to Crop Monitoring in Tropical Zones

5.1. Introduction: the case of tropical crops

5.1.1. *Remote sensing issues for monitoring tropical crops*

The benefits of remote sensing for observing crops have been widely shown through the success that it has had in top producing countries. It is omnipresent today in national and global systems for predicting large industrial crop harvests and in precision farming services. However, such systems are still not common enough to describe and quantify the traditional production of developing countries. The 2009 report on world agriculture by the International Assessment of Agricultural Science and Technology for Development [IAA 09] demonstrates the central role that smallholder farmers play in feeding the global population. Census counts indicate that there are nearly 500 million small farms around the world, and that in Africa 90% of agricultural production comes from small family farms [SPE 01]. Trend analysis further suggests that small farms are expected to continue dominating agricultural landscapes in developing countries, especially in Asia and Africa, for at least the next two or three decades [NAG 05]. Moreover, because of new land subdivisions and the cultivation of new areas, the number of these farms continues to increase in many countries.

Chapter written by Pierre TODOROFF and Jaco KEMP.

Presently, the international community recognizes the strategic role of economic, scientific and agricultural development in developing countries for solving global issues (food security, reduction of climate change, the fight against poverty, etc.) and offers support dedicated to the development of new information and communication technologies, including remote sensing. Monitoring tropical crops via satellite is therefore more relevant than ever. Traditional methods having reached their limits, new information technologies must now take over to allow tropical agriculture to play its part in key global issues. Just as remote sensing contributed significantly to the study of industrial crops since the 1990s, modern remote sensing offers extremely promising technological capabilities for providing crucial information about tropical agricultural production.

These new capabilities, adapted to the special characteristics of tropical crops, require different processing and analytical techniques than those initially developed for monitoring temperate crops using images produced by previous generations of satellites.

For this potential to be fully exploited, developing countries themselves must have the technical and scientific expertise necessary for the development of solutions corresponding to their specific needs. The training of skilled human capital is crucial for this development, and can be supported by institutions and research organizations in industrialized countries and with an ambitious scientific cooperation policy in this field.

5.1.2. *Specific contexts*

The remote sensing technologies used to observe agricultural production systems in industrialized countries are difficult to apply to systems found in developing countries, particularly those in tropical zones. This is not because the crops planted there are different; this difference alone does not justify a whole chapter in this work. The remote sensing methods and techniques developed for monitoring wheat or corn crops in Europe or North America could theoretically be applied in the same way to sugarcane, millet or rice. However, these techniques are inefficient in tropical developing countries because the production conditions are different.

The unique challenges that hinder the implementation of these solutions in tropical countries can be divided into three categories, described in detail in the sections that follow.

5.1.2.1. *Socio-economic conditions*

While intensive farming on a large scale exists and its development is encouraged in tropical countries, agriculture is still largely a family matter in these regions, practiced for self-subsistence. Producers have little revenue and therefore have limited access to means of improvement (inputs, mechanization, etc.). The production levels are low and the crops are sensitive to stress factors; sometimes harvests are devastated by bad climate conditions or attacks by pathogens or pests. These issues make it difficult firstly to make an inventory of cultivated surfaces and to identify crops using airborne or spaceborne imaging based on the recognition of the type of growing cycle, and secondly to estimate and predict production. Often, random or unobservable factors prevail over determining factors taken into account in classical yield development models (such as solar radiation, temperature, water).

Furthermore, the history, the social and familial rules, and the land policy of tropical countries usually lead to land fragmentation, divided into the hands of many smallholder farmers. Combined with the need to cover family food needs, this dispersal of production is expressed as small plots (sometimes a few hundred square meters), very diverse crops (grains, vegetables, tubers, etc.) within one farm and a large percentage of plots under rotation.

The mix of crop types within one plot is another characteristic inherited from traditional agricultural practices or those promoted by movements promoting agro-ecological development seeking to optimize land and natural resources while protecting soil. A secondary crop is also sometimes found between rows of the main crop or as permanent vegetation cover (in direct seeding mulch-based cropping systems).

5.1.2.2. *Climate conditions*

What is striking upon observation of tropical agricultural production by optical satellite imaging is the percentage of unusable images because of cloud cover. In fact, certain intertropical regions suffer from the presence of trade winds, monsoons, intertropical convergence zone oscillations, etc.

Cyclones, seasons with daily rainfall and clouds therefore prevent the observation of all or part of an area for a large part of the year, especially during the crop-growing season.

Furthermore, in dry tropical zones, sand particles suspended in the atmosphere sometimes make it difficult to use the images.

Moreover, the climate of tropical regions presents climatic seasons with much weaker temperature variations than in temperate zones, and much more so as one approaches the equator. Tropical crops thus demonstrate more or less continuous growth, with asynchronous cycles and therefore spread-out harvests (e.g. the harvest season of sugarcane takes place all year long in Kenya). In these conditions, techniques for analyzing temporal series of images used to map crops with annual vegetation cycle calendars are inefficient.

5.1.2.3. *Scientific and technological conditions*

Technological infrastructures and scientific means for researching tropical countries on the one hand and economic interests in their agricultural production on the other are disproportionate to those of industrial countries and their agricultural sectors. However, the delay in using satellite imaging technology for tropical agriculture is real and largely explains the gap that exists among industrialized countries in operational solutions for monitoring crops. In fact, the offer of satellite products for agriculture has been defined (specification of sensors and products) by and for the major users (governments and research institutions in developed countries, actors in high-revenue sectors). Therefore, this offer of satellite products has long sought to respond to the needs of industrial crops and economic conditions of their countries of production by privileging products with a moderate revisit time and a high spatial resolution, but which are commercial products.

Today, with the signing of cooperation treaties with and between emerging countries (China, Brazil and India in particular), the increase in regional remote sensing platforms for southern countries, space initiatives, and programs in certain countries (free access to Landsat archives, free MODIS data, Sentinel missions by the European Earth Observation Program Copernicus, World Heritage SPOT program to freely provide SPOT 1–SPOT 5 programs since 1986), the offer moves significantly towards free access to images, with sensors that are better adapted to crop monitoring. Recent satellites also offer free products with high revisit frequency at moderate and high spatial resolutions: MODIS (moderate resolution, revisit time of 1–2 days), in service since February 2000, and Sentinel-2 (high resolution, revisit time of 5 days with Sentinel-2A launched in June 2015 and 2B anticipated for 2016), which allow for the partial integration of tropical agricultural products in regional and global harvest forecasting tools.

Radar remote sensing presents a distinctive situation. In fact, agricultural research has relatively recently made use of this technology to monitor crops, thanks to the technological development of Synthetic Aperture Radar (SAR), which has given access to a spatial resolution near that of optical sensors, and compatible with the observation of crops [BAG 09, BAG 10]. Being equipped with active sensors that generate their own source of energy, radar satellites can capture images independently of cloud cover or daylight and are of significant interest for tropical zones, where frequent cloud cover is an obstacle for the use of optical imaging. However, the cost of data and the complexity of their processing have limited their application to date. Both launched in May 2014, the European mission Sentinel-1, providing free high spatial resolution radar images with high revisit frequency (6 days with satellites 1A and 1B), and the Japanese satellite ALOS-2/PALSAR in L-band, adapted to the monitoring of vegetation, should push this situation along. Today, technological and political conditions have come together so that emerging countries, with the assistance of research institutes in developed countries, can develop the technical capabilities necessary for monitoring tropical crops with radar imaging.

This chapter aims to contribute to the development of satellite image use for monitoring tropical crops by trying to demystify the discipline that is remote sensing and convince potential users in the agricultural world of its interest and accessibility. This is shown using several examples of how, using remote sensing data available today and with limited means and relatively simple methods, it is possible to characterize tropical crops. It also presents some techniques for mapping tropical crops using optical or radar imaging, for predicting yields, and for monitoring the harvest of sugarcane.

5.2. Crop mapping

Knowledge of land use, and agricultural use in particular, is crucial in tropical countries in order to orient development policies (where are the usable areas? What are the land dynamics?) and contribute to food security (what is the area growing which crops? Where are the fields located?).

In developed countries, mapping of land use is done by sophisticated image processing methods, supplemented by airborne or satellite image photointerpretation. This job is tedious, but nonetheless achievable, because land use types are rather easy to identify in images and land use is limited by regulations.

This is not the case in developing countries. Sources of information are sparse because of the absence of regular airborne photographic cover and difficulties in centralizing data, because of poor technological and financial means, and difficulties in field access. Remote sensing is therefore a means of collecting data that should be prioritized, particularly for the mapping of crops, which make up a large part of these still largely rural countries. It is a crucial tool for statistical studies and national and global predictions of food resources.

Mapping methods based on classification of satellite images have also been developed in temperate countries to move away from or complement administrative agricultural data. These classifications use either a pixel-based approach (this is the case of optical time series analysis-based methods) or an object-based approach, which has seen growing success with the multiplication of very high spatial resolution (VHSR) perspectives.

Very few studies have assessed the potential of the aforementioned methods for mapping tropical crops of family farming in developing countries, yet the fine resolution of VHSR images (submetric to metric resolution) is particularly well adapted to analyze very heterogeneous landscapes with small fields.

5.2.1. *Example of a method for analyzing optical image time series to map sugarcane plots in Kenya*

Certain methods are based on the analysis of an optical image time series at a high spatial resolution (Landsat, ASTER, SPOT 5 [MUR 01, THE 03, SER 08]) and are well adapted to agricultural landscapes whose crops show synchronized growth cycles.

Nevertheless, they can also be applied to tropical landscapes, as shown by a study carried out in Kenya to map sugarcane crops [MUL 15] at the heart of a very heterogeneous agricultural landscape, cultivated by more than 4,000 farmers, dominated by sugarcane farming, but where different types of vegetation (natural scrubland, pastureland) and cultivated plant species (sugarcane, subsistence crops), different cultural calendars (plantation of sugarcane plots throughout the year), and agricultural practices (weed control, harvesting cane green or after burning) blend together.

The developed method is based on a multi-date supervised classification of NDVI vegetation index (*Normalized Difference Vegetation Index*) time series, realized using free, high-resolution images (15–30 m) from the *Operational Land Imager* (OLI) sensor aboard the Landsat 8 satellite. Around 20 images have been used in this way to realize a supervised classification using a ground truth database representative of the crop diversity to be mapped.

The study area is the sugarcane plain of Kibos-Miwani, about 10 km northeast of Lake Victoria, extending over an area of 104 km^2 (0°03'01.6 " S, 34°54'47.6"E) (Figure 5.1).

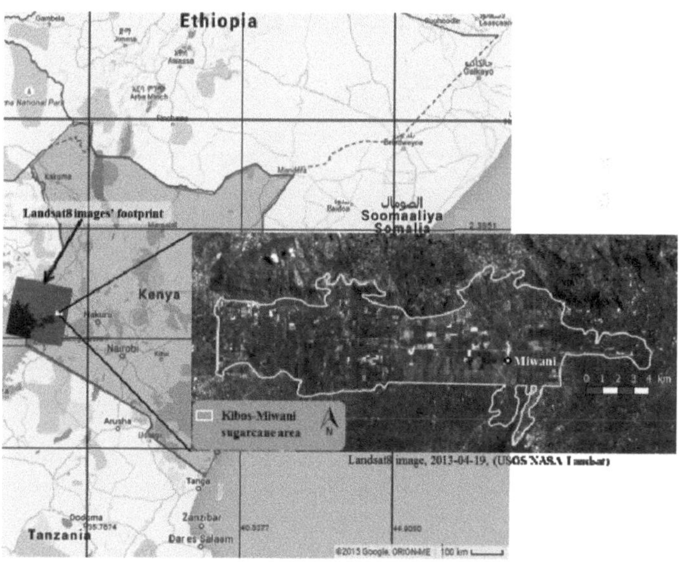

Figure 5.1. *Study area in the Kibos-Miwani region of Kenya (0°03'02"S, 34°54'48"E). For a color version of this figure, see www.iste.co.uk/baghdadi/3.zip*

In most southern countries, agricultural plots are not recorded on paper, let alone in digital form, and high spatial resolution, cloud-free images, which would allow for digitizing agricultural plots by segmentation, are very rare. Plots were therefore digitized through the photointerpretation of a cloud-free Landsat 8 image on which a fusion of resolution has been carried out between the multispectral bands (spatial resolution of 30 m) and the panchromatic band (spatial resolution of 15 m). This operation has also been applied to all Landsat 8 images from the time series in order to realize the classification at the same spatial resolution.

Each image, acquired in orthorectified level, has been translated into a reflectance image (*Top of Atmosphere* (TOA)), then into the NDVI index according to the formula:

$$\text{NDVI}=\frac{NIR\text{-}R}{NIR+R} \qquad [5.1]$$

where NIR and R are respectively the reflectances in the near infrared and the red.

A synthetic image has been created by superimposing each of the 20 NDVI images (Figure 5.2) into a single image.

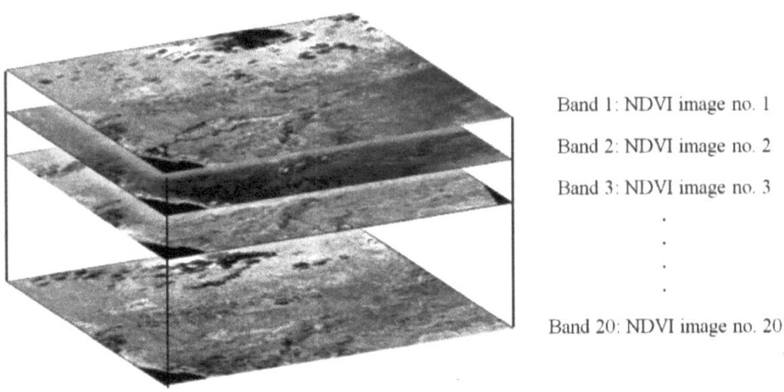

Figure 5.2. *Creating a multitemporal NDVI image*

Field surveys have allowed the cultivated species to be measured on around 10% of the plots in the study area (or around 1,200 observations), according to six classes (Table 5.1): five classes for sugarcane according to the age of the cane because the crop has a different spectral signature according to its stage of development, and one "other" class in which plots not planted with sugarcane will be classified.

Class	Age (months)
Sugarcane #1	0–2
Sugarcane #2	3–5
Sugarcane #3	6–8
Sugarcane #4	9–11
Sugarcane #5	≥12
Other	–

Table 5.1. *Definition of the crop classes*

The classification algorithm used is the is the maximum likelihood algorithm applied to a multitemporal synthetic image. The pixels classified in the five classes of sugarcane were then gathered into a single "sugarcane" class. Figure 5.3 shows the result of the classification.

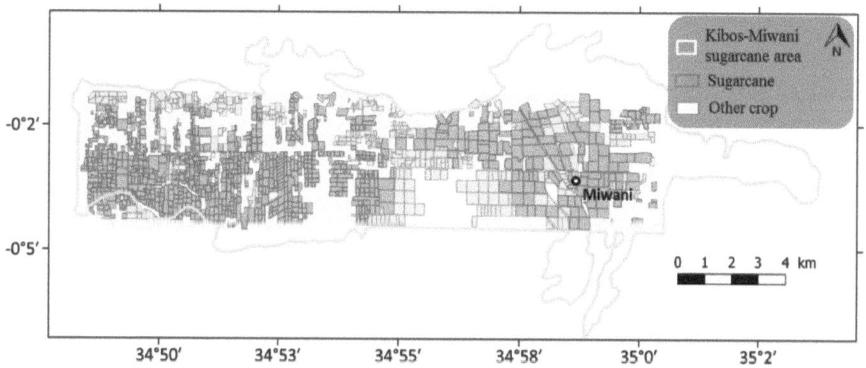

Figure 5.3. *Map of the sugarcane plots. For a color version of this figure, see www.iste.co.uk/baghdadi/3.zip*

The quality of the classification has been assessed using 25% of the observations not used in the classification process. The average error rate is 16%, which is very satisfactory considering the strong heterogeneity of the landscape. The result obtained establishes considerable progress in the knowledge of sugarcane production in Kenya.

This example shows that in situations where existing data are extremely limited (as is most often the case in developing countries), this "classic" method, relying entirely on open access data, works well despite the limited spatial resolution. It is indispensible for estimating agricultural resources and managing production. Future open-access images delivered by Sentinel-2 satellites at a spatial resolution of 10 m should provide even more precise results, particularly thanks to an increased acquisition of 5 days.

5.2.2. *Example of object-based image analysis of VHSR optical images to map land use and crops on Reunion Island*

Coupled with automatic classification algorithms, object-based image analysis (OBIA) has seen considerable development and growing success in recent years [BLA 10] for land cover and land use mapping [CAS 09, TUR 10, VAU 15].

The example below presents the technique applied for land use and crop mapping using a Pleiades image acquired in the southern part of Reunion Island (A French overseas department located in the southern Indian Ocean) (Figure 5.4).

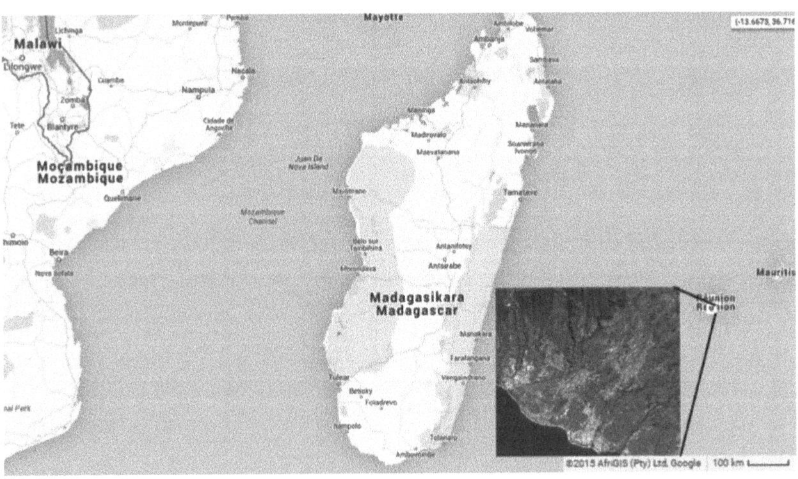

Figure 5.4. *Study area and Pleiades image from July 27, 2012*

5.2.2.1. *Presentation of the method*

Vegetation mapping studies using optical images at different spatial resolutions (10-m resolution SPOT 5 images, 15-m resolution ASTER images, 30-m resolution Landsat 7 images, 250-m resolution MODIS images) by pixel or object approach has shown that these spatial resolutions are insufficient for the accurate characterization of tropical vegetation [KIN 02, GAO 08, WHI 11].

A study was carried out using a VHSR (0.50 m) Pleiades image acquired on Reunion Island in 2012 in order to assess the potential of submetric resolution images for identifying the main land use trends and crops in tropical regions. The Pleiades image used covers a heterogeneous agricultural area made up of very diverse land covers (agricultural, natural and urban) and is representative of the main types of crops on Reunion Island.

Primarily aiming to address issues of food security and agricultural development, the method defines eight classes (Figure 5.5) instead of the 100, if not more, classes of land and soil use from reference land use maps that would then require hundreds of days of photointerpretation.

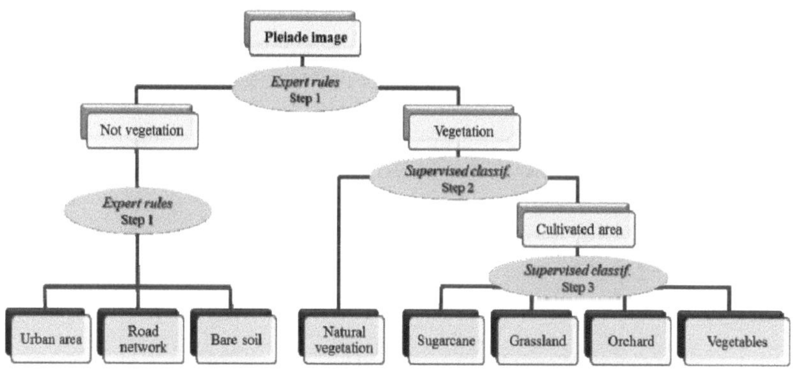

Figure 5.5. *Class typology used for land use and crop mapping by OBIA on Reunion Island*

5.2.2.2. *Building the training database*

5.2.2.2.1. Preprocessing

The image in question was obtained on July 27, 2012. To be able to reuse the classification models developed in this study on other images acquired in the same area on other days, with different satellite configurations, it is necessary to use standardized radiometric measurements. Pixel digital numbers have therefore been converted into *Top of Atmosphere* (TOA) reflectance values.

Additionally, to take advantage of the highest spatial resolution for calculating the texture of the different objects, radiometric bands B, G, R and NIR have been resampled at the panchromatic band's resolution (0.5 m) using resolution merge (by Brovey transform) (Figure 5.6).

The algorithms used in this preprocessing are available in Open Source software, such as Monteverdi2, which uses the open library Orfeo Toolbox (OTB)[1].

Lastly, the (few) pixels whose radiometry is saturated (i.e. with a TOA reflectance > 1) have been removed by filtering as well as the few cloudy areas cut out after manual digitizing.

Figure 5.6. *Resolution comparison between multispectral bands and the panchromatic band of a Pleiades image. For a color version of this figure, see www.iste.co.uk/baghdadi/3.zip*

1 https://www.orfeo-toolbox.org/ (accessed on August 12, 2015).

5.2.2.2.2. Segmentation

As a first step, object-based analysis consists of grouping continuous pixels and similar characteristics (radiometry, texture, structure, etc.) into objects (or segments) according to certain geometric criteria.

In this study, segmentation has been carried out on three resolution levels to identify the different objects of interest. The algorithm used is the eCognition® v9 software's "multiresolution" algorithm.

The first level's goal is to digitize the boundaries of urban surfaces. The criteria that distinguish artificial surfaces from surfaces with vegetation cover are the radiometry in the green band as well as the objects' texture (Table 5.2). The texture in the green band has also been added to the multispectral radiometric values as a segmentation criterion. Furthermore, segmentation has been constrained to follow road network lines. This first level of segmentation is rather basic. It produces objects at urban block level by reducing the detail of different urban areas.

A second level of segmentation is necessary to include isolated buildings. It builds on the objects not classified as urban, with the same criteria, but on a finer level.

The third level, independent of the first two, aims to digitize areas with vegetation cover. It starts with primitive objects at a sub-plot level until it reaches the plot level (Figure 5.7).

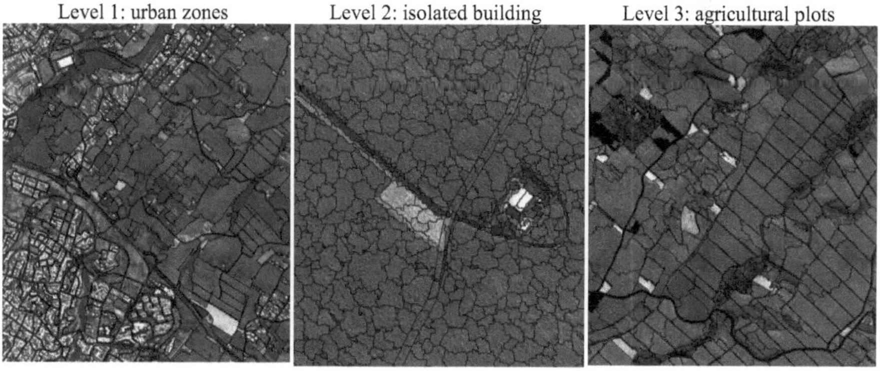

Figure 5.7. *Segmentation levels adapted to the objects to be classified in the study area. For a color version of this figure, see www.iste.co.uk/baghdadi/3.zip*

5.2.2.2.3. Classification

The observation (i.e. the class of the observed objects) inputs in the training database have been obtained primarily by photointerpretation. The great advantage of VHSR images is the ease for visually identifying the objects on the screen. A large part of the field survey time which is necessary when a supervised classification must be applied using a high spatial resolution image can thus be replaced by a few hours of photointerpretation.

A random selection of nearly 1,000 segments among all those generated was thus entered into the database.

Some field surveys have nonetheless been carried out to eliminate the ambiguities on 331 plots for which photointerpretation led to confusion.

The classification process has three steps, combining classification by expert rules and supervised classification (Figure 5.5):

Step 1:

Objects belonging to the general land cover classes are classified by Boolean expert rules because they can be efficiently differentiated with fixed thresholds of a reasonable number of indices (Table 5.2). The road network has been taken out with a buffer zone around the national road network's vector layer polylines (BD Topo 2009®, National Geographic Institute, France). These vector data are not always available for public services in developing countries, but numerous websites offer open access files (Open Street Map community[2], ArcGIS[3], non-governmental organizations, development projects, etc.).

Step 2:

The "vegetation" class has been divided into two daughter classes: "natural vegetation" and "cultivated area". These two classes present similar radiometric and textural characteristics. To differentiate them, 31 radiometric, vegetation and textural variables and indices (particularly those defined by Haralick [HAR 73]) have thus been calculated for each segment classified as "vegetation" by the expert rules. For the sake of efficiency and

2 http://www.arcgis.com/home/gallery.html.

3 www.openstreetmap.org.

reproducibility in an operational framework, the 5 most differentiating variables have been kept after analysis with the random forests algorithm [BRE 84]. A supervised classification by nearest neighbor algorithm has been run using these five variables on 80% of the observations in the training database.

Class	Indices used
Urban fabric	Mean Euclidean distance, Green band
Isolated building	Radiometric contrast Mean Euclidean distance, Green band Mean Red band Mean Green band NDVI Sum of reflectances/NIR
Vegetation	Slope Standard nearest neighbor (generated) Mean Red band Standard deviation Green band Standard deviation NIR band Sum or reflectances Vegetation's color index

Table 5.2. *Indices used in classification by expert rules*

Step 3:

The final goal of the study is to identify crops. The "cultivated area" class was therefore divided into four daughter classes: "sugarcane", "grassland", "orchard", and "vegetables". The radiometric and textural differences between these classes sometimes being very subtle, the 31 previously calculated indices have been synthesized in the form of their principal components (PC). The distribution of the first three PC values for 80% of the training database observations represents the standard profile of PC values of each class (Figure 5.8). It is noted that each class shows a relatively distinct profile, making it possible to separate the classes from their PC profile. However, the dispersion of the PC values makes it impossible to classify an observation whose profile corresponds to several classes using a model based on Boolean rules. A supervised classification based on fuzzy logic calculates the probability that the object belongs to each class as a function of the values of its three components (Figure 5.9). The object is assigned to the class that obtains the highest mean probability on the three PCs.

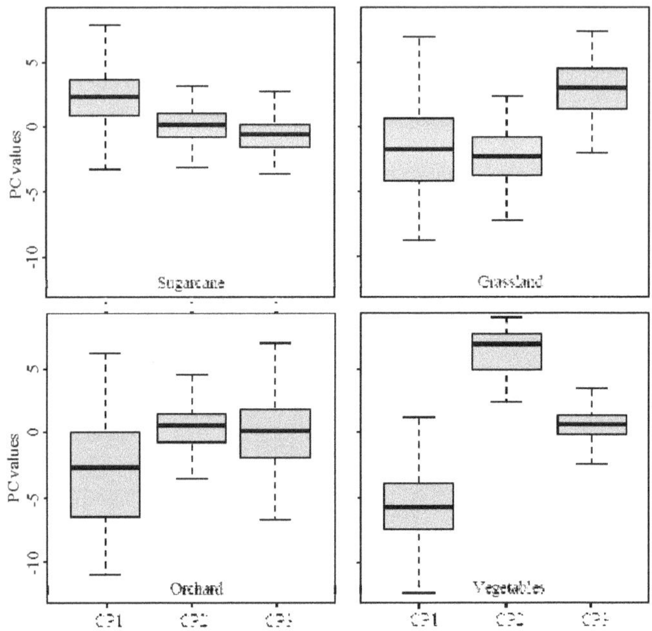

Figure 5.8. *Distribution of the key components' values according to crop classes*

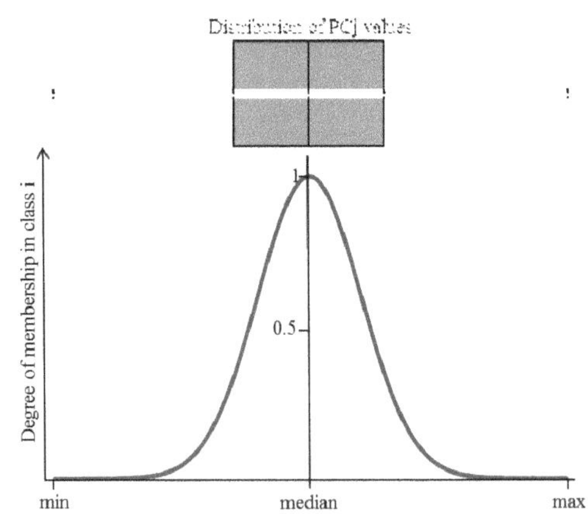

Figure 5.9. *Example of membership probability function to class i of an object as a function of its PCj*

5.2.2.3. *Results*

The classification model based on principal component analysis and fuzzy logic has been applied to the database's objects that were classified as "cultivated surfaces" by the previous supervised classification. The result is presented in Figure 5.10.

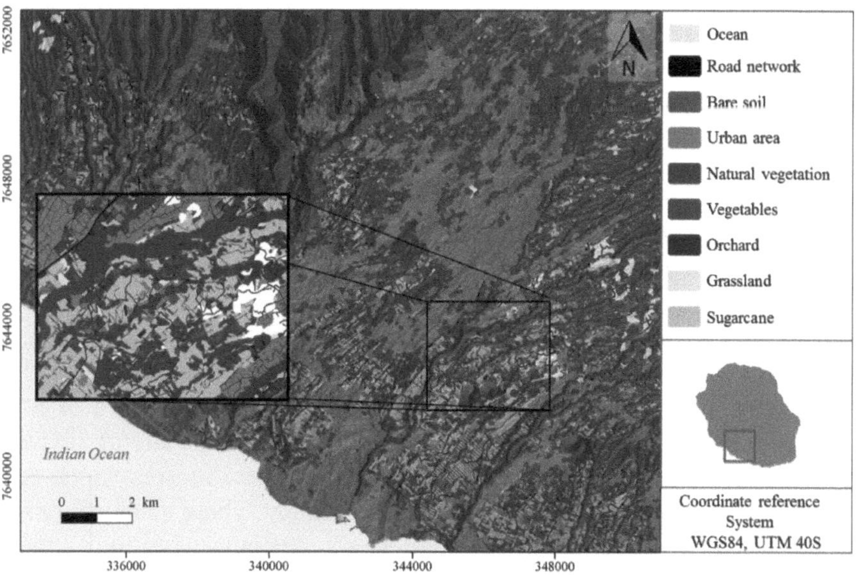

Figure 5.10. *Land use map of the southwestern part of Reunion Island (Lat.: 21°16'54"S, Long.: 55°29'46"E). For a color version of this figure, see www.iste.co.uk/baghdadi/3.zip*

The classification's quality has been assessed through the classification error rates computed for 20% of the training database's observations not used to calculate the classification models. The overall error for the classification "natural vegetation" compared with "cultivated area" is less than 13%. The error for the four crop classes is 19%. These results are extremely satisfying, despite confusions between the "sugarcane" and "vegetables" classes because of the coexistence of very different crops in the latter. These performances can be improved by dividing the "vegetables" class into several more specific classes of vegetables.

These results are similar to those of legal land use maps in industrialized countries: maximum of 10% error, but for a much larger number of classes. Thus, this method cannot replace the tedious photo interpretation work by operators, necessary for the elaboration of official maps, but it may provide a cheap calculation or updating of essential, but often inexistent, basic land and agricultural statistics which are central for the elaboration of economic and rural development and food security strategies in developing countries.

5.2.3. *Potential of radar polarimetry in South Africa*

SAR satellite imagery allows us to obtain information concerning the structure and orientation of crops through the analysis of the backscattered microwave radiation from the surface [FER 16]. These signals are not strongly affected by cloud cover, making SAR imagery an extremely powerful tool in tropical regions where persistent cloud cover is prevalent. SAR imagery records both the amplitude and phase of the backscattered waves and can differentiate between signals that were sent and received in different polarizations. Furthermore, the longer wavelengths used in SAR imagery means that backscattered waves can interact with more of the plant volume (and even the ground), in contrast with optical sensors which only measure reflectance from the top of the plant canopy. These combinations of parameters (amplitude, phase, polarization, wavelength) allow for a unique characterization and analysis of crops.

The use of SAR data for landcover and agriculture applications is rapidly developing, with several state-of-the-art SAR sensors recently launched and even more being planned, leading to what is being called "a golden age of SAR" [MOR 14a]. The availability of these new sensors, in turn, stimulates the development of new approaches, applications, algorithms, and models [QI 11].

The most evident advantage of SAR imagery for the mapping of crops in tropical settings is its capacity for image capture during poor weather conditions. This capacity means that observational gaps in a time series of image acquisitions are significantly less of a risk when using SAR imagery. For example, if the entire Landsat 8 image archive over Reunion Island is examined, there is not a single scene out of dozens (to date) for which at least some part of the island is not obscured by cloud. Unfortunately, in the domain of crop mapping using satellite imagery, the same level of scientific maturity has not been reached in the use of SAR data as has been reached with optical data.

5.2.3.1. *SAR image content for crop mapping*

The case study below presents some of the features of polarimetric SAR data (data containing the four polarizations VV, VH, HH, and HV) that can help in crop-type mapping and discrimination. The area of interest is situated in the Western Cape province of South Africa, and broadly coincides with the catchment area of the Berg River (Figure 5.11). A major agricultural region of the province and country, it is the main wheat-growing region in South Africa [DAF 10]. Major crop types include (in order of abundance): wheat, Lucerne (alfalfa), wine grapes, grazing and pastures, lupines and canola (colza). Secondary crops include tea, olives, vegetables, fruits and berries [WCD 13]. Most of these crops are not irrigated and rely on winter rainfalls, from May to August. Persistent cloud cover during this time of the year often precludes the use of optical imaging for monitoring the peak of the growing season. SAR therefore offers a very useful alternative. The data layers with the greatest potential for crop mapping based on polarimetric SAR can be separated into two broad groups: multipolarization backscattering and polarimetric decompositions.

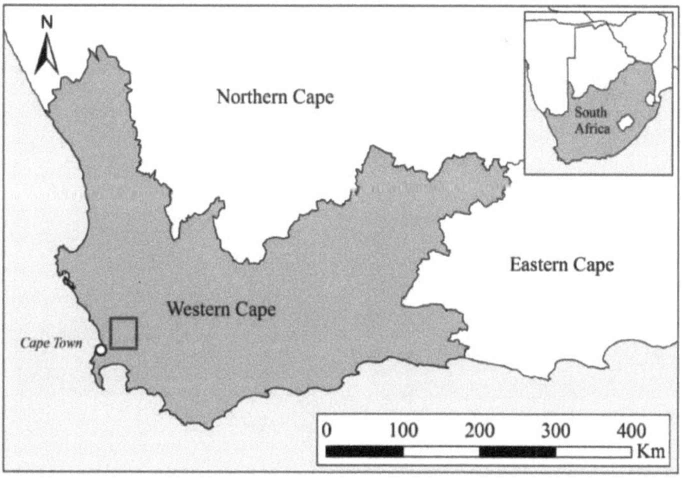

Figure 5.11. *Study area in the West Cape province (South Africa)*

5.2.3.1.1. Multipolarization backscatter

The microwave radiation backscattered by the Earth's surface is a factor of a variety of variables [BAG 16]. These include surface roughness (relative to the wavelength), the soil moisture content, the orientation of features and the degree of microwave penetration. These factors are all affected differentially in the different polarizations. For example, VV polarization is more sensitive than HH to surface roughness, allowing for the separation of roughness and soil moisture effects [MAT 97]. Co-polarized channels (HH and VV) are sensitive to double-bounce effects (the signal undergoes two reflections before returning to the sensor; this is often the case in urban zones), while cross-polarized channels (HV and VH) are not. Cross-polarized channels, however, are more affected by scattering in a volume of vegetation than the co-polarized channels [GHE 11]. Using this knowledge as established in the literature, the four backscatter channels can be used separately and in combination (through ratios etc.) in order to distinguish a variety of different ground parameters related to crop type.

Figure 5.12 shows examples of fully polarized X-band (with a wavelength of 31 mm, frequency of 9.65 GHz) TerraSAR-X backscatter in the Berg River's catchment. The image was acquired in December (summer) 2014 with an incidence angle of 21.7°.

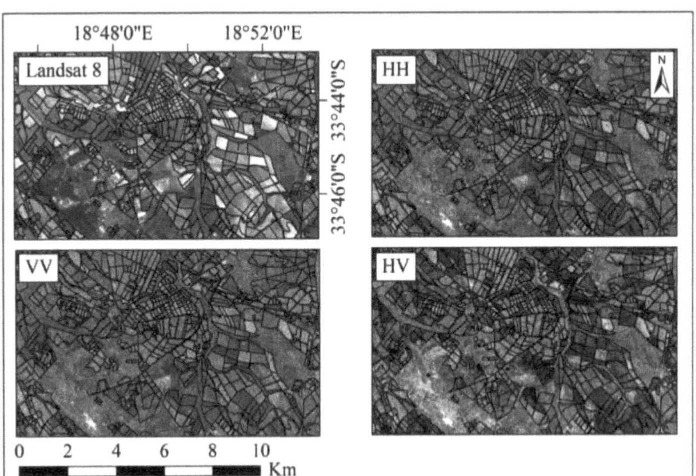

Figure 5.12. *Landsat 8 image (top left) and TerraSAR-X backscatter image (© DLR, 2014) in different polarizations, December 2014 (plot outlines are in black. For a color version of this figure, see www.iste.co.uk/baghdadi/3.zip*

In this figure, the Landsat 8 image (top left, combination of RGB bands 5-4-3) shows clear differences in the vegetation cover according to the crops and stages of growth. The co-polarized signal (HH and VV) in X-band shows differences according to the same parameters, but also according to differences in the bare soils' surface roughness. The nearly vertical incidence, however, reduces the sensitivity to surface roughness, and the backscatter variations visible in the images are moderate. The cross-polarized signal HV (bottom right) more clearly distinguishes the backscatter of bare surfaces (low backscatter level) from the volume scattering caused by vegetation (strong backscatter level).

In Figure 5.12, the dark areas in the HV-band generally reveal cereal crops or pastures that were bare in the summer, while the light areas reveal grapevines and natural vegetation. In all polarizations, however, the differences in soil moisture, surface roughness, texture and even the crop's orientation have an effect on the backscattered signal and should be taken into account.

5.2.3.1.2. Polarimetric decomposition

The S-matrix extracted from an image of the Earth's surface acquired in complete polarization describes the transformation of the incident wave's electrical field (E^i) into the scattered wave's electrical field (E^s):

$$\begin{bmatrix} E_H^s \\ E_V^s \end{bmatrix} = \begin{bmatrix} S_{HH} & S_{HV} \\ S_{VH} & S_{VV} \end{bmatrix} \begin{bmatrix} E_H^i \\ E_V^i \end{bmatrix} \qquad [5.2]$$

The purpose of polarimetric decomposition is to deconstruct the complex information contained in the scattering matrix (which encompasses both phase and amplitude for all polarizations) into estimates of various ground scattering mechanisms [FER 16]. Analogous to how a Tasseled Cap Transformation of optical data result in indices for brightness, greenness or wetness, polarimetric decompositions often aim to provide indications of scattering from surfaces, volumes, corners, etc.

Several polarimetric decomposition theorems have been developed, and can broadly be divided into large classes: coherent theorems (e.g. Pauli, Krogager, Cameron, Touzi), decompositions based on physical models (e.g. Freeman-Durden, Yamaguchi), eigenvector-based decompositions (e.g. Cloude-Pottier, van Zyl), and those based on the dichotomy of the Kennaugh matrix (e.g. Huynen, Holm and Barnes, Yang) [LEE 09a, CLO 96].

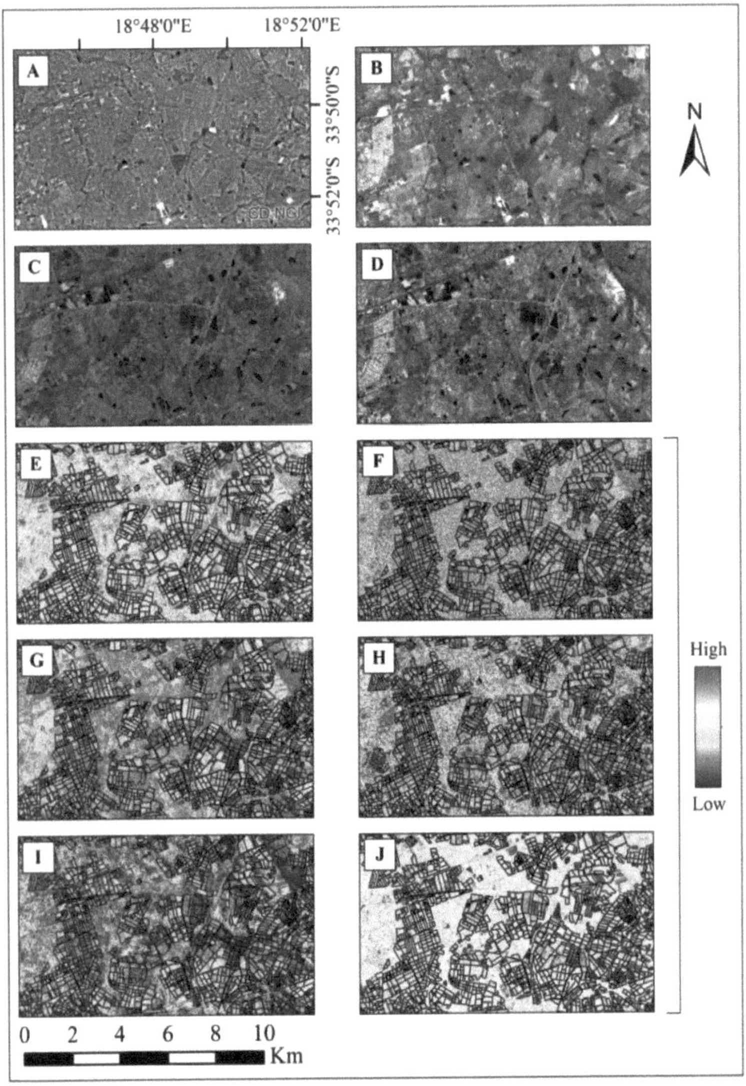

Figure 5.13. *Some radar variables measured in the Berg River's catchment (South Africa): a) Airborne photo; b) Landsat 8 (RGB: 5-4-3); c) HV scattering coefficient; d) Pauli decomposition; e) alpha angle, f) anisotropy; g) entropy; h) Freeman double-bounce; i) Freeman uneven bounces; j) Freeman volume scattering. For a color version of this figure, see www.iste.co.uk/baghdadi/3.zip*

The example in Figure 5.13, again from the Berg River's catchment in South Africa, presents some of the decomposition parameters from the Pauli,

Cloude-Pottier and Freemen-Durden target decompositions theorems based on TerraSAR-X data. For successful application in crop mapping the assumption is made that different crops at different stages of development will exhibit different scattering mechanisms to varying degrees.

Figure 5.13 shows an airborne photograph, a Landsat image and a backscatter image in HV polarization of the study site in thumbnails A, B and C, respectively. The Pauli decomposition is presented in thumbnail D. This representation shows the differences in scattering mechanisms, the blue indicating single-bounce scattering, the green volume scattering and the red double-bounce scattering. Thumbnails E, F, and G show the parameters alpha, anisotropy, and entropy of the Cloud–Pottier decomposition, respectively. Alpha (which provides a general indication of the dominant scattering mechanism) and entropy (which indicates the increasing amounts of depolarization and mix of scatterers) are particularly useful for characterizing agricultural landscapes. Thumbnails H, I and J show the Freeman–Durden decomposition's components for double-bounce, odd-bounce and volume scattering, respectively. This is a decomposition based on a physical model that aims to separate these three types of scattering. Rough surfaces demonstrate stronger odd-bounce scattering and lower volume scattering, while the interaction with vegetation cover produces greater volume and maybe even double-bounce scattering (according to the type, spacing, and orientation of the crop).

5.2.3.1.3. Challenges, opportunities and outlooks

SAR imaging, particularly fully polarimetric imagery, holds significant potential for agriculture mapping and monitoring applications. It is not, however, without limitations and its use without challenges.

SAR data suffer from a phenomenon known as speckle, which is a salt-and-pepper-like noise in the imagery (thought it is, in fact, not noise, but a measured quantity related to the coherent nature of the sensor). Speckle dramatically reduces the radiometric fidelity of pixels in an image and has the consequence that, unlike with optical data, a single pixel cannot be expected to be representative of the sampled area. For most applications, speckle reduction, either through multi-looking or speckle filtering is a requirement. Research on optimal filtering methods have been continuing for more than two decades and as SAR applications start to increase, this will remain an important problem [ARG 13].

The side-looking nature of SAR sensors also brings with it inherent terrain distortions known as foreshortening, layover and radar shadow. These geometric (and related radiometric) distortions become more pronounced with increasing slope and decreasing (steeper) incidence angle. This sometimes precludes the use of SAR in areas where crops are cultivated on steep slopes.

A further potential barrier to the use of SAR data is its relative complexity compared to optical image data. While freely available software is becoming increasingly available and mature, specialized commercial software is often required for quick and reliable SAR image processing. Acquiring, pre-processing and analyzing SAR data also require a comparatively high level of training and skill, and interpretation of the imagery and its derivatives is not always intuitive.

Moreover, data in complete polarization is only available via a few sensors (RADARSAT-2, PALSAR-2, TerraSAR-X), and is thus relatively expensive. Open-access SAR data of Sentinel-1 are not fully polarimetric and do not provide as much information content. Despite the existence of free data, access to this data on a regular time step and on some study areas is not guaranteed, such as programming, under control of the European Space Agency, which does not plan for regular acquisitions on all territories. It is therefore necessary for users and the scientific community to frequently communicate their needs in order to have satisfactory access to these data.

When the decision is made to acquire and use SAR imaging for agricultural applications, there are some considerations to take into account.

The first concerns the suitability of the imagery for the stated purpose. The luxury of simply choosing and acquiring the ideal data types is often not an option for developing countries. Choices generally focus on wavelength (C-band is commonly used in agricultural applications, but is not always the most appropriate), incidence angle (steeper incidence angles result in larger ground responses in row crops, for example), and the appropriate time of year. The latter is particularly crucial in crop monitoring studies where the phenological state of crops has a significant impact on the SAR signal. Research has shown that crop-type separability is often highest later in the growing season, after the ripening of crops [KAR 08, MCN 09]. An important consideration is, therefore, the crop calendars of the region of interest, especially if different crops mature and are harvested at different times of year.

When filtering is performed on SAR data for agricultural applications, consideration should also be given to the purpose for which the data is being prepared. While a suite of different filtering algorithms are available in free and commercial software, filtering generally improves radiometric quality at the cost of spatial resolution [LEE 09b]. If an area of interest consists of large contiguous fields, spatial resolution might be of lesser concern than if small parcels are present (as is common in tropical regions with steep slopes). Furthermore, a single filter type or filtering window size might not be appropriate for use on all polarimetric decomposition features. For example, it is known that varying degrees of filtering affect the entropy, alpha and anisotropy parameters of the Cloude–Pottier decomposition in different ways: larger windows result in higher entropy, but lower anisotropy [LEE 09b].

Despite these challenges, rapid development in sensor technology and image processing technology are leading to many opportunities for the use of SAR imagery in crop mapping and monitoring applications. These include new and more sensors, the use of OBIA, the opportunities of multi-date and multi-sensor data fusion, and the power of machine-learning algorithms.

5.3. Yield prediction

Yield prediction is a crucial operation for optimizing harvesting conditions in order to maximize farmers' revenue, anticipate the risks on food security, and elaborate import and commercialization strategies [EVE 05], but it is also a difficult operation, because there is no universal solution applicable at every level, in every culture and in every socioeconomic context. In developing countries agriculture rarely benefits from an efficient technical support. The heterogeneity of yields and the fragmentation of production make the exercise even more difficult: expert-based yield predictions at the level of each plot or farm are not very reliable or not always available, and the sugarcane cultivated areas are uncertain. The consolidation of predictions at the regional level is therefore particularly problematic.

Crop models used in this exercise have seen great success over the last 30 years by automatizing this task and increasing the results' reliability [SIN 95, BOU 97, LIS 05, BEZ 07]. However, they do not perform any miracles in very heterogeneous cropping conditions of low-income small-scale agriculture.

Remote sensing data offers a pertinent alternative. Indeed, satellite observation allows for the measurement of crops' vegetative and nutritional development in near real time. The assimilation of remote sensing data into crop models (using satellite data in crop models) allows for the initialization or readjustment of simulations with satellite observations [BOU 92, MAA 88, MOR 95]. This technique also compensates for the lack of climatic data on fine temporal and spatial resolutions often required by crop growth models.

5.3.1. *Example of yield prediction using optical remote sensing*

The example below shows how to predict the biomass yield and sugar content of sugarcane plots using a simple empirical yield-prediction model based on the NDVI vegetation index [BEG 10].

5.3.1.1. *Principle*

Temporal NDVI profiles from more than 120 plots in two French overseas departments (Guadeloupe, an island in the Antilles, and Reunion Island, in the Indian Ocean) have been calculated from a temporal series of around 20 SPOT 4 and SPOT 5 images covering three growth cycles (Figure 5.14). Similar measurements are now freely available via the European Sentinel-2 satellite.

The comparison of maximum NDVI-values with the yield in biomass (tons of fresh sugarcane harvested/ha) and the yield in sugar (in degrees BRIX) has shown that the maximum NDVI value strongly correlates to the yield in biomass. An exponential regression model allows the yield to be estimated with a root mean square error of 13.2 t/ha (Figure 5.15).

The relative error is around 20%, which is a high value. However, this is the model's error at the individual plot level. The farmer's estimations are generally more reliable, but the desired objective is to offer policymakers a harvest prediction tool on a larger scale (community, territory, region). By adding the predictions calculated at the plot scale to higher-scale levels (like the production basin), the error then drops by effect of the mean. To further reduce the error, the difference in estimated yield can be calculated between two consecutive years. The average yield variation is thus applied to the yield observed the year before in order to predict the yield for the following year. Thus, with this method applied on Reunion Island (Figure 5.16), the average error of the predictions observed is 5% at the production basin level, and 1.5% for the entire island [BOY 12], which is very satisfactory, taking into account the labor-saving effort made (no field survey necessary).

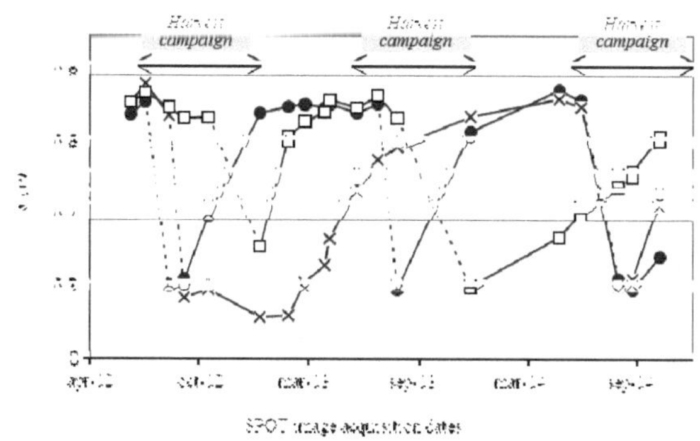

Figure 5.14. *Temporal NDVI profiles of three sugarcane plots on Reunion Island calculated using SPOT images. The dotted lines represent the plot's harvest date*

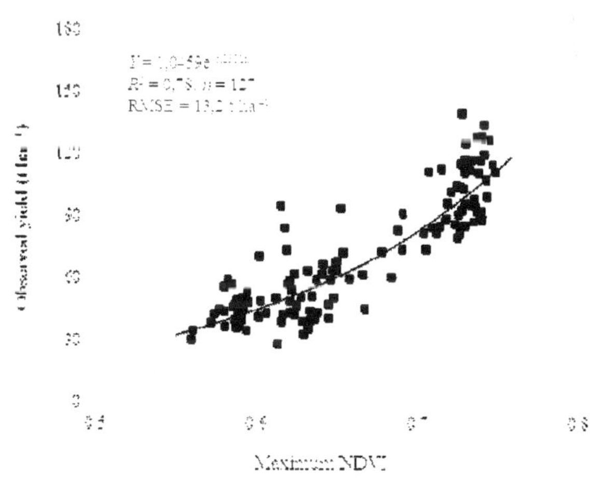

Figure 5.15. *Yield versus maximum NDVI*

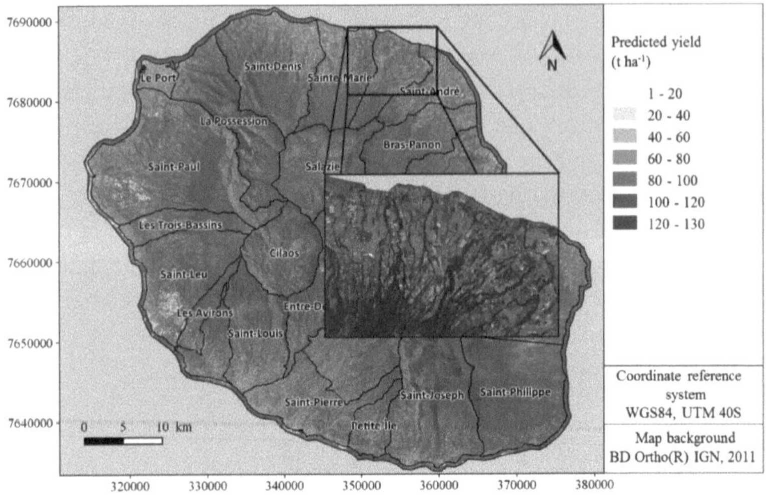

Figure 5.16. *Sugarcane yield-prediction map. For a color version of this figure, see www.iste.co.uk/baghdadi/3.zip*

5.3.1.2. *Practical implementation*

In practice, the maximum NDVI value is observed at the end of the growth period, which is to say 4–6 months after the previous harvest, according to climatic conditions. The NDVI value can remain stable (between 0.7 and 0.8) during the following ripening phase if there is no hydric stress, or decrease in the case of severe stress. Furthermore, leaf senescence causes a drop in the NDVI values, possibly dropping to 0.25.

The maximum NDVI can thus be extracted from a single image acquired at the end of the growth period, which is to say generally 1–2 months before the start of the harvest. When this is not possible (because of cloud cover, or when plot growth is not synchronized, or in the case of severe stress), the maximum NDVI of each plot can be estimated using their NDVI function modeled by two logistic curves [CLE 02] corresponding to the growth and ripening phases, respectively (Figure 5.17).

The NDVI equation is given by equation [5.3]:

$$NDVI(t) = G(t) - S(t) \qquad [5.3]$$

where G is the growth phase curve, and S is the senescence phase curve.

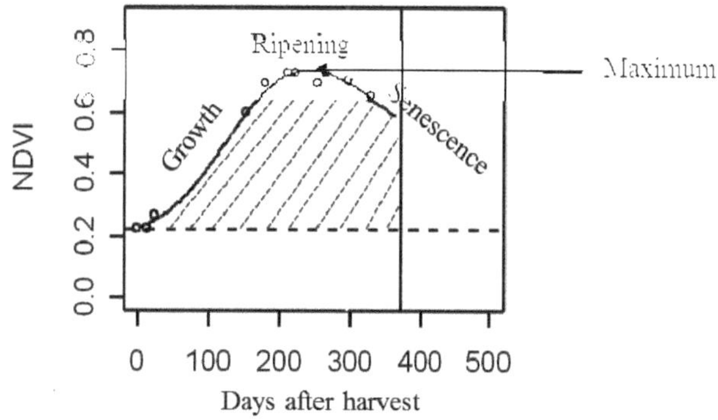

Figure 5.17. *NDVI evolution model*

The *G* and *S* functions are obtained by non-linear adjustment of equations [5.4] using measurements extracted from the image time series by minimizing the residual sum of squares:

$$\begin{aligned} G(t) &= M/\left(1+e^{(-a(t-tg))}\right) \\ S(t) &= M/\left(1+e^{(-b(t-ts))}\right) \end{aligned} \qquad [5.4]$$

M is the maximum value of the logistic curve, *a* and *b* are the slope values at the inflection points of functions *G* and *S*, *t* is the number of days after the harvest, *tg* and *ts* are the age of the crop at these inflection points. The ordinate at the origin of equation [5.3] is set equal to the average NDVI value observed on the plots in the two weeks that follow the harvest, that is to say, on the sugarcane residues (e.g. NDVI = 0.23 on Reunion Island).

The yield prediction in sugar is more delicate, as sugar is elaborated by the plant during the final ripening phase, that is to say, when the vegetative development has reached its maximum. The sugar content is largely determined by the climatic conditions during the weeks leading up to the harvest: heat and hydric stress favor the accumulation of saccharose in the stems. Strong rains can lead to a decrease in the sugar content. The relation between sugar content and NDVI can thus have quite a variable reliability (Figure 5.18). The yield prediction in sugar adds the yield-prediction model's error in biomass and the sugar content estimation model's error.

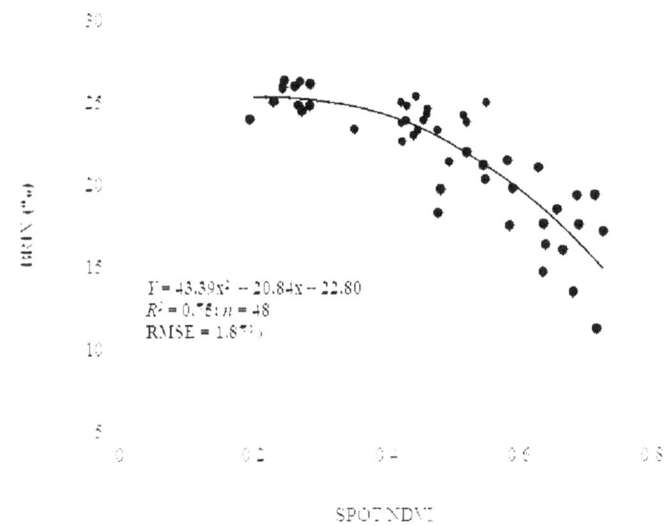

Figure 5.18. *Relation between maximum NDVI and BRIX index*

The yield prediction of grain or fruit crop is even more complex. Even if the yield is strongly correlated to the yield in biomass, numerous factors can modify the relationship between the maximum of NDVI and the yield in grain or fruit:

– poor fertilization perturbing blossoming or fructification;

– grain sterility because of the cold in the case of high-altitude rice.

– climatic events just before the harvest (cyclones, drought).

Biomass yield models based on satellite observations should, therefore, be completed using models that consider these factors. In certain cases, an ecophysiological model allows these factors to be considered (e.g. the model of rice grain sterility under the effect of cold temperatures). In developing countries, the variation factors are often due to suboptimal cultivation practices (an unsuitable sowing date, insufficient fertilization, etc.), which are not quantifiable and vary greatly from one farmer to the next. It therefore remains quite difficult to predict the yield of agricultural productions on small family farms with high precision. At the regional scale, remote sensing is nevertheless the most efficient technique. It brings considerable progress in the mapping and understanding of food resources in developing countries for which there is no affordable or reliable alternative.

5.4. Harvest monitoring

5.4.1. *Issues*

Knowing the progression of harvests, that is to say, localizing harvested fields and those to be harvested, particularly in the case of industrial crops, is not only crucial for maximizing the production chain's productivity (optimizing the allocation of harvest and transportation resources so that all the fields are harvested, reducing costs, ensuring an adequate supply to the processing units), but also for providing the dates of start and end of growth of the fields. These data are, indeed, extremely useful in agronomic studies aiming to simulate cultivation scenarios or to diagnose crop status. The models commonly used in these studies simulate vegetative growth and must be initialized with the date of start of growth on the fields. In the case of perennial crops, such as sugarcane, the crop cycle's start of growth for year n also corresponds to the crop cycle's harvest date for year $n - 1$ (the crop grows from ratoon from cycle $n - 1$). Remote sensing identification of harvested land at a high enough frequency thus allows us to determine the start date of harvesting activities.

In the tropics, frequent cloud cover does not allow for observation by optical remote sensing of fields in a region at a high enough frequency in order for:

– the information to be usable for policymakers;

– the harvest operation to be detected (observation must take place soon after the harvest to avoid confusing a field already regrowing with one that has not yet been harvested).

Radar remote sensing allows for these constraints to be eliminated. The examples presented in the following section show that it is an efficient solution for mapping the progression of the sugarcane harvest. We show a technique for detecting sugarcane harvests through radar backscattering analysis in this section.

5.4.2. *Example of sugarcane harvest monitoring by remote sensing on Reunion Island*

Harvesting sugarcane is a heavy duty operation in terms of workforce and equipments, requiring a complex organization and spread over several months. In many producing countries, sugarcane is produced by many small growers who harvest their fields by hand at a quite variable rate depending on the workforce available and weather conditions. They do not know the harvested area accurately and this information is rarely centralized. It is therefore very difficult for policymakers to be informed about the harvest's progress.

The radar signal is sensitive to the surface characteristics of the target observed. Baghdadi *et al.* [BAG 09] showed that the backscattered signal in VH polarization was sensitive to the height of the crop (Figure 5.19): the amplitude increases with the height of stalks, then saturates. The higher the frequency, the sooner saturation occurs. In L-band, the amplitude reaches its maximum at a height of around 2 m, and in C- and X-bands, before 1 m.

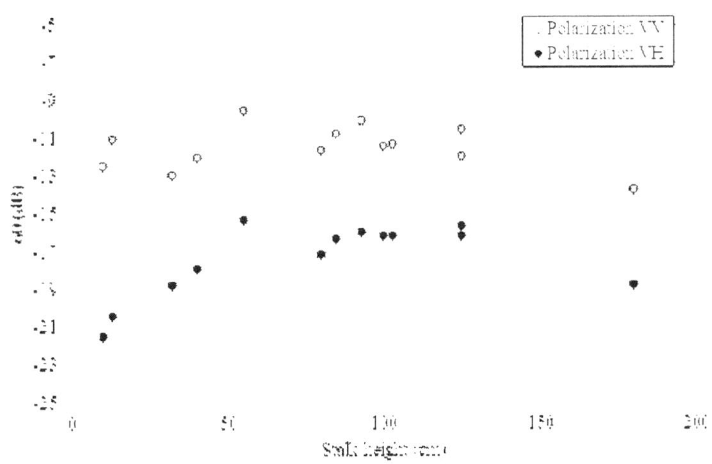

Figure 5.19. *Backscatter coefficient versus stalk height (signal in C-band, incidence angle = 44°)*

This property thus allows recently cut fields to be distinguished in radar images from standing crops using the backscatter coefficient value (Figure 5.20). Harvested fields appear in dark gray, meaning low backscatter. In fact, the ground of harvested fields is covered by mulch made up of the plant parts not used for sugar production (dry leaves, leaf sheaths, the tops of stalks). It has a low surface roughness and fosters the specular reflection phenomenon of the incident radar signal in the opposite direction. Unharvested fields appear in light gray: the cover's roughness and its heterogeneous structure scatter the incident signal in all directions; a significant part is scattered towards the satellite.

Radar satellites operating in X-band generally offer a higher (metric) spatial image resolution than that offered by satellites operating in the C-band (or L-band, *a fortiori*) (decametric resolution). However, radar signals in the C-band, with a longer wavelength, penetrate deeper into the cover than signals in the X-band. They are, therefore, theoretically better adapted to the characterization of the cover's structure. Furthermore, images acquired by the recent European Sentinel-1 satellites, in the C-band, are freely available, thus allowing developing countries access to spatial data for monitoring harvests.

Figure 5.20. *Field harvesting status by backscatter coefficient (TerraSAR-X (© DLR, 2014) image from October 6 2014, VV polarization, incidence angle = 37°, resolution = 2.5 m. For a color version of this figure, see www.iste.co.uk/baghdadi/3.zip)*

The backscatter coefficient values, measured in C-band, are indeed clearly contrasted depending on the field's harvest status. Figure 5.21 shows that the backscatter coefficient measured in cross-polarization (HV and VH) for an incidence angle of about 36° on fields located in the south of the island with the RADARSAT-2 satellite (C-band) varies between –19 dB and –25 dB for harvested fields. Note that these values increase rather quickly after the harvest, as the crop begins to grow again. After about 15 days, we run the risk of of not being able to distinguish between harvested and unharvested fields.

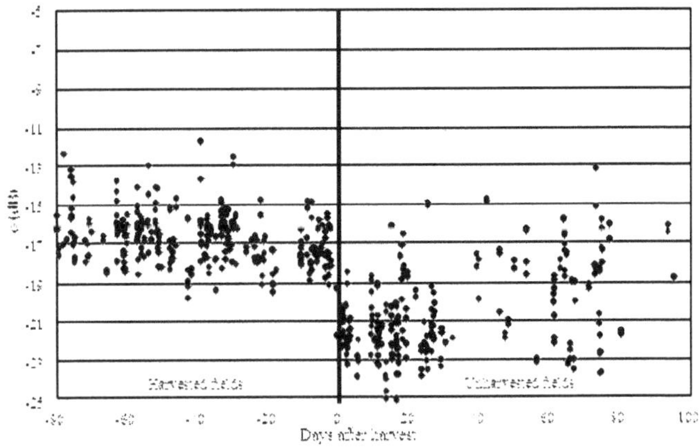

Figure 5.21. *Backscatter coefficient as a function of the age of the harvest (RADARSAT-2 satellite, HV and VH polarizations, incidence: 36°, resolution = 6 m)*

Moreover, the backscatter coefficient's sensitivity to the field's harvest status depends strongly on the radar acquisition configuration: wavelength, spatial resolution, polarization and incidence angle. It is necessary to determine the configuration that maximizes backscatter coefficient variations according to the status of the harvest. This optimal configuration depends on the satellite, the geography of the study area (topography, field size), and the crop. On Reunion Island, with very mountainous terrain, measurements in C-band with a moderate incidence angle (33°) show that the variations in the scattering coefficient are almost no longer visible in cross-polarization.

Operational implementation of this approach for monitoring sugarcane harvests therefore requires a preliminary study, either to determine the optimal radar configuration and thereby deduce the backscatter coefficient threshold that distinguishes harvested fields from unharvested ones in the region of interest, or, if using openly available images, to determine this threshold from images acquired by programmed acquisitions (in a pre-defined configuration) over the area of interest (e.g. Sentinel-1).

This method can be easily applied in an operational use, as it is relatively cheap and its processing is simple. Requiring only a single polarized image, the cost of data acquisition is remarkably low compared with multi-polarized images and the images can be provided by any radar satellite.

5.5. Conclusion and outlook

The few case studies presented in this chapter illustrate how recent technological advances in remote sensing are allowing for an efficient response to certain questions of agricultural development in developing countries.

Remote sensing data can solve certain problems which were until recently unsolvable for tropical developing countries. By coupling the wealth of satellite information with in-the-field knowledge and the ecophysiological mechanisms that drive agricultural production, more complex questions can be addressed, such as harvest prediction [MOR 14b] and the simulation of scenarios (climatic, cultural, etc.). These mechanisms are formalized and simulated by agronomic models (crop growth, yield prediction, diagnostic, etc.). Coupling of remote sensing and model data, often called "assimilation", "calibration", or "initialization", is the subject of numerous studies [LOS 95, MOU 98, LAU 05, YUP 08] and should see considerable progress in the near future thanks to the efforts of agronomic research to develop/adapt crop models to tropical crops and thanks to the technological evolution of satellite sensors.

New generations of satellites are regularly put into orbit and offer ever-increasing spatial, temporal and spectral resolutions. Agriculture, particularly in developing countries, will benefit most from this and is catching up in the application of this technique to produce more and better data. The optical satellite Sentinel-2, with its 13 high spatial resolution spectral bands and revisit time of 5 days, is predicted to break new ground in our ability to observe agricultural production. The constellations of SAR sensors (rather than single sensors) are becoming more and more numerous and promise to deliver almost continuous monitoring of the Earth's surface [MOR 13].

Despite the relative complexity of radar remote sensing in relation to optical remote sensing, the skills required for the analysis of radar data are expanding from the domain of signal processing and moving into the domain of end users and policymakers (agronomists, geologists, botanists, geographers, etc.). This is partly thanks to the multiplication of specialized and open-license software. While this move has been slow, it is currently taking place and should be supported.

However, with more and more sensors and data, the question arises as to "when is enough, enough?": which sensors should be combined, which combinations of dates yield the best results, which characteristics, ratios, decompositions, and indices must be used, and how many image dates over what time period are necessary for optimal results?

These questions are crucial, but are difficult to answer through systematic evaluation or even multivariate analysis. In this regard, the use of machine-learning algorithms present another significant opportunity for agricultural land use applications using remote sensing. However, the increase of data space dimensions (that stems from the large number of information layers used) in these applications raises issues of complexity in the management, storage and processing of this information. The Hughes effect describes this "curse of dimensionality", where the quantity of data, the storage capacity, and the calculation time increase exponentially with the number of dimensions, but the precision of the result diminishes beyond a certain quantity of information [HUG 68]. To illustrate this, let us imagine an image classification problem with data from several sensors on several dates. A simple optical image can produce dozens of radiometric bands, indices, ratios and texture measurements. With polarimetric SAR data, dozens of other variables can be added through polarimetric decompositions, associated textures and ratios. When several dates are used, such an exercise can quickly reach hundreds of data layers. Commonly, conventional image classification algorithms completely fail to separate crop or land cover classes when faced with such high dimensionalities, especially when training sets are small [MYB 13]. Machine-learning algorithms such as Classification and Regression Trees (CART), Random Forests (RF) and Support Vector Machines (SVM) allow for powerful levels of feature extraction, dimensionality reduction and classification [FER 14]. This allow us to not only sensibly utilize the tremendous amounts of data at our disposal, but also to gain insight into the relative importance of different sensors, image features, dates and polarization in the output classifications.

Given the abundance of Earth observation data and tools that are becoming more and more accessible to end users, the combined use of optical and SAR radar data is poised to make significant contributions to the way we map and monitor crops, particularly in tropical regions.

5.6. Key points

– The current context of satellite information opens new outlooks for the monitoring of tropical crops.

– The increase in optical sensors' spatial resolution allows for very heterogeneous cultivated areas of small-scale family farming in the tropics to be better characterized.

– The multiplication of freely available data gives access to very dense optical image datasets that partly compensate their limited spatial resolution and the risks of cloud cover.

– Radar remote sensing is reaching maturity and offers a competitive alternative to optical remote sensing for monitoring agriculture in zones with omnipresent cloud cover.

– The future looks even more promising with the advent of improvements in sensors, with shorter revisit times and ever more free-access data. Research is already developing multi-sensor data fusion methods. Data mining techniques for analyzing the tremendous masses of spatio-temporal data have been invading remote sensing laboratories for several years. However, questions concerning the ever-important management (storage, access and processing) of information even for developing countries, are still not resolved and will pave the way for future reaserch.

– Furthermore, the development of remote sensing applications for agriculture in developing countries will strongly depend upon international cooperation programs in this field. In fact, these programs will be decisive for these countries' ability to use their own technical and human skills to take ownership of these techniques and develop and implement new applications adapted to their needs.

5.7. Bibliography

[ARG 13] ARGENTI F., LAPINI A., BIANCHI T. *et al.*, "A tutorial on speckle reduction in synthetic aperture radar images", *IEEE Geoscience and Remote Sensing Magazine*, vol. 1, no. 3, pp. 6–35, 2013.

[BAG 09] BAGHDADI N., BOYER N., TODOROFF P. *et al.*, "Potential of SAR sensors TerraSAR-X, ASAR/ENVISAT and PALSAR/ALOS for monitoring sugarcane crops on Reunion Island", *Remote Sensing of Environment*, vol. 113, pp. 1724–1738, 2009.

[BAG 10] BAGHDADI N., CRESSON R., TODOROFF P. *et al.*, "Multitemporal observations of sugarcane by TerraSAR-X images", *Sensors*, vol. 10, pp. 8899-8919, 2010.

[BAG 16] BAGHDADI N., ZRIBI M., "Characterization of soil surface properties using radar remote systems", in BAGHDADI N., ZRIBI M. (eds), *Surface Remote Sensing in Continental Hydrology*, ISTE Press, London and Elsevier, Oxford, 2016.

[BEG 10] BÉGUÉ A., LEBOURGEOIS V., BAPPEL E. *et al.*, "Spatio-temporal variability of sugarcane fields and recommendations for yield forecast using NDVI", *International Journal of Remote Sensing*, vol. 31, pp. 5391–5407, 2010.

[BEZ 07] BEZUIDENHOUT C.N., SINGELS A., "Operational forecasting of South African sugarcane production: Part 1 – system description", *Agricultural Systems*, vol. 92, pp. 23–38, 2007.

[BLA 10] BLASCHKE T., "Object based image analysis for remote sensing", *ISPRS Journal of Photogrammetry and Remote Sensing*, vol. 65, no. 1, pp. 2–16, 2010.

[BOU 92] BOUMAN B.A.M., "Linking physical remote sensing models with crop growth simulation models, applied for sugar beet", *International Journal of Remote Sensing*, vol. 13, pp. 2565–2581, 1992.

[BOU 97] BOUMAN B.A.M., VAN DIEPEN C.A., VOSEN P. *et al.*, " Simulation and system analysis tools for crop yield forecasting ", in TENG P.S. *et al.* (eds)., *Application of Systems Approaches at the Farm and Regional Levels*, Kluwer Academic Publishers, 1997.

[BOY 12] BOYER N., TODOROFF P., PAULIN L. *et al.*, "Un modèle de prévision de rendement de la canne à sucre basé sur des images satellitaires SPOT - Exemple de La Réunion ", *Revue Française de Photogrammétrie et de Télédétection*, vol. 197, pp. 76–85, 2012.

[BRE 84] BREIMAN L., *Classification and Regression Trees*, Wadsworth International Group, 1984.

[CAS 09] CASTILLEJO-GONZALEZ I.L., LOPEZ-GRANADOS F., GARCIA-FERRER A. *et al.*, "Object- and pixel-based analysis for mapping crops and their agro-environmental associated measures using QuickBird imaging", *Computers and Electronics in Agriculture*, vol. 68, pp. 207–215, 2009.

[CLE 02] CLEVERS J.G.P.W., VONDER O.W., JONGSCHAAP R.E.E. *et al.*, "Using SPOT data for calibrating a wheat growth model under mediterranean conditions", *Agronomie*, vol. 22, pp. 687–694, 2002.

[CLO 96] CLOUDE S.R., POTTIER E., "A review of target decomposition theorems in radar polarimetry", *IEEE Transactions on Geoscience and Remote Sensing*, vol. 34, pp. 498–518, 1996.

[DAF 10] DAFF, Wheat Production Guideline, Department of Agriculture, Forestry and Fisheries, South Africa, p. 32, 2010.

[EVE 05] EVERINGHAM Y., INMAN-BAMBER G., TICEHURST C. *et al.*, "Yield forecasting for marketers", *Proceedings of the 27th Conference of the Australian Society of Sugarcane Technologists*, Bundaberg, QLD, Australia, pp. 51–60, 3–6 May 2005.

[FER 14] FERNANDEZ-DELGADO M., CERNADAS E., BARRO S., *et al.*, "Do we need hundreds of classifiers to solve real world classification problems?", *Journal of Machine Learning Research*, vol. 15, pp. 3133–3181, 2014.

[FER 16] FERRO-FAMIL L., POTTIER E., "Coherent diversity mode SAR imaging: SAR polarimetry, interferometry and tomography", in BAGHDADI N., ZRIBI M. (eds), *Microwave Remote Sensing of Land Surfaces*, ISTE Press, London and Elsevier, Oxford, 2016.

[GAO 08] GAO Y., MAS J.F., "A comparison of the performance of pixel-based and object-based classifications over images with various spatial resolutions", *GEOBIA 2008 – Pixels, Objects, Intelligence, GEOgraphic Object Based Image Analysis for the 21st Century*, Calgary, Canada, 2008.

[GHE 11] GHERBOUDJ I., MAGAGI R., BERG A.A. *et al.*, "Soil moisture retrieval over agricultural fields from multi-polarized and multi-angular RADARSAT-2 SAR data", *Remote Sensing of Environment*, vol. 115, pp. 33–43, 2011.

[HAR 73] HARALICK R.M., SHANMUGAM K., DINSTEIN I.H., "Textural features for image classification. Systems, man and cybernetics", *IEEE Transactions on Systems, Man, and Cybernetics*, vol. SMC-3, no. 6, pp. 610–621, 1973.

[HUG 68] HUGHES G., "On the mean accuracy of statistical pattern recognizers", *IEEE Transactions on Information Theory*, vol. 14, no. 1, pp. 55–63, 1968.

[IAA 09] IAASTD, Agriculture at a Crossroads: Global Report, International Assessment of Agricultural Science and Technology for Development, Washington, Island Press, DC, 2009.

[KAR 08] KARJALAINEN M., KAARTINEN H., HYYPPÄ J., "Agricultural monitoring using envisat alternating polarization SAR images", *Photogrammetric Engineering & Remote Sensing*, vol. 74, pp. 117–126, 2008.

[KIN 02] KING R.B., "Land cover mapping principles: a return to interpretation fundamentals", *International Journal of Remote Sensing*, vol. 23, pp. 3525–3545, 2002.

[LAU 05] LAUNAY M., GUERIF M., "Assimilating remote sensing data into a crop model to improve predictive performance for spatial applications", *Agriculture, Ecosystems & Environment*, vol. 111, pp. 321–339, 2005.

[LEE 09a] LEE J.S., POTTIER E., *Polarimetric Radar Imaging: From Basics to Applications*, CRC Press, Boca Raton, 2009.

[LEE 09b] LEE J.S., WEN J.H., AINSWORTH T.L. *et al.*, "Improved sigma filter for speckle filtering of SAR imaging", *IEEE Transactions on Geoscience and Remote Sensing*, vol. 47, pp. 202–213, 2009.

[LIS 05] LISSON S.N., INMAN-BAMBER N.G., ROBERTSON M.J. *et al.*, "The historical and future contribution of crop physiology and modelling research to sugarcane production systems", *Field Crops Research*, vol. 92, pp. 321–335, 2005.

[LOS 95] LO SEEN D., MOUGIN E., RAMBAL S. *et al.*, " A regional Sahelian grassland model to be coupled with multispectral satellite data. II: Toward the control of its simulations by remotely sensed indices", *Remote Sensing of Environment*, vol. 52, pp. 194–206, 1995.

[MAA 88] MAAS S.J., "Using satellite data to improve model estimates of crop yield", *Agronomy Journal*, vol. 80, pp. 655–662, 1988.

[MAT 97] MATTIA F., LE TOAN T., SOUYRIS J.C. *et al.*, "The effect of surface roughness on multifrequency polarimetric SAR data", *IEEE Transactions on Geoscience and Remote Sensing*, vol. 35, pp. 954–966, 1997.

[MCN 09] MCNAIRN H., CHAMPAGNE C., SHANG J. *et al.*, "Integration of optical and synthetic aperture radar (SAR) imaging for delivering operational annual crop inventories", *ISPRS Journal of Photogrammetry and Remote Sensing*, vol. 64, pp. 434–449, 2009.

[MOR 95] MORAN M.S., MAAS S.J., PINTER JR P.J., "Combining remote sensing and modeling for estimating surface evaporation and biomass production", *Remote Sensing Reviews*, vol. 12, pp. 335–353, 1995.

[MOR 13] MOREIRA A., PRATS-IRAOLA P., YOUNIS M. *et al.*, "A tutorial on synthetic aperture radar", *IEEE Geoscience and Remote Sensing Magazine*, vol. 1, no. 1, pp. 6–43, 2013.

[MOR 14a] MOREIRA A., "A golden age for spaceborne SAR systems", *20th International Conference on Microwaves, Radar, and Wireless Communication (MIKON)*, pp. 1–4, 2014.

[MOR 14b] MOREL J., TODOROFF P., BÉGUÉ A. *et al.*, "Toward a satellite-based system of sugarcane yield estimation and forecasting in smallholder Farming Conditions: a Case Study on Reunion Island", *Remote Sensing*, vol. 6, pp. 6620–6635, 2014.

[MOU 98] MOULIN S., BONDEAU A., DELECOLLE R., " Combining agricultural crop models and satellite observations: From field to regional scales", *International Journal of Remote Sensing*, pp. 1021–1036, 1998.

[MUL 15] MULIANGA B., BÉGUÉ A., CLOUVEL P. *et al.*, "Mapping cropping practices of a sugarcane-based cropping system in Kenya using remote sensing", *Remote Sensing*, vol. 7, pp. 14428–14444., 2015.

[MUR 01] MURAKAMI T., OGAWA S., ISHITSUKA N. *et al.*, "Crop discrimination with multitemporal SPOT/HRV data in the Saga Plains, Japan", *International Journal of Remote Sensing*, vol. 22, pp. 1335–1348, 2001.

[MYB 13] MYBURGH G., VAN NIEKERK A., "Effect of feature dimensionality on object-based land cover classification: a comparison of three classifiers", *South African Journal of Geomatics*, vol. 2, pp. 13–27, 2013.

[NAG 05] NAGAYETS O., "Small farms: current status and key trends", Information Brief, Prepared for the Future of Small Farms Research Workshop, Wye College, Ashford, 2005.

[QI 11] QI Z., YEH A.G.O., LI X. *et al.*, "A novel algorithm for land use and land cover classification using RADARSAT-2 polarimetric SAR data", *Remote Sensing of Environment*, vol. 118, pp. 21–39, 2011.

[SER 08] SERRA P., PONS X., "Monitoring farmers' decisions on Mediterranean irrigated crops using satellite image time series", *International Journal of Remote Sensing*, vol. 29, pp. 2293–2316, 2008.

[SIN 95] SINGHI G., PARIYAR M.P., "Crop yield forecasting, a review of methods used in developping countries of Asia", *Proceedings of the Seminar in Crop Yield Forecasting Methods*, Villefranche-sur-Mer, France, pp. 167–179, 1995.

[SPE 01] SPENCER D., "The future of agriculture in sub-Saharan Africa and South Asia: W(h)ither the small farm?" *Proceedings of an International Conference in Sustainable Food Security for all by 2020,* Bonn, Germany, pp. 107–114, September 4–6, 2001.

[THE 03] THENKABAIL P.S., "Biophysical and yield information for precision farming from near-real-time and historical Landsat TM images", *International Journal of Remote Sensing*, vol. 24, pp. 2879–2904, 2003.

[TUR 10] TURKER M., OZDARICI A., "Field-based crop classification using SPOT4, SPOT5, IKONOS and QuickBird imaging for agricultural areas: a comparison study", *International Journal of Remote Sensing*, vol. 32, pp. 9735–9768, 2010.

[VAU 15] VAUDOUR E., NOIROT-COSSON P.E., MEMBRIVE O., "Early-season mapping of crops and cultural operations using very high spatial resolution Pléïades images", *International Journal of Applied Earth Observation and Geoinformation*, vol. 42, no. 42, pp. 128–141, 2015.

[WHI 11] WHITESIDE T.G., BOGGS G.S., MAIER S.W., "Comparing object-based and pixel-based classifications for mapping savannas", *International Journal of Applied Earth Observation and Geoinformation*, vol. 13, pp. 884–893, 2011.

[YUP 08] YUPING M., SHILI W., LI Z. *et al.*, "Monitoring winter wheat growth in North China by combining a crop model and remote sensing data", *International Journal of Applied Earth Observation and Geoinformation*, vol. 10, pp. 426–437, 2008.

6

Monitoring of Agricultural Landscapes Using Remote Sensing Data

6.1. Introduction

In this chapter, agricultural landscapes are defined as ecosystems used by humans for agricultural purposes (crop/livestock productions). Within the framework of climate land use and land cover evolution, which are sometimes responsible of significant impacts on the environment and societies, represent a major challenge for the international community. Monitoring of agricultural landscapes using remote sensing images takes place within a larger scientific framework on the study of land cover and land use changes in general. The *Land Change Science* [TUR 08], which has emerged over the last 10 years as a full-fledged scientific discipline, aims at understanding land cover and land use dynamics as a complex system based on the humankind-environment relationships, by developing theories, notions and models investigated in the framework of social and environmental sciences. These studies [DEF 04, FOL 05] which focused in particular on the analysis of the interactions between land cover changes and global change in general (climate change, loss of biodiversity, impacts on natural resources…), and were developed within the framework of international research programs (LULCC: *Land Use and*

Chapter written by Samuel CORGNE, Laurence HUBERT-MOY and Julie BETBEDER.

Land Cover Change; GLP: *Global Land Project*, etc.), have four main objectives:

– improving the characterization of the main types of land cover;

– improving the identification of space-time dynamics that characterize land cover and land use changes (change dynamics, scales at which they occur, etc.);

– identifying the main factors (spatial, biophysical, climatic, socio-economic, etc.) that motivate land cover and land use changes;

– improving and developing new models able to assess the impacts of land use and land cover changes on societies, territories and the environment.

In this chapter, we focus on the first two objectives, i.e. the characterization of the main types of land cover and land use, and their monitoring within an agricultural landscape.

Knowledge land cover and land use dynamics in agricultural landscapes still represents a scientific challenge, since they are characterized by a "time and space impermanence" at different scales from the long- to the short-term and from the regional to the local level [COR 12a]. Driving factors of these dynamics are multiple and complex, whether they are economic, social, environmental or political factors. Agricultural landscapes are thus characterized by high diversity and a high space–time uncertainty, which represent an obstacle to the understanding of their functioning. Nevertheless, a fine knowledge of this type of landscape represent a prerequisite for the adaptation of agricultural practices to climate change. In this changing and uncertain context, a good knowledge of the space-time variations of land cover and land use represents, a key element, both for the *Land Change Science* scientific community and for the stakeholders in the agricultural sector.

For several decades, the monitoring of land cover and land use changes has generally been carried out using remote sensing data acquired in the optical domain. The high spatial resolution of the Landsat TM and SPOT images, with a decametric resolution, allows us to produce a map of the crop rotation patterns at a parcel scale in agricultural territories of various sizes, from the experimental site of a few km^2 to watersheds of several thousand km^2. If the number of satellites in orbit was limited in the 1990s and 2000s, when only few countries (Europe, United States, etc.) were developing spatial sensors for Earth Observation (SPOT, Landsat, etc.), their number has significantly increased since the 2000s, in particular with the arrival of emerging countries (India, China, Brazil, South Korea, etc.). Thus, in 2011, we could count nearly 120 operational satellites observing the Earth, and nearly 240 are projected for 2021, an increase of 30% compared with the last few decades [EUR 12]. The increase in the number of satellites was accompanied by an improvement in spatial, spectral and temporal resolutions, both in the optical and the microwave domains. Thus, the number and the quality of available optical and radar satellite images today make it possible to consider monitoring the agricultural landscape intra-annual dynamics at a parcel level.

The *Synthetic Aperture Radar* (SAR) imaging systems, long characterized by a decametric spatial resolution and radiometric information less rich than the optical imaging systems (sensors with one to two bands characterized by one or two polarizations) were very little used or not used at all for the study of agricultural landscapes. However, recently, new SAR sensors were developed with a substantial improvement in acquisition capacities with a higher spatial resolution, a higher revisit time and the possibility of acquiring data in polarimetric mode. SAR imaging systems have thus experienced a constant evolution over the last 25 years, since the first launches of SAR satellites characterized by decametric spatial resolutions (approximately 30 m × 30 m) and a single frequency (C-band, ~5.3 GHz). Thus, since the beginning of the 2000s, a substantial improvement in spatial resolution, which have nowadays a metric range for several satellite sensors (RADARSAT-2, ALOS-2, TerraSAR-X, COSMO-SkyMed, etc.) can be noticed. It especially notes a significant diversification of the frequencies provided by the SAR satellites (X-, C-, L-and P-bands in future years, etc.), as well as far richer radiometric information with the development of *Full Polar* or *Dual-Pol* SAR images.

Different types of information can be retrieved from SAR images in polarimetric mode, such as the intensity of the signal backscattered by the Earth's surface in different polarizations, the different backscattering mechanisms that prevail within an image pixel thanks to polarimetric decompositions, or even the elevation type measurements in case of images acquired in interferometric conditions.

Thus, space agencies, such as the European Space Agency (ESA), the Japanese Space Agency (JAXA), the Canadian Space Agency (CSA), the German Space Agency (DLR), or the Italian Space Agency (ASI), are developing ambitious research programs in order to use their SAR sensors to solve different environmental problems (deforestation, urban sprawl, agriculture, geology, glaciology…). However, radar data still remain under-exploited in programs studying agricultural landscapes, which generally only use optical images. However, cloud cover highly limits the characterization and the monitoring of land cover and land use changes with optical images, as a high revisit time is generally necessary to identify and map crops. On the other hand, as they produce their own energy source in the field of microwaves between 12 and 0.3 GHz (approximately at wavelength of 2–75 cm), SAR sensors have the great advantage of being able to acquire information day and night, whatever the cloud cover. In addition to their insensitivity to visibility conditions, radar images can characterize surface states through the acquisition of information on soil humidity and roughness, as well as on vegetation (leaf area index, water content, biomass, etc.). Generally, only backscattering coefficients and polarimetric variables are used to study agricultural landscapes. If some studies use radar images to analyze open agricultural landscapes with large parcels [MCN 02, MCN 09], few studies relate to landscapes characterized by small parcels [COR 14, BAG 09]. Furthermore, radar polarimetry, a technology developed on some sensors (RADARSAT-2, ALOS-2…) and which has a high potential to identify and monitor land cover changes [COR 14], has been very popular for the last few years when studying agricultural landscapes [BAG 09].

Based on the joint analysis of the intensity and the phase of the backscattered signal in different polarizations (HH, HV, VH and VV), polarimetric processing facilitates the characterization of the main backscattering mechanisms operating within a pixel, such as the scattering mechanisms of the surface, volume, etc.

Two examples of agricultural landscape studies using polarimetric SAR remote sensing are shown in this chapter. The first one relates to the identification and characterization of winter land cover from a high spatial resolution radar image in full polarimetric mode. The second one analyses the potential of a time series of very high spatial resolution radar images in partial polarimetric mode (two polarizations HH and VV) to identify crops. Those two applications of polarimetric radar imaging were carried out in a bocage landscape in Brittany (France), that can be defined as a rural landscape characterized by small agricultural parcels surrounded by different types of hedges (wooded, shrubby, etc.).

6.2. Identifying winter land cover within the framework of intensive agriculture

6.2.1. *Overall context*

In intensive agricultural landscapes, the characterization of winter land cover represents a significant environmental challenge, in particular for water quality. Indeed, during the generally rainy intercropping period, the presence of a vegetation cover on the soils can have a direct impact on polluting flow transfers (nitrates, pesticides, etc.) from agricultural fields towards rivers [DAB 01]. During the intercrop period, characterizing land cover and land use remains a major scientific challenge, because agricultural practices are numerous and difficult to identify (plowing, crop residues, winter crops, intercropping, etc.). With high spatial resolution optical satellite data such as SPOT or Landsat images, it is possible to discriminate soils covered with vegetation and bare soils or soils with little vegetation cover. Those two categories can be determined from a vegetation index, the NDVI (*Normalized Difference Vegetation Index*), thresholded and validated by means of ground measurements carried out on the study area, [LAM 94, COR 02]. However, characterizing and monitoring agricultural landscapes in winter is still not often studied with optical images, because of the fact that, in many regions, the cloud cover is generally too prevalent in winter to register usable images. Furthermore, optical data cannot be used to identify some farming practices on bare soil fields or on fields with little vegetation cover (deep plowing, shallow plowing, crop residues, etc.), because the lack of chlorophyll activity limits their extraction [DAU 01]. SAR images in polarimetric mode have already been used to differentiate some farming practices on bare soils (plowing, crop residues, etc.) in open agricultural landscapes with large fields [MCN 02, AUB 11], but few have been used in landscapes with small agricultural fields surrounded by hedges and embankments. We are proposing here to evaluate the contribution of radar polarimetry to identify bare soils and soils covered by vegetation on a watershed scale, and to characterize the roughness of bare soils in fragmented agricultural landscapes [COR 12b, COR 12c].

6.2.2. *Study site and data*

The study site is the Yar watershed, located in Brittany (France). This coastal watershed, measuring 61.5 km^2, is responsible for more than 60% of the polluting flows feeding the Saint-Michel-en-Grève bay (Figure 6.1). The Yar watershed is characterized by cattle farms and a landscape composed of small parcels with an average size of 1 ha (2.5 acres), mostly surrounded by wooded hedges. The significant feed of nutrients from agriculture to rivers results in an eutrophication[1] phenomenon on the coast, commonly known as "green tides". Different agro-environmental actions were carried out on the watershed to understand and halt that phenomenon, and have resulted in land cover modifications, in particular through production mode changes (extensification with an increase of meadows, conversion to organic agriculture, etc.) and an extension of the winter parcel cover through the implantation of intercropping. In fact, the Nitrates European Directive, applicable since July 2001, imposes the presence of vegetation covers in winter in all areas subject to nitrate transfer. Actions aiming at promoting intercropping implantation of the CIPAN type (intermediate crops which trap nitrates, such as phacelia or mustard) are also implemented after the annual harvest of wheat, rapeseed or even corn. The identification through satellite imaging of the land cover thus remains complex and often requires the use of multi-sensor satellite data [CLA 12, FIE 13], because the sowing season is relatively short (between September and November according to annual crops).

A polarimetric SAR radar image (with all the HH, HV, VH and VV polarizations) of the RADARSAT-2 sensor was acquired on February 10th, 2011 on the study site with an incidence angle of 38° and a 5 m range image spatial resolution. Field campaigns were carried out on 123 agricultural plots to identify bare soils and soils covered by vegetation (half of the parcels are used for the analysis of radar variables and image classification, the other half are used to validate the classification created). In this study, a soil is considered as bare when its vegetation cover rate is less than 10%. The information gathered on the parcels (Figure 6.2) concerns land use, vegetation cover rate, if a parcel is covered by vegetation, and bare soil

1 Eutrophication can be considered as a pollution form that results in aquatic ecosystem degradation. It occurs when an aquatic environment receives too many nutrients (nitrogen, phosphorus) that can be assimilated by algae, which then proliferate.

roughness: shallow plowing, deep plowing, bare soils with crop residues, regrowth after a corn crop, winter young cereal, etc.

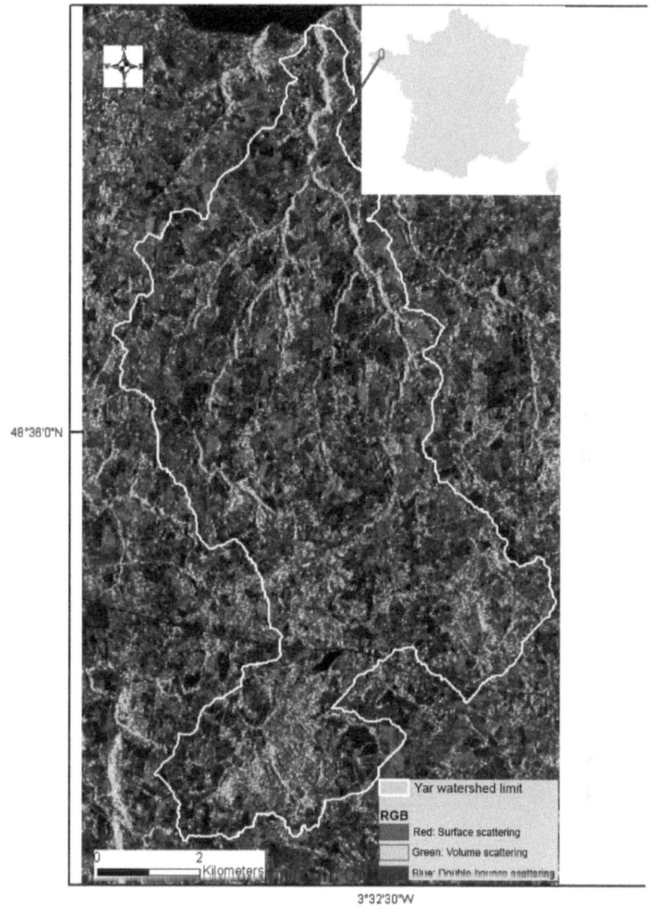

Figure 6.1. *Yar watershed location. Composite image derived from the polarimetric decomposition of Durden-Freeman on the Yar watershed (RADARSAT-2, on February 10th, 2011). RGB (Red, Green, Blue channels) = surface, volume and double-bounce. For a color version of this figure, see www.iste.co.uk/baghdadi/3.zip*

The surface states observed on the agricultural parcels during the winter period are varied (plowing, presence of a vegetation cover, winter crop) and are linked to numerous factors of an economic (grants for certain crops), climatic (pluviometry), and environmental (agro-environmental measures) nature. From an environmental point of view, it is important to differentiate bare soils with a "low roughness", that are homogeneous and smooth, from bare soils with "high roughness", that include crop residues and/ or deep plowing (as the first ones can accelerate the polluting flow transfer towards rivers or encourage the formation of slaking crusts during winter rains).

Figure 6.2. *Different types of bare soil parcels in the Yar watershed*

6.2.3. *Methodology*

6.2.3.1. *RADARSAT-2 image pre-processing*

First, the RADARSAT-2 image was radiometrically calibrated with the following formula:

$$\sigma^0_{\text{linear}} = \frac{(\text{DN}^2 + \text{A0})}{\text{Aj}} \sin(\theta) \qquad [6.1]$$

with: Aj, the sensor gain; A0, the offset; Θ, the incidence angle; DN, the image digital number.

Then, a filter adapted to the polarimetric speckle was applied [LEE 99] on the scattering matrix in order to reduce noise while preserving the spatial and spectral information of the image. The size of the sliding window was established at 7 × 7 pixels, in order to mitigate the speckle present on the image while keeping as many details as possible. The image was then geometrically corrected and geo-referenced in order to correct the topographic deformations, using the Digital Elevation Model *Shuttle Radar Topography Mission* (SRTM) and the control points on the ground selected from aerial pictures. The geometric correction was accurate to within less than one pixel.

6.2.3.2. *Extraction of radar variables from the RADARSAT-2 image*

In full polarimetric mode, it is possible to use the complex matrix S (Sinclair matrix), which contains all the polarimetric properties of a pixel, to first calculate the backscattering coefficients σ^0 in HH, VV and HV polarizations (average of HV and VH, given that HV and VH are considered as equal in case of a monostatic acquisition). Those backscattering coefficients are then calculated in decibels (dB) by applying the following equation:

$$\sigma 0\ (\mathrm{db}) = 10\log_{10}\left(\sigma^0_{\mathrm{linear}}\right) \qquad [6.2]$$

Secondly, the use of the matrix S by means of different matrix vectorization methods [CLO 96] allows us to determine the covariance matrix C_3 and the coherence matrix T_3 [LEE 09]:

$$C_3 = \begin{bmatrix} \langle |S_{hh}|^2\rangle & \sqrt{2}\langle S_{hh}S_{hv}^*\rangle & \langle S_{hh}S_{vv}^*\rangle \\ \sqrt{2}\langle S_{hv}S_{hh}^*\rangle & 2\langle |S_{hv}|^2\rangle & \sqrt{2}\langle S_{hv}S_{vv}^*\rangle \\ S_{vv}S_{hh}^* & \sqrt{2}\langle S_{vv}S_{hv}^*\rangle & \langle |S_{vv}|^2\rangle \end{bmatrix} \qquad [6.3]$$

$$T_3 = \begin{bmatrix} \langle |S_{\mathrm{hh}}+S_{\mathrm{vv}}|^2\rangle & \langle (S_{\mathrm{hh}}+S_{\mathrm{vv}})(S_{\mathrm{hh}}+S_{\mathrm{vv}})^*\rangle & 2\langle (S_{hh}+S_{vv})S_{hv}^*\rangle \\ \langle (S_{\mathrm{hh}}-S_{\mathrm{vv}})(S_{\mathrm{hh}}+S_{\mathrm{vv}})^*\rangle & \langle |S_{\mathrm{hh}}-S_{\mathrm{vv}}|^2\rangle & 2\langle (S_{hh}-S_{vv})S_{hv}^*\rangle \\ 2\langle S_{\mathrm{hv}}(S_{\mathrm{hh}}+S_{\mathrm{vv}})^*\rangle & 2\langle S_{\mathrm{hv}}(S_{\mathrm{hh}}+S_{\mathrm{vv}})^{2*}\rangle & 4\langle |S_{hv}|^2\rangle \end{bmatrix} \qquad [6.4]$$

From those two matrixes [LEE 09], different incoherent polarimetric decompositions can be applied in order to extract polarimetric variables allowing the analysis of the main scattering mechanisms linked to a pixel. From T_3, we can thus apply the decomposition based on the eigenvectors of

Cloude–Pottier [CLO 96] and of Touzi [TOU 92]. For C_3, the decompositions are based on a physical model [FRE 98, VAN 90]. From those different polarimetric decompositions, different radar variables are extracted (Table 6.1).

Radar variable description	Designation
Backscattering coefficients	
Backscattering coefficient HH	σ^0_{HH}
Backscattering coefficient VV	σ^0_{VV}
Backscattering coefficient VH	σ^0_{VH}
Span (total power)	$\lvert S_{\text{hh}}\rvert^2 + \lvert S_{\text{vv}}\rvert^2 + \lvert S_{\text{vh}}\rvert^2$
Polarimetric decompositions	
Cloude–Pottier: entropy; alpha angle; anisotropy	H; α; A
Durden–Freeman: volume; surface; double-bounce	$Vol; S; Dr$
Touzy: orientation angle; dominant eigenvalue; phase, anisotropy	Ψ; λ; ϕ; A
Van Zyl: surface; dihedral angle; volume	S; Ad; Vol
Polarimetric discriminators	
Variation coefficient, maximum and minimum of the fully polarized component; polarization degree; minimum and maximum of the power received, pedestal height, etc.	

Table 6.1. *Description of the radar variables extracted from a fully polarized radar image*

As an example, Figure 6.1 shows a colored composition with the three predominant scattering mechanisms derived from the Freeman–Durden decomposition. The double scattering is associated with the blue color, the volume scattering with the green color and the surface scattering with the red color. Through that colored composition, we can interpret the following:

– Surface scattering: associated with the red color in the colored composition, surface scattering appears in that color when the surface scattering prevails and in orange when other scattering mechanisms intervene. It usually corresponds to bare soil parcels or parcels with low vegetation cover, whatever the roughness (plowing, crop residues, regrowth, etc.);

– Volume scattering: as it is sensitive to the random scattering of targets, volume scattering is usually associated with wooded vegetation. In green on the colored composition, it corresponds here to river bank vegetation and the forest located in the south of the Yar watershed;

– Double-bounce scattering: it is associated with the blue color in the colored composition. It mainly corresponds here to hedgerow and urban centers favorable to that type of scattering.

6.2.3.3. *Identification and classification of significant radar variables*

The analysis of radar variables (backscattering coefficients, variables derived from polarimetric decompositions, etc.) with data derived from the field campaign was first performed from a principal component analysis (PCA) in order to determine the most significant radar variables for the study of winter land cover. An automatic *Classification Tree Analysis* (CTA) was then applied in order to:

– map bare soils and soils covered with vegetation;

– map bare soils with a low or high roughness.

The automatic classification tree analysis was applied because it is an efficient non parametric supervised classification technique useful when few classes need to be identified [ZAM 06].

6.2.4. *Results*

6.2.4.1. *Classification of bare soils and soils covered by vegetation*

The principal component analysis of the different radar variables extracted from the RADARSAT-2 image with the test parcels allowed us to extract 10 statistically significant variables to identify bare soils and soils covered by vegetation (Table 6.2). Those 10 variables include different types of radar information: σ^0 in polarization VV, which is sensitive to the vertical elements of the objects studied; volume scattering and entropy, derived from the polarimetric decompositions, attest to the random scattering processes, such as those created by vegetation; the "minimum power received" and the *pedestal height* indicate the levels of intensity and depolarization of the signal.

Radar variables
1) Maximum of the fully polarized component: polarimetric discriminator derived from the scattering matrix. It describes the maximum intensity of the polarized component for each image pixel
2) Polarization degree: polarimetric discriminator derived from the scattering matrix. It describes the ratio between the power in the polarized part of the electromagnetic wave and its total power
3) Entropy (*H*): polarimetric parameter derived from the Cloude–Pottier decomposition. It describes the scattering random nature of each pixel
4) Contribution of volume scattering (Vol): polarimetric parameter derived from the Freeman-Durden decomposition. It describes the contribution of the volume scattering of each pixel compared with the surface scattering and the double-bounce scattering
5) Contribution of surface scattering (*S*): polarimetric parameter derived from the Freeman-Durden decomposition. It describes the contribution of the surface scattering of each pixel compared with the volume scattering and the double-bounce scattering
6) Pedestal height: polarimetric discriminator derived from the scattering matrix. It indicates the depolarization level of each pixel
7) Touzi phase (ϕ): polarimetric parameter derived from the Touzi decomposition. It characterizes the magnitude and phase of each pixel
8) Touzi anisotropy (*A*): polarimetric parameter derived from the Touzi decomposition. It characterizes the importance of secondary mechanisms compared with the principal mechanism for each pixel
9) Backscattering coefficient VV (σ^0_{VV}): it measures the intensity backscattered in the VV polarization in dBs for each pixel
10) Minimum of the power received: polarimetric discriminator derived from the scattering matrix. It characterizes smooth surfaces with a specular scattering

Table 6.2. *Identification of the 10 statistically significant variables for mapping bare soils and soils covered with vegetation*

Table 6.3 shows the confusion matrix of the classification tree analysis created from the 10 radar variables of Table 6.2. It shows a satisfying classification rate, with a Kappa coefficient of 0.86. We can note that a significant part of the watershed surface was left with bare soil (28% of the total area and 46% of the utilized agricultural area). Part of the soils that look bare on the radar on February 10th, 2011 can in fact correspond to young winter cereal crops which, at that time of the year, have a low covering of vegetation.

KAPPA: 0.86		Site	
		Bare soils	Covered soils
Classification	Bare soils	94.0%	6.0%
	Soils covered with vegetation	8.0%	92.0%

Table 6.3. *Confusion matrix for the classification of the RADARSAT-2 image of February 10th, 2011*

6.2.4.2. *Classification of bare soils with a low and high roughness*

Only bare soils identified during the previous classification are selected here. From the field data, two classes are studied:

– low roughness, corresponding to parcels with a shallow plowing with no vegetation or a young crop;

– high roughness, corresponding to parcels with crop residues, significant regrowth, or deep plowing.

The PCA facilitated the identification of nine statistically significant radar variables in order to differentiate the two roughness types associated with bare soil parcels: entropy, variation coefficient, Touzi phase, minimum power received, pedestal height, maximum scattered intensity, maximum of the fully polarized component, surface roughness and dominant eigenvalue (derived from the Cloude–Pottier decomposition). The backscattering coefficients do not contribute to the differentiation of the roughness types. The F1 axis, which explains 63.32% of the variance, reflects the variable scattering properties with, for the negative values, a high level of random scattering and, for positive values, a single predominant scattering mechanism. The F2 axis with 35.51% of the variance explained indicates the depolarization level of the signal (high for positive values and low for negative values).

The bare soil classification in two classes (low and high roughness) is carried out, like the previous classification, with an automatic tree analysis. The classification accuracy for those two classes is 100%. At the watershed level (Figure 6.3), 29% of the utilized agricultural area (UAA) corresponds to bare soils "with a low roughness", which usually correspond to parcels where a crop is planted in autumn. That class can then indirectly allow us to identify winter cereals in the watershed. Bare soils with a high roughness (17% of the UAA) correspond to bare soils before a spring crop (corn, temporary meadow, etc.).

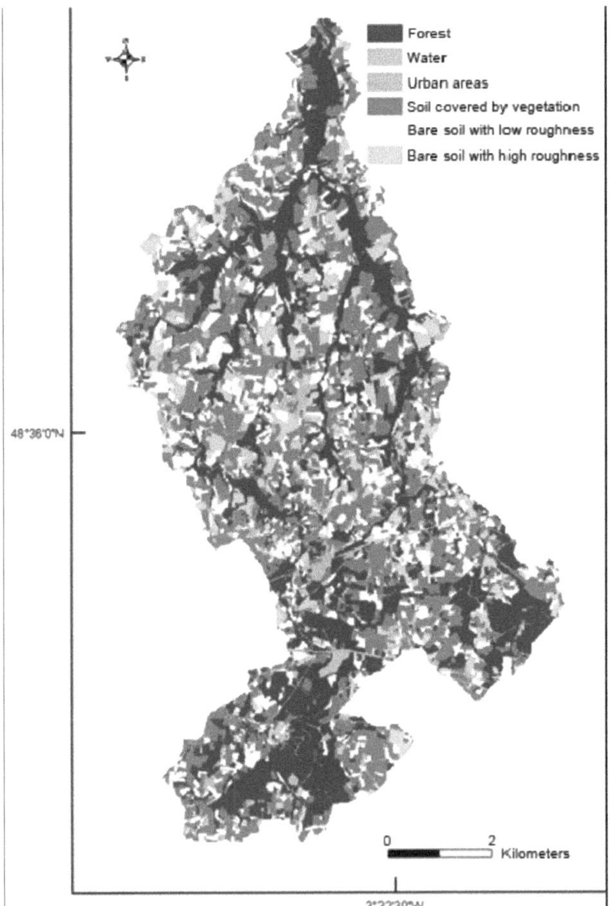

Figure 6.3. *Winter land cover from the RADARSAT-2 image of the Yar watershed, France (February 10th, 2011). For a color version of this figure, see www.iste.co.uk/baghdadi/3.zip*

6.3. Phenology monitoring and crop characterization from a series of radar images

6.3.1. *Overall context*

Although crop identification in intensive agricultural landscapes represents a significant challenge from an environmental point of view, their identification at a parcel level still remains critical [LOB 07]. The high space–time variability of the crops involves a multi-date acquisition at an intra-annual level that can be freed from meteorological conditions, which can currently only be systematically provided by data with a high or very high spatial resolution [COR 14, BAG 09, BET 14]. Here, we are proposing to analyze the potentials of a SAR time series with high spatial resolution in order to identify annual crops within an agricultural bocage framework.

6.3.2. *Study site and data*

The study site is located in Brittany (France), in the south of the Mont-Saint-Michel bay (Figure 6.4). It covers 9400 ha (approximately 23,000 acres) and is characterized by a temperate oceanic climate with an average rainfall of 600 mm/year, the average temperature in January is 6 °C and 18.2 °C in August. The agricultural landscape of this site is dominated by production systems of the "crop-livestock farming" type, which are characterized by a high layout variability (size of the parcels, density of hedges), which is explained by the soil and subsoil characteristics. At the north-eastern limit of the site, vegetable crop farms are established on soils rich in alluvium; the far northwest as well as half of the south have subsoils in granite and the center has shaly subsoils, on which meadows and crops are found. Thus, the landscape layout diversity is organized along a north-south gradient of 15 km with heavy hedgerow in the south containing small parcels surrounded by hedges; bocage in the north is far less dense. The crop systems and the practices to manage the edges of wooded fields implemented by farmers are closely linked to that hedgerow gradient [THE 02, LOT 10].

A series of radar satellite images was acquired between February and June 2013. The radar time series includes four TerraSAR-X images (February 16th, 2013; March 10th, 2013; April 15th, 2013 and June 17th,

2013) acquired in double polarization HH and VV, with a 37° incidence angle, in high *Spotlight* (1.5 m × 1.5 m) resolution mode. As the VV polarization is sensitive to vertical structures, it allows us to identify the different types of vegetation as their vertical structures vary, and so indirectly, to identify the vegetation phenology. On the contrary, the HH polarization, which tends to more easily penetrate vertical structures, can provide interesting information on the underlying soil state, like the type of crop practices (types of plowing, seedling, etc.). The image dates were chosen in order to identify the crop phenological stages: at the beginning of the year during seedling (February, March), in spring during bolting and corn seedling (April) and at the beginning of the summer during crop growth and reaping (June). Surveys were conducted in the field according to each radar image acquisition date on 26 parcels spread over the whole study site. For each parcel, the land cover type, as well as the vegetation height, were recorded. Numerous SAR sensors facilitate the acquisition of data in double polarization with different configurations (HH, VV; HH/HV or VV/HV). Even though the radar data acquired in full polarization can be used to identify several types of polarimetric mechanisms, they are not readily available. In fact, most of the SAR sensors, such as TerraSAR-X and ALOS-2, mainly acquire images in single or double polarization modes. In addition, the full polarization mode has a ground footprint generally lower than the single or double mode. The new embarked radar sensor Sentinel-1 (launched in 2014) is also a double polarization radar sensor.

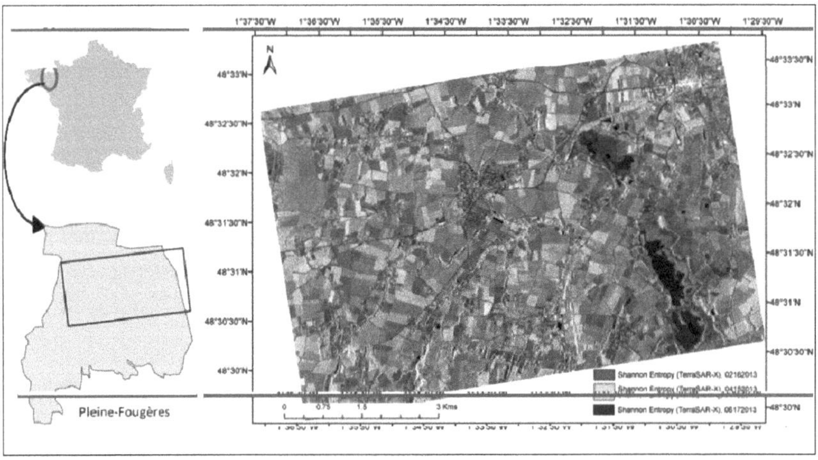

Figure 6.4. *Localization of the study site in Brittany, France. For a color version of this figure, see www.iste.co.uk/baghdadi/3.zip*

6.3.3. *Methodology*

6.3.3.1. *Extracting TerraSAR-X image backscattering coefficients*

The TerraSAR-X images were first radiometrically calibrated according to the following equation [FRI 07]:

$$\sigma^0_{linear} = (Ks \cdot DN^2 - NEBN) \cdot \sin(\theta) \qquad [6.5]$$

Ks represents the calibration constant, *DN* is the digital number of each pixel (magnitude of the backscattering signal), *NEBN* (*Noise Equivalent Beta Naught*) corresponds to the sensor noise and θ is the radar incidence angle. This equation allows us to transform the digital number of each pixel in backscattering coefficient on a linear scale.

A *Lee Refined* [LEE 81] filter was then applied by means of a sliding window in order to reduce speckle, because it allows preservation of details on the heterogeneous areas and averaging of information on homogeneous areas [LEE 09]. The size of the sliding window was fixed at 7 × 7 pixels in order to reduce the speckle present on the image, while preserving as many details as possible. The image was then orthorectified using a Digital Elevation Model (SRTM, 90 m × 90 m spatial resolution) and aerial photographs. The accuracy of the geometric correction is to within one pixel (~1 m).

6.3.3.2. *Extraction of TerraSAR-X image radar variables*

In dual polarization mode, it is possible to reduce the covariance matrix C_2 which will then have two dimensions (the covariance matrix is presented here in the case of data acquired in HH, VV mode):

$$C_2 = \begin{bmatrix} \langle |S_{hh}|^2 \rangle & \langle S_{hh} S_{vv}^* \rangle \\ \langle S_{vv} S_{hh}^* \rangle & \langle |S_{vv}|^2 \rangle \end{bmatrix} \qquad [6.6]$$

S_{hh} and S_{vv} represent the backscattering matrix elements corresponding to HH and VV polarizations. The radar variables that can be extracted from those data are shown in Table 6.4.

From those different radar variables, two analyses are presented: the first one is based on the time profile analysis of radar variables according to the different crops studied, the second one concerns the study area classification.

The classification developed here is based on the merging of radar variables through the Dempster–Shafer rule of combination [BET 14].

Description of the parameters extracted from the data acquired in HH, VV double polarization mode	Designation
Backscattering coefficient HH	σ^0_{HH}
Backscattering coefficient VV	σ^0_{VV}
Ratio HH/VV	$\sigma^0_{HH}/\sigma^0_{VV}$
Shannon entropy	SE
Single-bounce (Pauli decomposition)	$\lvert S_{hh}+S_{vv}\rvert^2$
Double-bounce (Pauli decomposition)	$\lvert S_{hh}-S_{vv}\rvert^2$
Span	$\lvert S_{hh}\rvert^2+\lvert S_{vv}\rvert^2$

Table 6.4. *Radar variables derived from data acquired in double polarization (HH, VV)*

6.3.4. *Results*

The time evolution of the radar variables for each type of crop is shown in Figure 6.5. It is analyzed by the following variables.

6.3.4.1. *Radar variable analysis*

– *Single-bounce and double-bounce*: We can note that, whatever the class, the values of the single and double-bounce are high in winter and at the beginning of spring. They are reduced in April. For the "meadows" and "rapeseed" land cover classes, the double-bounce parameter (and single-bounce for just meadows) increases again in June.

– *Backscattering coefficient:* σ^0_{HH}, σ^0_{VV}. The evolution of the backscattering coefficients HH and VV has the same overall appearance as the one of the single and double-bounce. Profiles are rather stable for all of the classes until April (between –5 dB and –15 dB according to the crop types). The backscattering coefficients then greatly decrease in June for the barley and wheat classes, and slightly decrease for the corn and rapeseed classes.

– *Shannon entropy span*: Whatever the class, the span and the Shannon entropy show similar, time evolutions. We note that the two variables appear to be correlated with each other, because they often have the same behavior over time. Furthermore, we can note that the differences in values between the dates for the different crop types seem to be more significant for the Shannon entropy than for the other polarimetric indicators or backscattering coefficients.

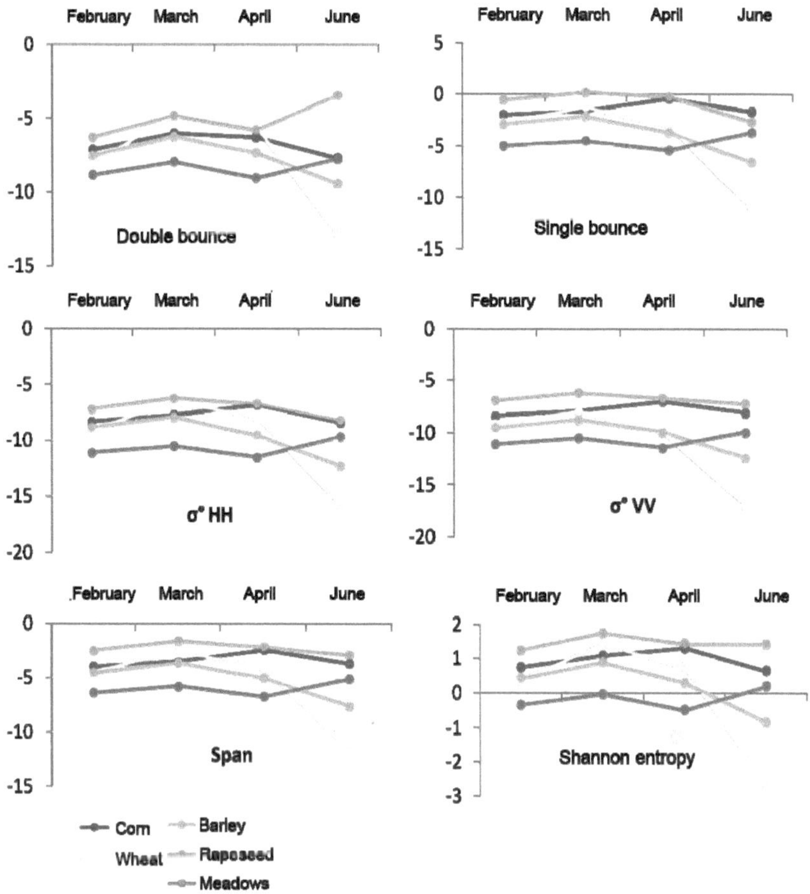

Figure 6.5. *Time evolution of the crop indicators derived from TerraSAR-X images. For a color version of this figure, see www.iste.co.uk/baghdadi/3.zip*

6.3.4.2. *Radar variable analysis according to crops*

– *Corn*: the time profiles derived from the obtained satellite images correspond with corn phenology (seedling in March–April, vegetation growth in June–July). The evolution of backscattering intensity in horizontal (HH) and in vertical (VV) allows us to retrace the corn changes with backscattering coefficients between –6 and –10 dB, when soils are bare (seedling of the April month). That range corresponds with the values described by Aubert et al. [AUB 11]. It confirms the sensitivity of the X- band to soil roughness. Regarding the change over time of the double and single-bounce, the values are very high in winter and at the beginning of spring because of the presence of winter intercropping, then the signal decreases during plowing and seedling in April and June, when the vegetation cover is still relatively little developed. The Shannon entropy is a polarimetric indicator, which reflects the diversity of the mechanisms that take place in the soil during the interaction between the signal and the object observed, and thus reflects the diversity in intensity of the backscattered signal. The latter shows high values (around 1) at the beginning of the year (February) and increases until April (up to approximately 2), because of the plowing which causes a high disturbance. In June, during the beginning of the corn growth, entropy highly decreases because of the homogeneity of the surfaces. The Shannon entropy values in June are, for that class, approximately 1.5. This is mainly because of corn growth, which leads to an increase of the backscattered mechanisms (mainly of the double-bounce) and an increase of the backscattered signal intensity. The span corresponds to the total intensity (in horizontal and in vertical) sent back to the sensor. That indicator is highly correlated with the Shannon entropy (86% correlation). The increase of span values is mainly because of corn growth and plowing, which generate an increase of the backscattered signal towards the sensor.

– *Wheat, barley*: those two types of cereals are generally sowed in October for winter cereals and in March for spring cereals. The time profiles of the indicators derived from the radar are highly similar for wheat and barley at the beginning of the year. Those two types of cereals indeed show very similar physiognomies and phenologies. However, we can differentiate those crops thanks to their harvest date, which is earlier for the barley (end of June, beginning of July) than for the wheat (mid-June or end of July). Unfortunately, no satellite image was acquired during that period, which would have allowed us to properly differentiate them. Nevertheless, we can note that the values of all the indicators derived from SAR images are higher for wheat in June. This is certainly linked to the barley specific structure. In fact, barley is a faster growing crop than wheat and has long ears covered with barbs. That cereal thus covers the entire soil. On the other hand, wheat is less covering and more "vertical". The scattering mechanisms that occur between the signal and those two crops thus seem different, with prevailing double-bounces for wheat and prevailing volume scattering for barley in June. The profiles observed for σ_{VV}^{0} are similar to those described by Ferazzoli [FER 02] (ERS, C- band) and so confirm the relevance of VV polarization use in order to identify wheat and barley.

– *Rapeseed*: rapeseed is generally sowed in October–November, then shows a rather low growth in winter. Its growth accelerates in April, when it reaches a height of approximately 50 cm. In May and June, rapeseed parcels have very high plants (around 150 cm) and are very covering. That vegetation cover degree can be seen through radar variables, which all have high values (for example Shannon entropy values are between 0 and 2) on nearly all the time series. From June, radar signal decline is explained by the rapeseed harvest. Those results confirm those of Ferazzoli [FER 02], who has demonstrated that intensity variation of the backscattered signal in C- band.

– *Meadow*: characterizing meadows on agricultural landscapes of the "dairy" type remains extremely complex. In fact, several management modes characterize meadows in Brittany (mowing, grazing, etc.), which make characterizing and monitoring them over time uncertain [DUS 12]. We can thus see in the field a high variability between the different samples, which show very different time profiles from each other. However, the whole of the time profiles for meadows is characterized by low values, which are typical of a relatively homogeneous covering vegetation.

6.3.4.3. *Classification of the study area*

The classification is carried out from different multi-date radar variables. The classification result is shown in Figure 6.6. The classification quality was estimated through a Kappa index calculated from validation samples representing 2% of the study area and which were not used for the classification. The kappa thus obtained is equal to 0.85 and the global accuracy is equal to 89% (Table 6.5). The different crop types are globally well detected, even if we note some confusions between the corn and meadow classes. On the other hand, the differentiation between crops, like wheat and barley which are very close from a phenological point of view, appears to be very good, which is a very interesting result for monitoring agricultural landscapes within an intensive agriculture framework.

Classification	Wheat	Rapeseed	Barley	Corn	Meadows
Wheat	92.0	0	3.7	4.3	0
Rapeseed	0	100	0	0	0
Barley	0	0	97.0	0	3.0
Corn	9.5	1.1	8.0	81.4	0
Meadows	0	0	0	24.5	75.5

Table 6.5. *Confusion matrix (%) for land use classification in 2013*

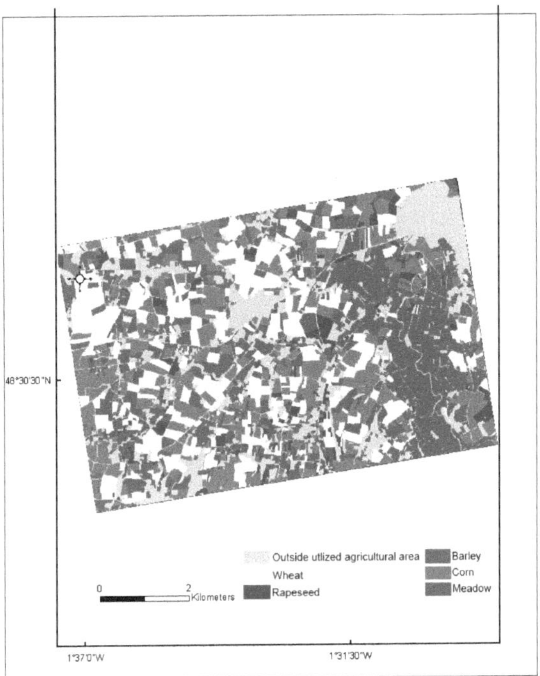

Figure 6.6. *Crop map issue of the TerraSAR-X image series in 2013 on the study area in Brittany, France. For a color version of this figure, see www.iste.co.uk/baghdadi/3.zip*

6.4. Prospects

A better knowledge of agricultural landscapes, which are characterized by high space and time dynamics, requires high temporal frequency data acquisition, whatever the atmospheric conditions. Therefore, space agencies have developed at short and medium term ambitious space programs. Thus, the European Space Agency recently launched, the Sentinel-1A radar satellite (April 3rd, 2014) which will be followed in 2016 by the Sentinel-1B satellite within the framework of the Copernicus program (European Earth monitoring program). Sentinel-1A and 1B are equipped with SAR systems in C band, which will allow them, when the two satellites will be operational, to observe the same area every 6 days, with a spatial resolution between 5 and 40 m, and with multiple acquisition modes (single polarization and double polarization). The high resolution optical component of Copernicus Sentinel-2, whose first satellite (Sentinel-2A) was launched on June 23rd, 2015, is equipped with a multispectral imager including 13 spectral bands (from the visible to mid-infrared) with a spatial resolution between 10, 20 and 60 m according to the spectral bands, and a swath width of 290 km. Associated with sentinel-2B (launch planned in 2016), the revisit time of the area will be 5 days. The radar/optic synergy of sentinel-1 and 2 will thus provide very interesting opportunities to identify and monitor crops "in real time" on a fine scale, thanks to their high spatial and time resolutions, as well as their complementarities: radar variables for Sentinel-1 (roughness, surface soil moisture…) and optical variables for Sentinel-2 (biophysical variables, vegetation indexes…).

At the same time, other space programs developed by other agencies, such as the Canadian Space Agency with RADARSAT-Constellation (first launch planned from 2018), and the German Space Agency with the future TerraSAR-X2 (planned in 2016), will allow us to evaluate the input of multi-frequency images for a better analysis of crops and intercrops.

Those future radar satellites, combined with optical sensors, will thus provide high precision and high quality spatial information, which is crucial to better characterize and monitor crops. Assimilated into agronomical and hydrological models, that information will allow us to develop decision support tools for the actors and managers of the agricultural sector (yield maps, irrigation management, precision agriculture, etc.). Within a framework of global change (climate, demographic, economical, etc.), the rich and diverse data provided by radar satellite thus represents a major challenge and opportunity for better management of agricultural territories and their resources.

6.5. Key points

High and very high spatial resolution radar polarimetry is of significant interest when studying agricultural landscapes characterized by heterogeneous small parcels.

In full polarimetric mode, radar variables derived from different polarimetric decompositions help to differentiate bare soils from soils covered with vegetation, and the roughness levels of the bare soils (low and high roughness).

A series of very high spatial resolution images acquired on the key dates of the crop season (seedling, growth and harvest) allow us to monitor crop phenology at a parcel level. The radar variables (Shannon entropy, span, etc.), derived from the TerraSAR-X image series acquired in double polarization (HH, VV), allow us to identify and monitor crops within an agricultural framework through differentiation.

6.6. Bibliography

[AUB 11] AUBERT M., BAGHDADI N., ZRIBI M. *et al.*, "Analysis of TerraSAR-X data sensitivity to bare soil moisture, roughness, composition and soil crust", *Remote Sensing of Environment*, vol. 115, no. 8, pp. 1801–1810, 2011.

[BAG 09] BAGHDADI N., BOYER N., TODOROFF P. *et al.*, "Potential of SAR sensors TerraSAR-X, ASAR/ENVISAT and PALSAR/ALOS for monitoring sugarcane crops on Reunion Island", *Remote Sensing of Environment*, vol. 113, no. 8, pp. 1724–1738, 2009.

[BET 14] BETBEDER J., CORGNE S., POTTIER E. *et al.*, "Multi-temporal optical and radar data fusion for crop monitoring: application to an intensive agricultural area in Brittany (France)", *IGARSS*, pp. 1493–1496, 2014.

[CLA 12] CLAVERIE M., DEMAREZ V., DUCHEMIN B. *et al.*, "Maize and sunflower biomass estimation in southwest France using high spatial and temporal resolution remote sensing data", *Remote Sensing of Environment*, vol. 124, pp. 844–857, 2012.

[CLO 96] CLOUDE S.R., POTTIER E.,"A review of target decomposition theorems in radar polarimetry", *IEEE Transactions on Geoscience and Remote Sensing*, vol. 34, pp. 498–518, 1996.

[COR 02] CORGNE S., HUBERT-MOY L., BARBIER J. *et al.*, "Follow-up and modeling of the land use in an intensive agricultural watershed in France", *9th International Symposium on Remote Sensing*, pp. 342–351, 2002.

[COR 12a] CORGNE S., HUBERT-MOY L., LECERF R. *et al.*, "Observation des changements d'occupation et d'utilisation des sols par télédétection dans le Grand Ouest", in MEROT P., DUBREUIL V., DELAHAYE D. *et al.* (eds), *Evaluation, Impacts, Perceptions. Espace et Territoires*, PUR, 2012.

[COR 12b] CORGNE S., BETBEDER J., RAPINEL S. *et al.*, "Dempster-Shafer fusion rule of optical and polarimetric data for winter land cover mapping", *International Geoscience and Remote Sensing Symposium* (IGARSS), Munich, Germany, July 22–27, 2012.

[COR 12c] CORGNE S., SEIGNE P., MERCIER G. *et al.*, "Bare soils extraction and characterization in an intensive agricultural area with polarimetric data", *International Geoscience and Remote Sensing Symposium* (IGARSS), Munich, Germany, July 22–27, 2012.

[COR 14] CORGNE S., DALLON D., MERCIER G., "Land use and land cover monitoring with multitemporal and multipolarization RADARSAT data: Application to an intensive agricultural area (France)", *International Geoscience and Remote Sensing Symposium (IGARSS)*, Québec, Canada, pp. 4239–4242, 13–18 July 2014.

[DAB 01] DABNEY S.M., DELGADO J.A., REEVES D.W. "Using winter crops to improve soil and water quality", *Communications in Soil Science Plant Annals*, vol. 32, pp. 1221–1250, 2001.

[DAU 01] DAUGHTRY C.S.T., "Discriminating crop residues from soil by shortwave infrared reflectance", *Agronomy Journal*, vol. 93, pp. 125–131, 2001.

[DEF 04] DEFRIES R.S., ASNER G.P., HOUGHTON R.A. (eds), "Ecosystems and land use change", *American Geophysical Union*, Washington, DC, 2004.

[DUS 12] DUSSEUX P., GONG X., CORPETTI T. *et al.*, "Contribution of radar images for grassland management identification", *Remote Sensing*, vol. 6, no. 7, pp. 6163–6182, 2012.

[EUR 12] EUROCONSULT, Satellites to be Built & Launched by 2021, World Market Survey, 2012.

[FER 02] FERRAZOLLI P., "SAR for agriculture: advances, problems and prospects", *Retrieval of Bio-and Geo-Physical Parameters from SAR Data for Land Applications*, vol. 475, pp. 47–56, 2002.

[FIE 13] FIEUZAL E., BAUP F., MARAIS-SICRE C., "Monitoring wheat and rapeseed by using synchronous optical and radar satellite data - from temporal signatures to crop parameters estimation", *Advances in Remote Sensing*, vol. 2, pp. 162–180, 2013.

[FOL 05] FOLEY J.A., DEFRIES R., ASNER G.P. *et al.* , "Global consequences of land use", *Science*, vol. 309, pp. 570–574, 2005.

[FRE 98] FREEMAN A., DURDEN S.L., "A three-component scattering model for polarimetric SAR data", *IEEE Transactions on Geoscience and Remote Sensing*, vol. 36, pp. 963–973, 1998.

[FRI 07] FRITZ T., *TerraSAR-X Level 1b Product Format Specification*, TX-GS-DD-3307, 2007.

[LAM 94] LAMBIN E.F., STRAHLER A.H., "Indicators of land cover change for change vectors analysis in multitemporal space at coarse spatial scale", *International Journal of Remote Sensing*, vol. 15, pp. 2099–2119, 1994.

[LEE 81] LEE J.S., "Speckle analysis and smoothing of synthetic aperture radar images", *Computer Graphics and Image Processing*, vol. 17, no. 1, pp. 24–32, 1981.

[LEE 99] LEE J.S., GRUNES M.R., DE GRANDI G., "Polarimetric SAR speckle filtering and its implication for classification", *IEEE Transactions on Geoscience and Remote Sensing*, vol. 37, pp. 2363–2373, 1999.

[LEE 09] LEE J.S., POTTIER E., *Polarimetric Radar Imaging: From Basics to Applications*, CRC Press, 2009.

[LOB 07] LOBELL D.B., FIELD C.B., "Global scale climate–crop yield relationships and the impacts of recent warming", *Environmental Research Letters*, vol. 2, no. 1, p. 7, 2007.

[LOF 10] LOFTI A., JAVELLE A., BAUDRY J. *et al.*, "Interdisciplinary analysis of Hedgerow network landscapes' sustainability", *Landscape Research*, vol. 35, no. 4, pp. 415–426, 2010.

[MCN 02] MCNAIRN H., DUGUAY C., BRISCO B. *et al.*, "The effect of soil and crop residue characteristics on polarimetric radar response", *Remote Sensing of Environment*, vol. 80, pp. 308–320, 2002.

[MCN 09] MCNAIRN H., CHAMPAGNE C., SHANG J. *et al.*, "Integration of optical and Synthetic Aperture Radar (SAR) imagery for delivering operational annual crop inventories", *ISPRS Journal of Photogrammetry and Remote Sensing*, vol. 64, no. 5, pp. 434–449, 2009.

[THE 02] THENAIL C., "Relationships between farm characteristics and the variation of the density of hedgerows at the level of a micro-region of bocage landscape. Study case in Brittany, France", *Agricultural Systems*, vol. 71, no. 3, pp. 207–230, 2002.

[TOU 92] TOUZI R., GOZE S., LE TOAN T. *et al.*, "Polarimetric discriminators for SAR images", *IEEE Transactions on Geoscience and Remote Sensing*, vol. 30, pp. 973–980, 1992.

[TUR 08] TURNER B.L., LAMBIN E., REENBERG A., "The emergence of land change science for global environmental change and sustainability", *Proceedings of the National Academy of Sciences of the United States of America*, vol. 104, no. 52, pp. 20666–20671, 2008.

[VAN 90] VAN ZYL J., "Calibration of polarimetric radar images using only image parameters and trihedral corner reflector responses", *IEEE Transactions on Geoscience and Remote Sensing*, vol. 28, pp. 337–348, 1990.

[YAN 11] YANG C., EVERITT J. H., MURDEN D., "Evaluating high resolution SPOT 5 satellite imagery for crop identification", *Computers and Electronics in Agriculture*, vol. 75, no. 2, pp. 347–354, 2011.

[ZAM 06] ZAMBON M., LAWRENCE R., BUNN A. *et al.*, Powell S., "Effect of alternative splitting rules on image processing using classification tree analysis", *Photogrammetric Engineering & Remote Sensing*, vol. 72, pp. 25–30, 2006.

7

Applications of Multispectral Optical Satellite Imaging in Forestry

7.1. Introduction

Optical imaging, which makes use of the reflective properties of short wavelength solar radiation (400 to 2,500 nm), is the oldest source of spatial data on which its applications in forestry have been developed.

For a long time aerial photography has been the main source of remote sensing data for operational forestry applications, these being forest mapping, inventories, resource estimation (volume of wood, biomass), forest management planning or observation of health condition of the trees. Thanks to technological advances such as the move from film to digital shooting, three-dimensional digital reconstruction of the forest canopy by photogrammetry using stereoscopic shots, the routine production of orthophotographs, and the use of sensors on board drones, its use in observation at tree or stand level is still relevant. Aerial photography also shows the potential of sub-meter (<1 m) spatial resolution satellite imaging.

Over several decades, a wealth of experience of routine observation of forest ecosystems has been acquired from numerous multispectral satellite sensors of varying spatial resolutions (~50 cm to 1 km) and revisit capabilities (from 1 day to several weeks). The expectations are not limited

Chapter written by Dominique GUYON and Nathalie BRÉDA.

to the applications of aerial photography previously mentioned. It is not a matter of simply gathering spatial information on the conditions at any given time, but also monitoring the development and the dynamics of forest covers, as well as any disruptions over time, ranging from the growing season to several decades, on the basis of remote sensing data time series. This monitoring allows the areas of application to be extended to current environmental concerns regarding the response of forest ecosystems to global changes (climate, deforestation) and hazards (extreme climatic events, pests, fires), and thus provides information that is useful in diagnosing the response and the resilience of forest ecosystems to environmental constraints.

Over the past few years, monitoring potential has increased due to technological innovations, methodological advances and also through the accumulation of sets of images acquired over more than 30 years. It is also worth mentioning the recent increase in very high spatial resolution satellite observation sources (VHR ~50 cm to 5 m), i.e., a pixel size less than or close to the size of tree crowns, with sensors such as PLEIADES, SPOT 6/7, IKONOS, QUICKBIRD or WORLDVIEW, for example. There is also a supply of long-term temporal archives systematically covering all of the Earth's surface, well-suited to the monitoring of forests with long life-cycles: on the one hand using sensors with medium resolution (MR ~250 to 1,000 m) and daily revisits, such as AVHRR (resolution 1.1 km at nadir) since 1981, VEGETATION (1 km) from 1998 to 2014, and its successors with higher spatial resolution, PROBA-V (2013 to 2018) and Sentinel-3 (since 2016), and MODIS (250 m to 1 km) since 2000, as well as using sensors with higher resolution and lower revisit frequency (several weeks) such as TM, ETM+ and OLI (30 m resolution) taken on board the Landsat 4,5,7 and 8 satellites since 1982. Most satellite systems with high resolution (HR ~5 to 30 m), for example SPOT 1,2,3,4 and 5, in orbit between 1986 and 2014, and all those with very high resolution (VHR, ~< 5 m) have the potential for a high revisit frequency, but this occurs to the detriment of a periodic coverage of the whole forest area, and at the price of considerable variation in viewing angles from one date to another, or one view to another. Thus, in the case of a major incident such as a storm, while it is easy to program the acquisition of images after the event, the potential lack of prior data hinders the detection of forest damage. However, for several years, a policy of regular acquisition of HR or VHR data over entire areas has been established in France, for example with the GEOSUD/THEIA initiative, or the KALIDEOS program of the CNES. In the future, the Sentinel-2 sensors (the first being launched on June 23 2015, with the second planned for 2016), which offer a high revisit frequency (3 to 5 days, depending on latitude) of the whole of the land area, at high resolution (10 m) with consistent viewing angles, should facilitate the monitoring of all forest surfaces at the stand level over the course of either seasons or years.

Finally, nowadays, the use of satellite data is facilitated by the provision by the scientific community of time series produced with homogeneous methods and with a known quality. The pre-treatment of data consists primarily of the following: geo-referencing, cloud masking, correction of atmospheric and topographical effects, modeling and normalization of directional effects, estimation of surface reflectance (also known as top of canopy reflectance), calculation of vegetation indices (which are combinations of the reflectance of a number of spectral bands), such as the Normalized Difference Vegetation Index[1] (NDVI). Along with the main products of these processing chains (reflectances, vegetation indices) there are quality indicators, including the detected presence of cloud or snow. These facilities are already well-established for the MR (MODIS, VEGETATION, and AVHHR in particular). This is more recent for the HR, especially with the products of the United States Geological Survey (USGS) and NASA in the US (Landsat in particular), as well as those from the MUSCATE chain of the thematic center of continental surfaces, THEIA, in France (Landsat, Formosat2, SPOT4 and SPOT5 Take Five data, simulating those of Sentinel-2), resulting from work done by the French research laboratory CESBIO and the CNES.

The possible uses of satellite imagery produced by multispectral radiometers are examined here, i.e. those with a limited number of relatively broad spectral bands (width > 20 nm) in the wavelengths of visible (blue, green, red), near infrared (NIR) and middle infrared (MIR, also known as SWIR, for *Short Wave Infrared)*. Aerial imaging (drone, ultra-light aircraft, and plane) will not therefore be treated as such. The use of hyperspectral data will also not be addressed, because, with a continuous sampling of the

1 The NDVI is the normalized difference in reflectance between the red and the near infrared spectral bands (NIR): NDVI = (NIR-Red)/(NIR+ Red).

whole range of wavelengths from 400 to 2500 nm, with many narrow bands (width ~10 nm), specific methods are required.

After an initial section presenting the fundamental points specific to forest cover, two types of applications will be presented and illustrated by various examples: estimating and monitoring the forest resource, and monitoring the responses of forests to global changes and hazards.

7.2. Specific key points of the forest cover

This section is not intended to be an exhaustive account of the physical bases of optical remote sensing and the modeling of reflectance. A prior basic general knowledge is assumed (see [BRI 16] and also Chapter 2 in this book). The fundamental issues specific to forest canopy are highlighted, together with, where necessary, a brief reminder of the basics, with a view to providing elements to help frame existing or potential applications in forestry.

A first sub-section relates to the specificity of forest canopy in terms of structure in comparison to the other types of plant communities. The second deals with the nature of the forest information that can be expected from multispectral satellite imagery.

7.2.1. *Structure of forest cover and reflectance*

The surface reflectance measured by satellite is a result of radiation exchanges (or radiative transfer) within the ensemble consisting of the tree canopy (one or more strata of trees), the vegetation from the understorey (low woody and herbaceous plants, as well as mosses and lichens) and the underlying soil (surface and sub-surface). These exchanges depend on the quantity of plant material, its spatial organization and its optical properties (reflection, transmission, absorption) in the wavelengths in question.

The quantity of plant material and its spatial organization (vertical and horizontal) form the cover structure. In the case of forests, this may be variously defined, according to the intended applications and the space and time scales under consideration.

7.2.1.1. *The forest cover: a collection of trees and stands*

The cover structure can be described with variables that are observable at tree and stand scale and that are useful in terms of forest management: the species composition, the stage of development or the age, the spatial distribution of the trees and their dimensions (density – number of trees per unit of surface, stem diameter, height, crown diameter, etc.), derived dendrometric characteristics, such as the basal area[2] and the dominant height[3] [PAR 09] and the crown cover fraction. Their spatio-temporal variations are a result of their development, growth, seasonality, as well forest management practices and silvicultural techniques (cutting in particular, but also modes of regeneration and forestry – high forest, coppice, presence of understorey, etc.). These depend on soil and climatic conditions and hazards (pests, droughts, storms for example). On the scale of the landscape and beyond, forest cover appears as a more or less complex mosaic of forest stands that are spatially homogeneous in relation to the above-mentioned variables.

The dendrometric variables measurable *in situ* are useful to the forestry practitioner in estimating timber volumes, aerial standing biomass and carbon stocks. To this end, allometric models, which are defined on the scale of the tree or the stand and are generally species-specific, are applied, with the diameter of the trunk at 1.3 m above ground, the age or the height often being used as input variables.

7.2.1.2. *Vertical structure of forest cover and possible presence of vegetation in the understorey*

The vertical structure of the forest cover can be quite complex, particularly in connection with the diversity of species, the natural processes of competition, mortality and regeneration, and also the methods of tree-felling and stand renewal. Trees rarely form a vertically homogeneous layer. The vertical profiles encountered in single-species or even-aged plantations, where the trees form a single stratum, are among the most homogeneous. The tree crowns can also be layered into several highly differentiated strata,

---

2 Sum of the surface of the stem cross-sections of all trees in the stand, at 1.30 m above the ground (i.e. at breast height), reduced to one hectare.

3 Average height of the sub-population of trees with access to light, which, in practice, is generally defined as that of the hundred largest trees in the hectare.

as in the case of coppice with standards where the tall trees from seed-origin overlook those regenerated by coppicing. In the most heterogeneous structures, the vertical distribution of the tree vegetation is continuous, with a broad range of ages, sizes and species. The forests treated according to the selection system in temperate regions and the primary forest of the equatorial region are included amongst these.

However, forest cover is not simply a more or less complex set of trees planted on bare soil. The soil can be carpeted by mosses or lichens. The understorey can also be populated by low woody or herbaceous vegetation, whose species composition, height development and cover fraction are dependent on climatic conditions, competition for water and light with the tree strata, and possible silvicultural operations (clearing, cutting). Moreover, understorey vegetation does not always form a storey that is clearly distinguishable from the trees, especially when the height of their lower level is low, as occurs during the first stages of regeneration, or in the first years following reforestation by planting.

7.2.1.3. *The forest cover: a collection of leaves*

From a functional point of view, it is the quantity of foliage, its spatial organization (distribution, orientation) and its biochemical composition, as well as the seasonal dynamics of these, which play the predominant roles in the ecophysiological functioning of the canopy [BRE 08]. They are the main factors in the radiation, water and carbon exchanges between the canopy and the atmosphere by photosynthesis and evapotranspiration. Thus, the leaf area index, or LAI (*Leaf Area Index*, surface of foliage by unit of surface of horizontal ground), is a key variable in process-based models - i.e. simulating the ecophysiological processes - for estimating water and carbon balance and productivity. The LAI is a relevant structure variable irrespective of the scale of observation (for a pixel size from several meters to several km) and of the study (from the individual stand to the whole region and beyond). Furthermore, in optical remote sensing, given the physical mechanisms involved, the LAI, but also the fraction of cover (Fcover, which is equal to 1 – the gap fraction at the nadir) and the fraction of photosynthetically active radiation absorbed (fAPAR[4]) that depend on it, are obviously related to the reflectance measured [GOW 99].

---

4 Field of wavelengths of photosynthetically active radiation, or PAR: 400 to 700 nm.

7.2.1.4. *Modeling the reflectance of forest cover*

Much specific work has been carried out, over the last thirty years or so, on interpreting the remote sensing signal of forest cover, either by modeling the radiative transfer for simulating this signal, or by using semi-empirical approaches often based on vegetation indices. The radiative transfer models rely on a parametrized mathematical description of the two (2D) or three-dimensional (3D) structure of the tree canopy and understorey, and on information on the optical properties of soil and leaves (measured using spectrometry or modeled from the biochemical characteristics – pigments, water, organic compounds – and the cellular structure). To gain a global vision of forest models, one should refer, in particular, to Stenberg *et al.* [STE 08] and Widlowski *et al.* [WID 07]. The hybrid geometric-optical reflectance models [STE 08] are particularly interesting. The tree canopy is viewed as an assemblage of tree crown envelopes whose location is not explicitly stated. The spatial distribution of the crowns is described statistically so as to take into account mutual shading and the distribution of gaps between crowns. Even if the representation of the shape of the trees and of the distribution of their dimensions is simplified, the non-random spatial organization of the components of the cover at different scales is taken into account: distribution of leaves in branches or shoots, distribution of shoots and branches in the crown, distribution of the trees in the stand.

7.2.1.5. *Reference data on the structure variables of foliage and their temporal variations*

The *in situ* measurements of the LAI and Fcover, and of parameters describing the organization of the foliage inherent to the architecture, the development and the spatial distribution of the trees, such as the tilt angle of the leaves or aggregation (also called clumping) parameters, provide essential reference data.

These variables are generally measured indirectly, using optical methods based on the *in situ* measurement of the directional gap fraction. Thus, taking a stochastic approach, in the case of a random distribution of leaves, the directional gap fraction $P_0(\theta,\phi)$ in the direction specified by the zenith angle θ and the azimuth angle ϕ can be related to the LAI using a Poisson model:

$$P_0(\theta,\phi) = \exp\left(-G(\theta,\varphi)\mathrm{LAI}/\cos(\theta)\right) \quad [7.1]$$

where $G(\theta, \varphi)$ is the mean coefficient of projection of the leaves, depending on their angular distribution (tilt, orientation).

If the foliage forms randomly distributed aggregates, the Poisson model can be generalized using an aggregation parameter $\lambda(\theta)$ which is less than 1:

$$P_0(\theta, \phi) = \exp\left(-G(\theta, \varphi)\lambda(\theta)LAI/\cos(\theta)\right) \qquad [7.2]$$

These optical methods have been extensively described and studied [WEI 04, JON 04a]: instruments (hemispherical photography, LAI2000, TRAC, DEMON and others), experimental measurement conditions, and radiation interception models have been used. These have been widely evaluated for forests [CHE 97] and, more rarely, compared with direct measurements taken from litter collections [DUF 95, BRE 03]. These methods provide the LAI, the Fcover, the angular distribution of the foliage. Some even provide information on the spatial aggregation of the foliage (hemispherical photography, TRAC, DEMON). If this is not the case, it is the "effective" LAI (LAIe) that is obtained (LAIe $(\theta) = \lambda(\theta)$ LAI); this can lead to a marked under-estimation of the LAI in conifers, where the distribution of the needles is generally highly aggregated within the crowns.

Contrary to estimations from allometric relationships or from litter collection, these optical methods do not, strictly speaking, allow for an estimate of the LAI, but rather the Plant Area Index (PAI), all plant components that intercept radiation, whether chlorophyll-containing (leaves and other green organs) or not (trunks, branches, senescent leaves...) being taken into account. The measurements of gap fraction are more complex in the understorey. They are carried out upwards, both below the understorey and below the tree crowns; the LAI and the Fcover of the understorey are calculated by assuming that the two strata are independent. Hemispherical photographs can be acquired by downwards viewing, the fraction of seen soil being a proxy of the gap fraction, if the top of the understorey is not too high.

The collection of fallen leaves using a litter box is complementary and provides access to the true LAI (and not the PAI), but it is cumbersome to implement in terms of spatial and temporal sampling. For deciduous hardwood forests, it enables a measurement of the contribution of the various tree species in the LAI produced during the entire growing season. Among evergreen conifers, it is interesting to know the intra-annual dynamics of needle fall, which depends on their age, their longevity and the climatic conditions in previous years, regardless of the unfolding of the new generation of needles in the next spring.

Estimating the LAI by allometric relationship to the dendrometric variables that are easy to measure *in situ*, such as stem diameter or basal area, is equally possible due to the functional relationship between the sapwood area on the stem section (sap conduction) and the expanded leaf area (transpiring surface) in terms of water exchange at equilibrium. However, the established equations have limitations. They are specific to the species, the silvicultural history of the stand, and are particularly dependent on the season and the year of observation in which they were calibrated [BRE 03]. They represent a compromise for evergreen species and in cases of high LAI, where the optical measure is saturated.

In terms of timing, relative illumination or radiation interception measurements in various wavelengths (global radiation throughout the solar spectrum, or photosynthetically active radiation alone, PAR) have long been used to continuously monitor the dynamics of the LAI throughout the annual cycle of leaf development [BRE 99], in particular to link these dynamics to the canopy evapotranspiration or carbon assimilation. This approach, initially used the Beer-Lambert law of extinction of radiation, requires the establishment of the vertical radiation extinction profile in the canopy to determine the extinction coefficient, which varies according to the wavelength involved [BRE 02].

More recently, *near surface remote sensing* techniques are re-examining these methods of continuous measurements of reflected, absorbed or intercepted radiation using low cost directional and/or multispectral sensors that are relatively easy to use on instrumented site networks: NDVI [SOU 12], PAR intercepted [GUY 14], or even Webcam images [RIC 07]. They have an obvious importance in the calibration and validation of current or future products from remote sensing concerning dating leaf phenology events from satellite time series data: the beginning of leaf expansion, date of maximum expansion, senescence and fall, duration of the growing period.

In addition, on the matter of seasonal leaf dynamics, as Nilson *et al* emphasize [NIL 12], there is a lack of ground based data for the validation of simulations of the seasonal trajectory of forest cover reflectance, even though reflectance models have long been available. Quantitative links with the seasonal variations in the spatial structure of the cover features (leaf tilt angle, leaf aggregation, for example) and of the biochemical (evolution of composition and pigment content, [MON 09]) and structural properties of the leaves have yet to be brought into widespread use on a functional basis.

7.2.1.6. *Spatial sampling*

7.2.1.6.1. Scale of spatial heterogeneity observable by sensors

Roughly speaking, observation scales for forest cover fall into three categories according to the spatial resolution of the sensor (Figure 7.1):

– the tree, or small group of trees, using a very high resolution (<5 m), the pixels being of a smaller or similar size to that of the crowns. The intra-stand variability of the canopy structure is observable by means of a large number of pixels;

– the stand, using a high resolution (between 10 and 30 m). The pixel observed, smaller than the stand, generally contains up to several hundred trees, depending on the density and distribution of the trees. Within the same pixel there may be different species of trees, different dimensions or different health conditions, the presence of clearings, areas of understorey in sunlight or shade, etc.;

– the group of stands, using a medium resolution (250 to 1,000 m, equivalent to a surface of 6.25 to 100 ha). The measurement includes information about one or more forest stands, or even about different types of land use, including some non-forest uses. Within the same pixel, environmental conditions can show considerable spatial variation: soil, understorey vegetation, topography, for example.

The reflectance measured in a pixel depends on the tree canopy structure in neighboring pixels. This is a result of the combined effects of the spatial response of the sensor (or *Point Spread Function)* and the pixel size in relation to the dimension of the trees (total height, diameter and length of crown). The maximum height of a tree generally being several tens of meters, this latter effect is not insignificant at high or very high spatial resolutions. It depends on the position of the sun (azimuth, zenith angles) and the viewing angles, which define the length and orientation of the cast shadows and the length and orientation of the portions of crowns as viewed by the sensor. For example, the length of shadow cast by a tree whose height is 25 m is 9.1, 25 and 43.3 m respectively, for a sun zenith angle of 20°, 45° and 60°. These effects are illustrated in Figure 7.1.

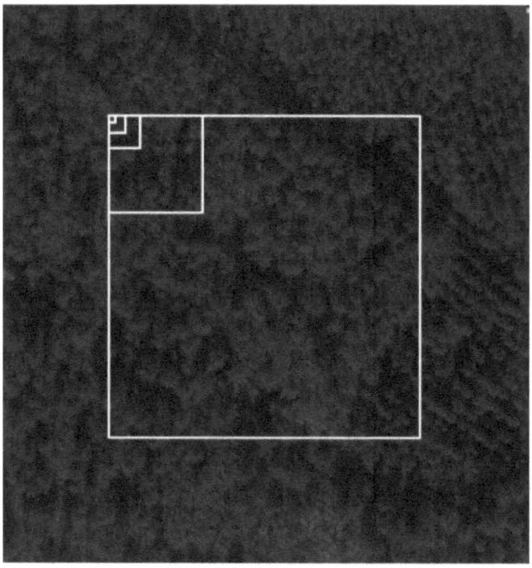

Figure 7.1. *Spatial sampling of the forest canopy for square surfaces of 2, 5, 10, 30 and 100 meters wide. In the background: aerial photograph in color-infrared (spatial resolution = 15 cm) taken in winter (sun zenith angle=52°) in a maritime pine forest. For a color version of this figure, see www.iste.co.uk.baghdadi3.zip*

7.2.1.6.2. Sampling strategies for error of character size reference data

All of these factors must be taken into consideration in the sampling strategies of the spatial variation of the forest cover for ground based collection of reference data: number and distribution of samples in the *in situ* plots, size and distribution of the plots within the stand and the landscape. The ground based measurements of variables of the tree canopy structure relate to surface elements and not points: LAI, stem diameter distribution, distribution of total height, species, tree density, basal area, biomass, etc. The spatial sampling method is defined on statistical criteria, with a possible stratification of forest areas into previously delineated categories of cover type. For dendrometry, the sampling procedures are those used for the forest inventories [PAR 09]. In the case of the LAI measurement of the tree layer using optical methods, the area sampled by the instrument depends on the height of the trees and the angular range of zenith angles investigated

[BRE 02]. For example, with a maximal angle of 60°, the measurement range at ground level, on flat ground, reaches 17 and 43 m for a uniform tree height of 10 and 25 m respectively, this being a maximum circular sampled surface of ~900 and 6 000 m². The procedure for relating the satellite signal measured (reflectance, vegetation index) to the ground based forest cover features, or to the satellite signal simulation which could be derived from the latter ones, is therefore essential. Its high quality is crucial to the spatial estimation of the biophysical variables of interest (class of cover, structure variable, etc.) and an appropriate quantization of errors.

7.2.1.6.3. Accessing intra-pixel forest information

Most strategies used for extracting information assume that the content of the area covered by the pixel is spatially homogeneous, which is rarely the case when using high or medium resolution. There are approaches for retrieving intra-pixel information. In particular, there are spectral or temporal disaggregation *(or unmixing)* methods for:

– finding the composition (area fraction) of the pixel in cover types (or *endmembers)*, by means of a prior information of the spectral or temporal variation of the reflectance of the types present;

– conversely, characterizing the spectral signatures or the temporal trajectory of the reflectance for a given cover type, using prior information about the intra-pixel composition.

The methods are mathematically complex, particularly if one is trying to model the spatial variability (that is, pixel by pixel) of the temporal trajectory [CAR 08] or the spectral signature [SOM 11] of each cover type. Spectral disaggregation is commonly used in hyperspectral data exploitation, but less so for multispectral data; a typical example of its application in the detection of forest change (logging, regrowth) using high resolution images (Landsat-5 TM) is provided in Adams *et al.* [ADA 95]. Temporal disaggregation can be applied in the characterization of the seasonal trajectory of the reflectance or of the vegetation index of a given forest cover type, using medium resolution time series (such as from AVHRR, MODIS, VEGETATION). An illustration is provided in Guyon *et al.* [GUY 11] who use the Cardot *et al* method [CAR 08] in a study of the altitudinal gradient of leaf phenology in deciduous broadleaf forests using a time series of

VEGETATION images in a region where broadleaf, coniferous or mixed stands, or even other types of land use, co-exist within the same pixel. Moreover, some spatio-temporal interpolation methods for seasonal trajectories acquired at high spatial resolution and low temporal frequency (for example, time series of Landsat images), with some trajectories acquired during the same period at medium resolution and high frequency (for example, MODIS), have been developed for the precise spatialization of seasonal information. We refer to the approach of Cardot *et al.* [CAR 08] or that of Gao *et al.* [GAO 06] which Hilker *et al.* [HIL 09] used for forests.

7.2.2. *What information about forest cover can be expected?*

7.2.2.1. *Exploitation of spectral variation*

The spectral properties (reflection, absorption, transmission) of leaves and needles depend on their biochemical compound content (pigments, water, cellulose, lignin, etc.), and their structure (cuticle, wax, hairiness, cell turgidity, number, arrangement and thickness of the various cell layers, etc.). These more or less interrelated elements are determined by the species, the stage of development of the leaf or its physiological condition (which can deteriorate due to water stress, disease, etc.) (Figures 7.2 and 7.3).

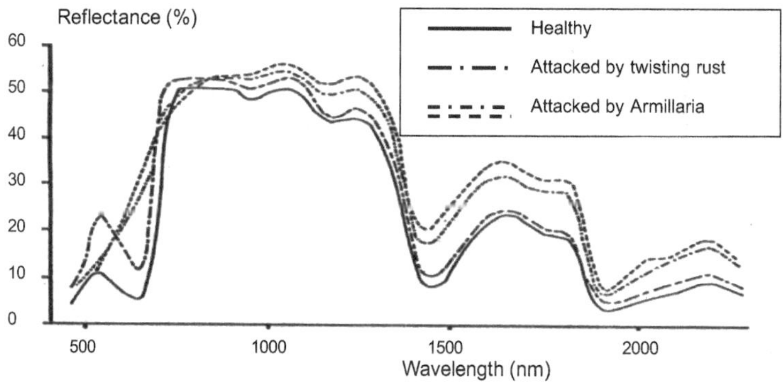

Figure 7.2. *Reflectance spectra of maritime pine needles (Pinus pinaster) taken from 4 trees in contrasted health condition: one healthy, one attacked by twisting rust (Melampsora pinitorqua, a fungus affecting leaf shoots), two in advanced stages of decline (foliage entirely red to yellow) resulting from an Armillaria attack (Armillaria ostoyae, a root fungus that impedes water supply) [GUYO 89]*

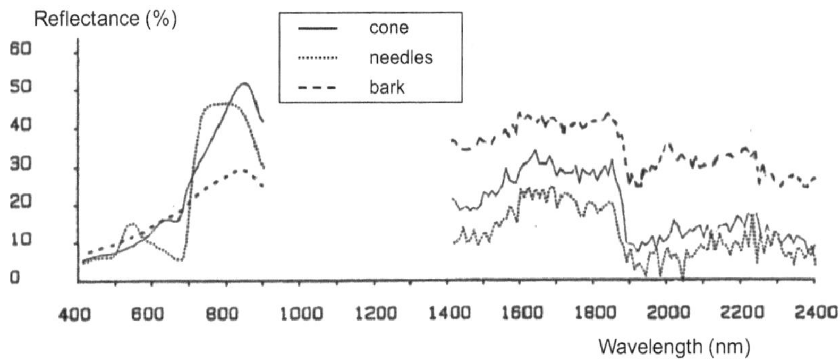

Figure 7.3. *Reflectance spectra of the various organs of a spruce (Picea abies): needles of 1 year, twig bark, cone scales [GUYO 89]*

When considered on the scale of forest cover, chlorophyll concentrations (CHL µg cm^{-2} leaf) and leaf Nitrogen (N, g nitrogen. g^{-1} of dry matter), which drive the uptake of Carbon by photosynthesis, are determined by the soil fertility and the physiological condition of the trees. These change during the course of leaf development (thus the date in the season) and also with the age of the needles in the case of evergreen conifers. The leaf mass per area, the ratio between the leaf mass and the leaf area (LMA g dry matter m^{-2} leaf), should also be taken into consideration. The LMA characterizes the adaptation of species to their microclimate environment and light in particular. It is determined by the development stage of the leaves during the season and by the radiation received, determined by their position within the canopy: the LMA decreases from the leaves in the light to the leaves in the shade; it affects leaf photosynthesis and thus Carbon assimilation in the canopy. Finally, the leaf area index of the cover (LAI m^2 of leaf by m^2 of ground) integrates these properties and controls the fluxes of energy and exchanges of matter: interception of radiation and precipitation, evapotranspiration, and Carbon assimilation. It is determined by the number of leaves or needles, as well as their surface area. The leaf area index varies according to the silvicultural practices (stand density, type and intensity of thinning), species composition, climatic and biotic conditions (for current and previous years), etc. [BRE 03]

One study, carried out on deciduous species of the Fontainebleau forest (beech, oak and hornbeam in particular) by Le Maire *et al.* [LEM 08] focused on the possible use of hyperspectral satellite data (Hyperion) in estimating the functional parameters of forest cover, which drive the exchanges of energy fluxes with the atmosphere. The interesting thing about this study is that it highlights the fact that the wavelengths providing information about these parameters vary according to the scale of measurement of reflectance: the leaf or the canopy. These results are based on the simulation of numerous reflectance spectra for leaves and forest canopies with the help of radiative transfer models (PROSPECT simulating leaf reflectance and SAIL coupled with PROSPECT simulating that of the cover [JAC 09]) with a broad range of variation in the functional parameters of interest, such as CHL, LMA and LAI. For all wavelengths taken in pairs (per bandwidth of 5 nm) a normalized difference index is calculated (the difference between the two reflectances divided by their sum), denoted ND, in order to separately predict each functional parameter with the help of a previously calibrated empirical model. Figure 7.4 shows the spectral regions of the ND index that provide the best RMSE (root mean square error) for the estimate of each parameter. It is found that scaling from the leaf to the canopy changes the layout of the effective regions. The CHL and LMA regions remain fundamentally the same on both scales, which means that there are no other biophysical variables to be taken into account in changing the scale of the indices in these regions. By contrast, regions that are suitable for estimating Nstruct (the leaf mesophyll structure parameter of PROSPECT) undergo substantial changes from one scale to another, due to the LAI, which has a very similar effect to Nstruct and which only occurs on the scale of the canopy. The Nstruct regions on the leaf scale are replaced by those of the Cw (leaf water content) on the scale of the forest canopy. On the scale of the canopy, regions that are suitable for estimating the LAI are under-represented, since the representation in this case is based on a normalized difference index (ND) and not a difference index, which is more appropriate for LAI determinations when one is dealing with very narrow spectral bands such as those used in this study.

The intrinsic characteristics of the species, related to the optical properties of the foliage and tree architecture, can impart spectral variations in forest canopy reflectance, which can be put to good use in the differentiation of species using hyperspectral sensors. Some differentiation is possible with multispectral broadband sensors. Thus, coniferous and deciduous broadleaf canopies generally have different reflectance spectra (Figure 7.5), which is significant in the near infrared where coniferous cover has the weakest values. There are several reasons for this, one being that the reflectance of the foliar organs is generally higher in deciduous broadleaf species than in conifers in the near infrared, the reverse being observed for absorbance. Furthermore, aggregation of the foliage around the branches of conifers also allows for a "better view" of the intra-crown shadows and branches which have very different spectral properties to those of needles (figure 7.3).

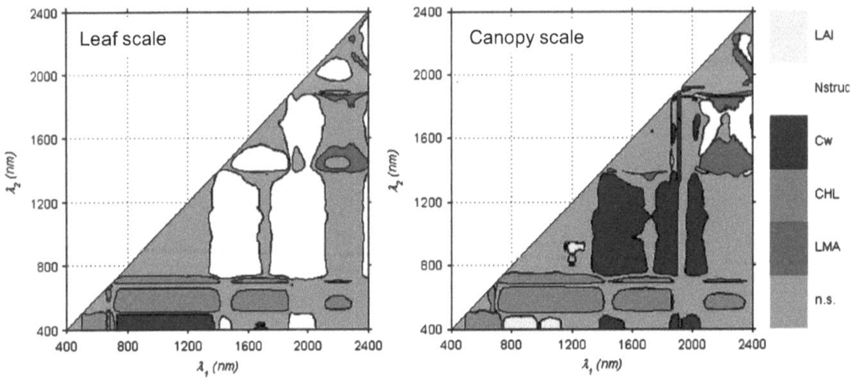

Figure 7.4. *Representation of wavelength pairs of ND indices (ND= (ρλ1−ρλ2)/(ρλ1+ρλ2)) enabling the prediction of five functional parameters with the smallest error. Comparison between (left) the scale of the leaf (reflectance simulated using PROSPECT) and (right) that of the canopy (reflectance simulated using SAIL+ PROSPECT) [LEM 08]. LAI: leaf area index of the cover, NStruct: indicator of leaf cellular structure, Cw: leaf water content, CHL: leaf chlorophyll content, LMA: leaf mass per area. These predictions are made parameter by parameter and the regions of insignificant indices (n.s) correspond to a high RMSE (root mean square error) for all parameters. For a color version of this figure, see www.iste.co.uk.baghdadi3.zip*

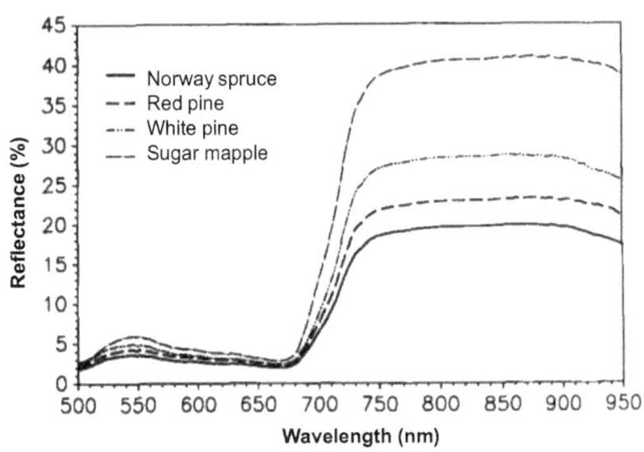

Figure 7.5. *Reflectance spectra in the visible and near infrared of forest canopies of various species [WIL 91]. Coniferous: spruce (Picea abies), two species of pine (red pine, Pinus resinosa; white pine, Pinus strobus); deciduous broadleaf: sugar maple (Acer saccharum)*

7.2.2.2. *Quantitative relationship between reflectance and structure variables of the forest cover*

As for other types of vegetation cover, the relationship between the cover reflectance and the LAI is asymptotic from a certain level. This saturation effect occurs in the visible wavelengths for lower LAI values than in the middle infrared, due to the strong absorption of visible radiation by pigments ([BRO 00], Figures 7.6 and 7.7); this underlines the importance of middle infrared in the study of forests. Vegetation indices such as the NDVI ([SPA 90], Figure 7.8) or the SR[5] (Figure 7.6), and even those based on the middle infrared (such as the RSR[6], Figure 7.6), have their limitations in the estimation of LAI when its values exceed the saturation level.

The marked dispersion around the relationship of the reflectance and vegetation indices [SPA 90, BRO 00] with the tree cover LAI is multifactorial. Among the major factors are: structure characteristics and optical properties specific to the species, fraction of tree cover, spatial variability of vegetation in the understorey, underlying soil surface properties, and conditions in which remote sensed data were acquired, such as solar and viewing angles.

Figures 7.6, 7.7 and 7.8 illustrate the difficulties involved in estimating LAI accurately, whether for the full cover or the tree layer only, from a single multispectral reflectance measurement in a unique geometry of illumination and viewing. One can easily understand the weakness of the relationship with a specific dendrometrical variable, given no additional information, as shown in the example in Figure 7.9.

Vegetation indices based on middle infrared reflectance (MIR) are used extensively on forests, since they are more sensitive to the structure of the cover and its modifications than indices combining reflectances in the red (R) and the near infrared (NIR), such as the simple ratio SR = NIR/R , or the normalized difference NDVI = (NIR-R)/ (NIR+R). There is the RSR mentioned above, and shown in Figure 7.6. The normalized difference (NIR-MIR) / (NIR + MIR), built on the model of NDVI, replacing red by the middle infrared, is widely used. Its name varies according to the application, and sometimes according to the MIR spectral band used and, the source

5 SR = NIR/R.

6 RSR = SR (1 - (MIR –MIRmin)/ (MIRmax –MIRmin)).

sensor: NDWI, NDII, LSWI, NBR for example, where the W in NDWI and LSWI refers to the information provided regarding the water content of the soil and vegetation, and the B in Normalized Burn Ratio (NBR) refers to information on the biomass burned during fires.

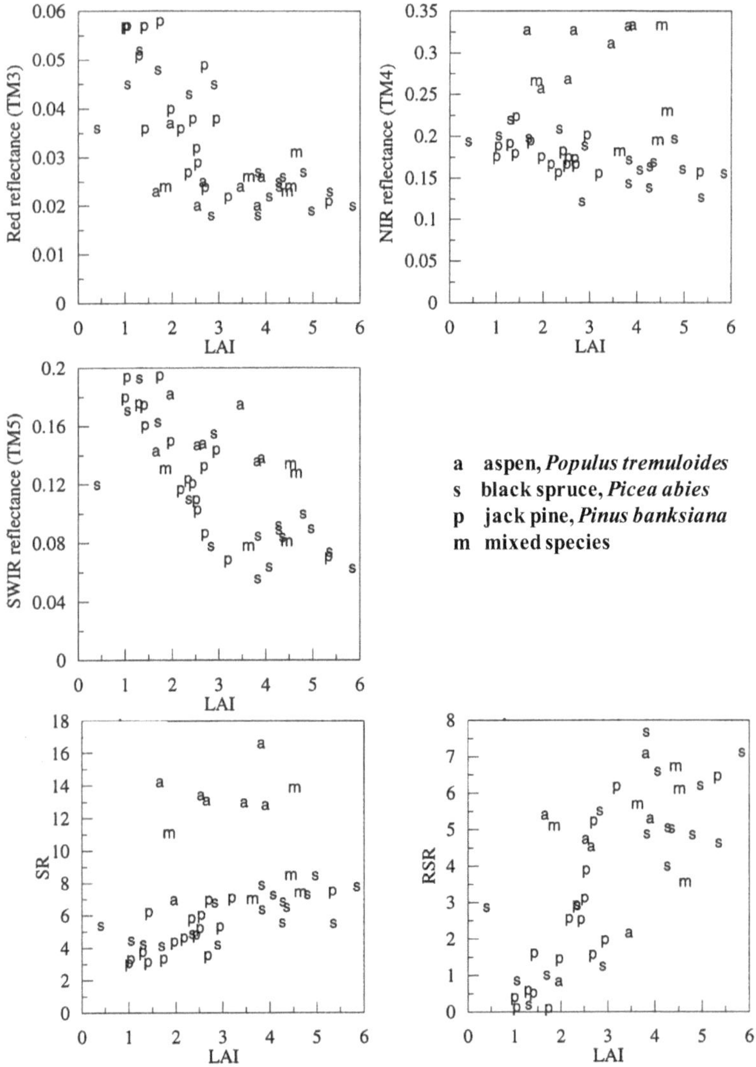

Figure 7.6. *Reflectance in the red, near infrared (NIR) and middle infrared (SWIR), and vegetation indices SR and RSR of forest covers of various species of the Canadian boreal forest as a function of the tree layer LAI measured in situ: observations made using the Landsat TM satellite sensor [BRO 00]*

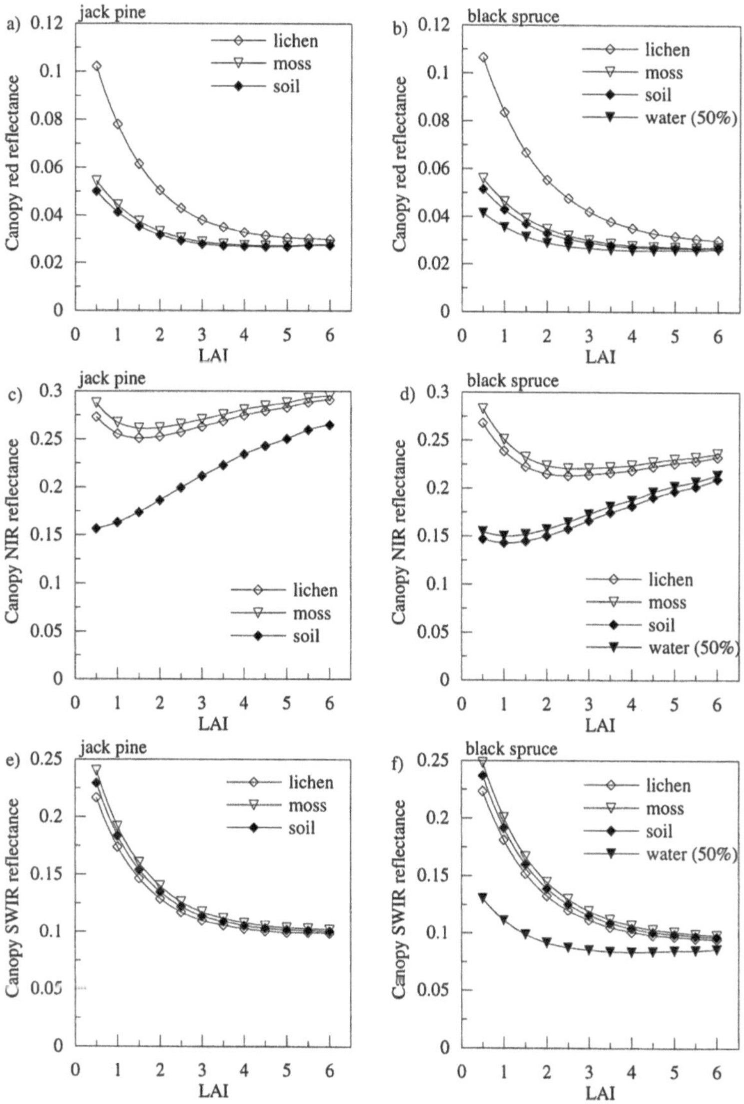

Figure 7.7. *Red, near infrared (NIR) and middle infrared (SWIR) reflectance of spruce (black spruce, Picea abies) and pine canopies (jack pine, Pinus banksiana) of the Canadian boreal forest as a function of the tree layer LAI measured in situ and the composition of the understorey: simulation using the 4-scale model for 4 types of understorey (lichen, mosses, bare soil, 50% water / 50% mosses) [BRO 00]*

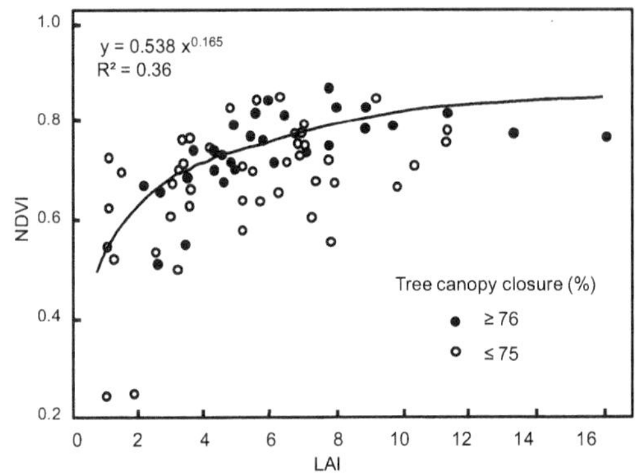

Figure 7.8. *NDVI of coniferous stands of various species (belonging to the genera: pine "pinus", fir "abies", spruce "picea", larch "larix") observed using the Landsat TM sensor, as a function of the tree layer LAI measured in situ [SPA 90]. The relationship saturates for LAI > 5, and depends on the closure of the cover. The tree canopy closure (= crown cover fraction) described in two classes; 3 sites in 3 American states; 4 TM images acquired in summer at various dates (July 7 to August 19 1984); LAI estimated by allometric relationship between leaf biomass and DBH (stem diameter at 1.3m above ground) for each tree*

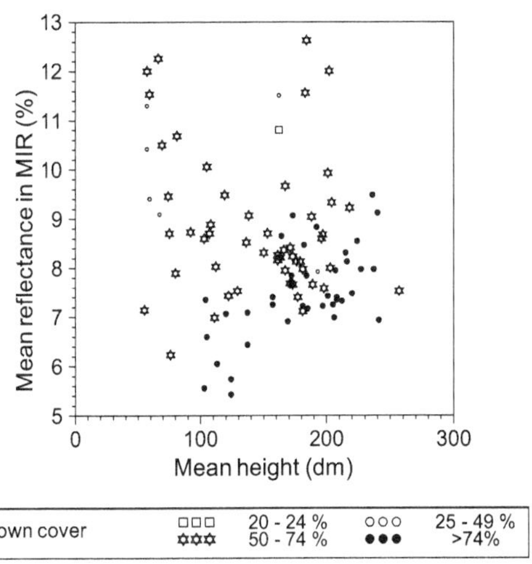

Figure 7.9. *Reflectance in the middle infrared "MIR" (TM5 band) for maritime pine stands measured using the Landsat TM sensor (August 20 1987), according to the average height of the trees and the percentage of crown cover, measured in situ [JOL 93]. 130 inventory plots from the French National Forest Inventory distributed across the Gironde (South West France); percentage of cover by crowns is described in four categories; high diversity of pedoclimatic conditions (dry to wet), composition and development of understorey vegetation*

7.2.2.3. *Taking the understorey into account*

Generally speaking, the contribution of the understorey to the remote sensing signal is especially important when the tree canopy is open, or when its LAI is low. It is dependent on the species composition and the structure of the understorey, which can significantly spatially vary depending on soil and climatic conditions, fertility, microclimate or under the influence of silvicultural operations.

It has been particularly studied in boreal forest where the trees are quite sparse and their cover fraction is low. The understorey can then represent a significant part of the total LAI, and show a marked spatio-temporal variation in reflectance, also influenced by the dynamics of the snow cover [JON 10]. The impact of the nature of the understorey on the sensitivity of the stand reflectance to the tree layer LAI is clearly observable in the example given in Figure 7.7.

The questions of understorey contribution to reflectance and vegetation indices also arises for other forest biomes. For example, the tree LAI of the maritime pine stands of south-western France is relatively low (<4), and can approach that of the understorey (Figure 7.10). Guyon *et al.* [GUY 01] have studied the effects of the understorey on the reflectance of these stands based on simulations of radiative transfers and data acquired by the Landsat TM satellite sensor in various seasons. They have shown that the variability of understorey reflectance represents a large share of that of the total cover. This contribution made by the understorey is significant, even when the pine crowns provide a closed canopy. Moreover, a reduction of the LAI due to clearing of the understorey can lead to a change in reflectance higher than that caused by removal of pine trees by thinning operations (generally cutting dominant trees, with less than 25% of the trees being removed).

The influence of the understorey can cause problems when one is trying to obtain separate biophysical information (structure variables, fAPAR, and phenology) on the two layers. Decoupling solutions are possible: either when one is focusing on only one of the two layers, and biophysical information about the other is provided (LAI, spectral reflectance, etc.), or when the effect of the two layers is decoupled, repeating the observations in different conditions, for example under multiple solar or viewing angles (section 7.2.2.4).

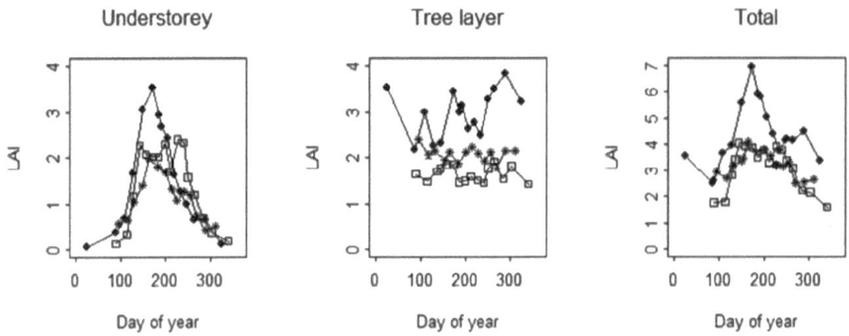

Figure 7.10. *Comparison of seasonal dynamics of LAI between the understorey and tree layer for three adult stands of maritime pine (Pinus pinaster) located in various common soil and climatic conditions in the Landes forest (South-western France) [YAU 08, YAU 12]. The understorey LAI shows a marked annual cycle, which is not the case for the tree layer, whose intra-annual dynamics of LAI is lower. The seasonal variations in the LAI of the forest cover as a whole are dominated by those of the understorey. The LAI estimated from the ground by optical measurements (hemispherical photography) is actually a PAI. ♦ wet conditions, understorey dominated by moor grass (Molinia caerulea) * dry, heather (Calluna vulgaris); □ intermediate, fern (Pteridium aquilinum)*

The effects of the understorey on the seasonal trajectory of the reflectance, which must be taken into consideration in phenological studies, are poorly documented [NIL 08, RAU 11]. The seasonal amplitude of the LAI is less pronounced in evergreen coniferous forests than in deciduous forests, with a non-zero winter value due to the partial renewal of the needles in spring and fall of the oldest generations of needles, which can occur throughout the year. The seasonal variability of the total LAI of the cover can thus be strongly determined by that of the understorey vegetation. This applies to the pine maritime forest of south-western France, where the understorey generally has a very weak LAI in winter and can reach a similar value to that of the trees in summer (Figure 7.10). Broadly speaking, the LAI for deciduous species has a simple cycle (increasing in the spring and decreasing in the autumn), but the possible leaf flushing in the understorey vegetation before that of the tree layer can complicate the phenological interpretation of the satellite observations [AHL 06].

7.2.2.4. *Exploitation of directional variations of reflectance*

The effects on reflectance from the direction of the sun and viewing by the sensor (zenith and azimuth angles), plus a possible anisotropy of the cover (row plantations), have been extensively studied for forest covers by means of modeling of radiative transfers and airborne sensor measurements ([LEB 99]. This cannot be ignored, the length of the shadows seen by the sensor and the amplitude of the hot-spot[7] being dependent on the height of the trees. An illustration on two pine maritime stands, highlighting the difficulty of interpreting spatial and temporal variations of measurements under different solar or viewing angles, is provided in Figure 7.11. Moreover, it is worth mentioning the influence of topography on the directional variations of reflectance, since steep slopes are often encountered in forested regions.

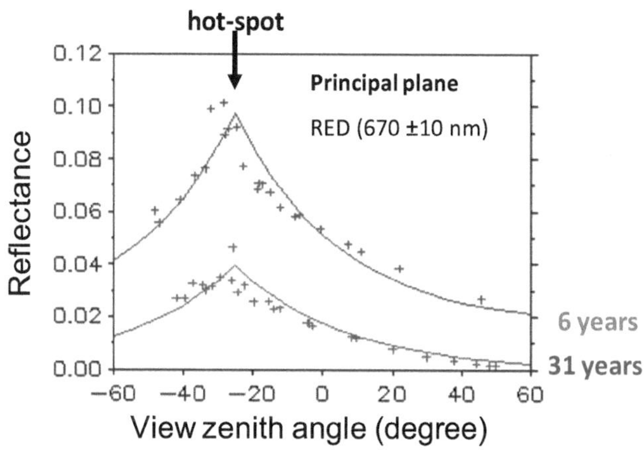

Figure 7.11. *Directional variations in reflectance in the red wavelengths of two maritime pine stands, aged 6 and 31 years, in the principal plane (sensor viewing and sun in the same plane) for a solar zenith angle θs ~25°. The dominant height of the pines is 4.2 m and 21 m, their optically measured LAI 1.3 and 1.96, for the two stands of 6 and 31 years respectively. Experimental data obtained from the airborne sensor POLDER, acquired in summer (July 21 2001) on the Landes forest (South-western France); – directional signature by statistical fitting of the model of Bréon et al. [BRE 02]*

These directional effects can be corrected, or, on the other hand, put to good use, when the satellite sensor is able to measure the reflectance in

7 The hot-spot corresponds to the maximum reflection observed on an uneven surface (soil, vegetation...) in the direction of backscattering (where the illumination and viewing directions are collinear). This is related to the fact that the fraction of shadows seen is zero. As soon as one deviates from this direction, the fraction of shadows seen increases and the radiation reflected in this direction is correspondingly decreased.

various solar and viewing angular configurations during a relatively short time lapse, where the structure of the cover can be assumed to be invariant. The distribution function of bidirectional reflectance (bidirectional reflectance distribution function (BRDF)) can actually be estimated using semi-empirical models. It is on this principle that the directional effects on satellite time series acquired on a daily basis with sensors such as MODIS, VEGETATION or AVHRR are corrected [BAR07], and the BRDF products obtained from MODIS measurements are developed. Some satellite instruments, such as MISR (aboard the Terra system, also carrying the MODIS sensor, resolution: 275 m, 1.1 km)), or POLDER, are designed to acquire measurements in several viewing angles simultaneously (9 in the case of MISR).

Multiangular data makes a decoupling between the understorey and tree layers possible. The reflectance of the entire soil plus vegetation (that is, the understorey) below the tree layers can thus be reconstructed (for example [PIS 12] with the BRDF produced from the MODIS sensors). Harnessing the anisotropy of the reflectance also provides better information about the LAI, height and percentage of cover of the tree layers, and thus the volume of wood and biomass, by means of allometric models, than the use of mono-angular observations (for example [CHO 08] with MISR measurements).

Knowledge of the BRDF can also be put to use in mapping the foliage aggregation parameter (section 7.2.1.5). We refer to the work of de He *et al.* [HE 12] on this subject, based on MODIS data.

7.2.2.5. *Importance of stereoscopy and digital photogrammetry*

Stereoscopic observation (i.e., two images acquired simultaneously in different geometrical configurations of viewing) enables, using digital photogrammetry techniques, a three dimensional reconstruction of the forest cover. The digital surface model (DSM), that is, of the roof of the tree cover, is generated by *image matching* techniques. It gives a digital canopy height model (CHM) by subtraction of a digital elevation model (DEM) provided elsewhere. Stereoscopy is possible with some satellite sensors VHR (PLEIADES, IKONOS, for example). Because of the relatively large size of trees to that of the pixels (50 cm panchromatic with PLEIADES), a quantification of the height of the tree cover can be achieved, and thus of the volume and biomass if the forest structure is not complex (in the case of purely coniferous forests, for example, see [STO 08, BEL 15]). The main difficulties are the parameter settings of the image matching and the identification of areal metrics of the CHM in modeling the key height variables (dominant height, for example), volume or biomass per stand surface element.

7.2.2.6. *Analysis of textural information*

The image texture, which may be defined as the interpixel spatial variation of the signal in the near vicinity, is also rich in information about the structure of the forest cover. This wealth of information is notable with the VHR. In fact, when the spatial resolution (for example 50 cm in the panchromatic band and 2 m in the multispectral bands with PLEIADES) is close to or less than the size of the tree crowns, the texture is particularly related to the structure of the tree cover as defined by forest managers, and in particular to the diameter of the crowns, the height of trees, their density and their spatial distribution (Figure 7.1). This relationship is complicated to describe mathematically. It is dependent upon the conditions in which the images are acquired [BRU 98]: solar and viewing angles (which determine, in particular, the length of shadows seen within and between the trees, the proportion of understorey or tree crowns seen in shade or in sunlight), atmospheric illumination conditions (proportion of scattered radiation in the incident radiation, which affects the contrast between surfaces in shade and those in sunlight). To the seasonal variations in texture relating to the sun's position should be added those relating to phenology.

The idea of using the texture of an image to describe the spatial structure of the forest cover (heterogeneities of the canopy, row plantations) or to estimate dendrometrical variables or biomass, is not new. There has been a renewed interest in it during the past few years, with the launch of numerous very high spatial resolution satellite sensors (such as PLEIADES, QUICKBIRD or WORLDVIEW). The increase in information processing capabilities of computers has facilitated the development of applications of mathematical methods in describing texture, both old (co-occurrence matrices of Haralick *et al.* [HAR 73], Fourier analysis, variograms for example), and new (e.g. wavelet decomposition).

7.2.2.7. *Wealth of temporal variation*

The value of reflectance in different wavelengths (or vegetation indices), measured in a single geometric configuration of viewing on a given date, provides little quantitative information on the forest cover structure. By contrast, a comparison between two dates, or temporal monitoring on the growing season or over the years, provides a wealth of information on its dynamics and its changes: seasonal variations related to the phenological cycle, forest cutting, reforestation, changes in land use, climatic and biotic perturbations (fires, storms, extreme drought events, pest outbreaks, etc.), in particular. A simple illustration over a forest landscape dominated by maritime pine, with the presence of deciduous trees along river banks and large agricultural fields for cultivation of corn, is provided in Figure 7.12. The temporal information there (inter – season, inter – years) seems more extensive than the spatial information.

Figure 7.12. *Wealth of temporal information. Illustration with multi annual series of Landsat TM images (7 years, various seasons) on an area of 5000 ha in south-western France. Color composition of the surface reflectances (corrected for atmospheric effects): red = TM5 "middle infrared", green = TM4 "near infrared", blue = TM3 "red". Silvicultural changes in maritime pine forest: clear cutting (especially between 1984 and 1985, due to frost damage), reforestation (particularly from 1986 to 1988); seasonal changes: phenology (deciduous: spring, summer, autumn), crop calendar (corn: plowing and sowing in April, maximum LAI in July, maturing in September, post-harvest in October); changes in land use: deforestation for conversion of pine maritime surfaces to corn crops from 1987 to 1989*

7.2.2.7.1. Sudden and major changes

The possibilities of detecting sudden and major changes (clear cutting, fires, heavy storms, for example) by comparison between two images acquired successively (before and after the event) have been studied and implemented for a long time, in particular using high spatial resolution data. But they are dependent on the variable availability of images that are totally free of clouds and snow, and the time interval between the two images involved. One cannot, therefore, put a precise date on the change within this period, which may be relatively long (from several weeks to a year or more, according to the climate, the meteorological conditions and the orbital cycle of the sensor field of view). Neither can one pin down the event, when, during this period, the change in the cover structure is mitigated by the dynamics of recovery of the vegetation (regrowth of understorey, regeneration of tree cover, spreading of the remaining tree cover, for example). Finally, differences in the illumination and viewing configurations, or in the phenological stages between the two dates, also affect the precision of detection [TOW 12].

Low temporal sampling is also an obstacle to the detection of more subtle changes, that is to say less intense or more spatially diffuse, or more gradual over time, such as seasonal variation in green leaf area due to the phenology, or some disruption in the health condition of the trees (drought, pests and diseases). But, having said that, a more regular temporal sampling will not guarantee successful identification of the phenomena of interest if the maximum variation of reflectance is low.

7.2.2.7.2. Surface phenology

Phenology is an indicator of the impact of climate change, and plays a key part in the seasonality of water and carbon fluxes. Methods for dating the key-stages of leaf phenology, or the extraction of phenological metrics (leaf expansion in spring, senescence of leaves - that is, change in color and fall - in autumn, duration of growing cycle, integral of the satellite signal during the cycle, for example) have been developed since the 1990s and applied to forest covers. They are based on modeling of the seasonal trajectory (i.e. intra-annual) of the satellite signal (multispectral reflectance or vegetation indices, biophysical variables derived thereby, such as the LAI) observed at high temporal frequency. This is the trajectories observed year after year with the medium spatial resolution time series produced by daily revisit sensors (of the MODIS, AVHRR and VEGETATION types, see

[ZHA 03]) for example), or that resulting from the compilation over several consecutive years of series acquired with high spatial resolution and lower frequency (such as Landsat, see [FIS 06]). A filtering of cloud data and a normalization or correction of directional and atmospheric effects are prerequisites.

Figure 7.13 shows a typical seasonal trajectory for a whole year of the vegetation index NDVI on deciduous forest, and illustrates the principles of its modeling for the extraction of phenological information. The modeling may be parametric with a statistical fitting of a preset curve (as in Figure 7.13), the hypothesis being that the shape of the seasonal dynamics of the foliage are known and fixed beforehand. Hence, the effect of possible interferences is ignored. The curve usually combines two sigmoid functions, an upward one in the spring and a downward one in the autumn, which mimic the leaf expansion and leaf fall. An asymmetrical double logistic function is often used ([FIS 06], see example in Figure 7.13). Non parametric approaches do not have these constraints: various methods of temporal smoothing are used, such as the Fourier analysis [BEC 06], analysis of the covariance temporal structure [GUY 11], and the use of filters such as that of Savitzky-Golay [JON 04b] (Figure 7.14). The dating of the phenological transitions involved, (which are not necessarily limited to those corresponding to the phenophases conventionally observed *in situ*), is inferred, either by thresholding on the modeled trajectory with a preset value, or by identification of characteristic points on the curve (for example, the inflection point, illustrated in Figure 7.13, or the point where x% of the inter season amplitude is reached, illustrated in Figure 7.15). Often it is the NDVI that is treated, but it can be less informative on the vegetation dynamics than linear indices (that is, resulting from a linear combination of reflectance in the red and the near-infrared, such as the PVI, for the *Perpendicular Vegetation Index*) which have a higher signal to noise ratio, or fewer saturation problems. Other ratio type indices can be used, such as the EVI (*Enhanced Vegetation Index*) or the NDWI (*Normalized Difference Water Index*, see section 7.2.2.2.). The latter uses middle infrared reflectance which is accessible using VEGETATION or MODIS. The middle infrared in particular is valuable when separating the effect of snowmelts from that of the beginnings of vegetation season at the end of the winter [DEL 05].

The main difficulty in modeling the seasonal trajectory of the remote sensing signal is the lack of temporal continuity of the data used, because of the data missing due to cloudiness in particular.

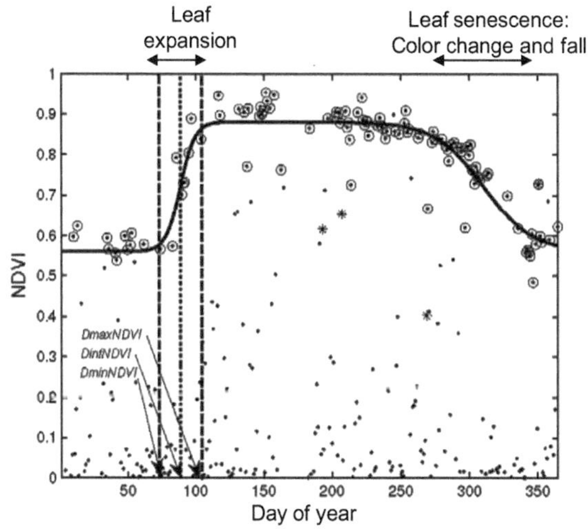

Figure 7.13. *Seasonal trajectory of the NDVI after daily measurements by the MODIS sensor (MOD09GQK product) during a given year, on a deciduous stand (oak, Quercus sp) and modeled using a double asymmetric sigmoid function [SOU 08]. • Measurements eliminated according MODIS indication of quality criteria, provided pixel by pixel (presence of clouds in particular); *.measurements eliminated by filtering out on a moving window. Circled points: measurements retained and used to adjust the model (continuous solid line). Definition of the three metrics of the expansion phase of the leaves indicated: DinfNDVI is the date of the inflection point, DminNDVI and DmaxNDVI correspond to characteristic curvature points [ZHA 03] at the beginning and end of the NDVI increase*

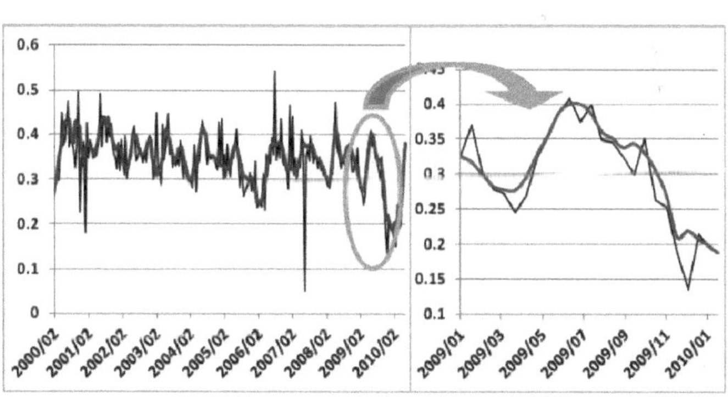

Figure 7.14. *Non-parametric modeling of the multi-annual seasonal trajectory of a vegetation index: an example of use of the TIMESAT algorithm that makes use of the Savitzky-Golay filter [JON 04b]. Illustration for the vegetation index EVI, provided by the MOD13Q1 product (16 day frequency) on one pixel, in a maritime pine forest. In black: before filtering, in red: after filtering. For a color version of this figure, see www.iste.co.uk.baghdadi3.zip*

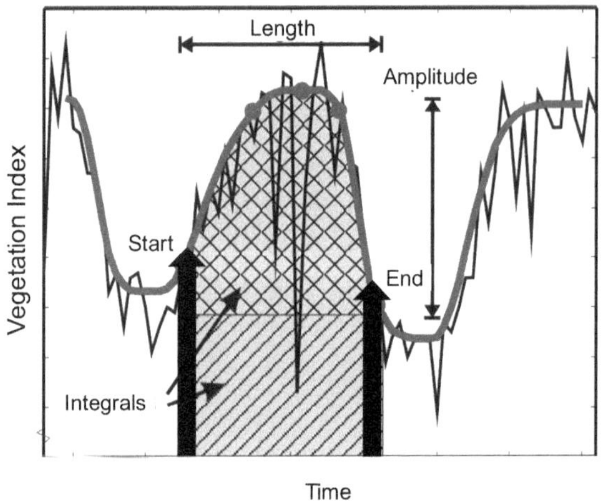

Figure 7.15. *Illustration of the extraction of seasonal metrics of the temporal trajectory with the TIMESAT algorithm: start, end, length of vegetation season, integral of the vegetation index over the vegetation season (shading to the right), that scaled to the amplitude (shading to the left) [JON 04b]. The points "start of vegetation season" and "end of vegetation season" indicated here = the dates on which 20% of the yearly amplitude is observed*

In reality, the information obtained by satellite is not strictly phenological. It is a "surface phenology" (*land surface phenology).* It combines the seasonal change of the leaf area due to climatic forcing, and its interaction with potential hazards (drought, pest insects, for example), with more drastic changes due to accidents (storm, fire, and outbreak of pest insects, etc.), or human intervention (forest cutting, clearing, land use changes, etc.), or with other seasonal changes in surface conditions (snow for example) which do not necessarily correlate with the phenology itself. It is indicative of the seasonal dynamics of all the covers that make up the pixel [BAD 04], a combination of tree layers and understorey in the case of forest cover [NIL 12]. But the mixture, in the same pixel, of forest species with differing phenological behaviors makes it difficult to understand the relationship between the phenological information as observed by satellite and the spatio-temporal variability of air temperature if the species composition is unknown [FIS 07]. Some disaggregation techniques enable the information to be refined, but the lack of knowledge of the species composition, particularly in mixed stands, and the need for a limited number of parameters in the disaggregation model, are generally limiting. For example, Guyon *et al.* [GUY 11] reconstruct the seasonal trajectory of the vegetation index Perpendicular Vegetation Index (PVI) of 3 forest types only: deciduous broadleaf, coniferous and mixed, for each 1 km^2 pixel of the time series studied using VEGETATION data.

The phenological information contained within the satellite data acquired during various seasons can also be put to good use in the mapping of forest formations made up of species that differ in their phenological cycle, in the tree strata [ZHU 14] and even in the understorey. The difficulties reside, in particular, in variations in reflectance due to topographical variations in illumination and the existence of phenological environmental gradients (for example, those due to elevation) in the area being examined.

7.2.2.7.3. Spatio-temporal dynamics of ecosystems

The accumulation of satellite data over several decades has enabled studies to be carried out on the spatio-temporal dynamics of forest ecosystems using a retrospective analysis of the multi-annual trajectory of reflectance or vegetation indices. It provides some characterization and monitoring elements of the dynamic processes in response to environmental constraints (climate, soil), and anthropological activities (forest management, urbanization, clearing, etc.) in conjunction with hazards (storms, fires, pests and diseases): changes or disruptions in leaf development, tree development or spatial structure of the cover, the ensuing responses in terms of recovery (regrowth, natural or artificial regeneration) and forest succession (succession of the stages of forest development or age increment, for example), and long-term trends.

Various methods have been developed and applied to forest covers using multi-annual data series acquired with very high temporal frequency and medium spatial resolution sensors (such as MODIS, AVHRR or VEGETATION). Thus, a decomposition of the time series into different time scales (seasonal, inter-annual, long term) can be applied to separate cyclical seasonal variations of a phenological nature, change or sudden accidental disruptions and long-term trends (for example, see the BFAST algorithm of Verbesselt *et al.* [VER 10]). Indicators of seasonal anomalies of leaf development (e.g. as a result of defoliation by insects or early senescence due to extreme drought event) can also be established for this type of image series, based on statistical criteria built up during previous reference years that have had no disturbance to the foliar phenological cycle. But these reference years can be difficult to identify, particularly in cases where there are successions of disruptions over several consecutive years, and in complex dieback cases, whose consequences may last for quite lengthy periods.

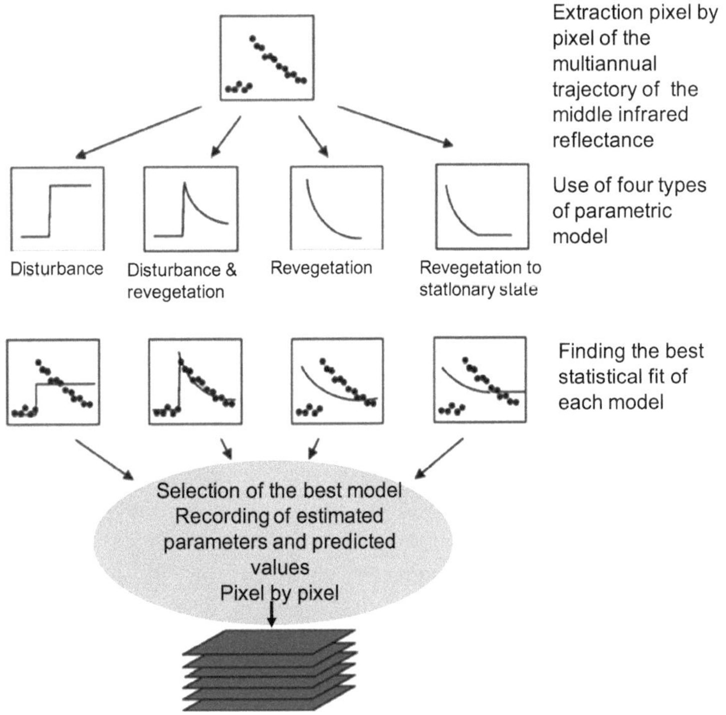

Figure 7.16. *Reconstruction of the long-term dynamics of forest cover by comparison of the multi-annual trajectory of reflectance observed from Landsat TM time series (from 1984 to the present day) with preset curves corresponding to typical successions of disruptions and recovery phases. Conceptual diagram of the modeling method proposed by Kennedy et al. [KEN 07]*

Long-term analysis of multi-year trajectories of the satellite signal may also be carried out, in order to reconstruct, in time and intensity, the succession of disruptions or changes (abrupt or gradual, natural in origin or otherwise) and recovery phases (natural or via silvicultural practices). Two kinds of methods may be distinguished. The one illustrated in Figure 7.16 is based on modeling or identification of temporal patterns, which can be non-monotonic or non-linear, corresponding to typical types of disruption and recovery phases [KEN 07, HUA 10]). The other is a breakdown into a succession of short-term trends (several years) by line segments: up, down, and stationary trends [KEN 10]. The multi-annual metrics (date change, phase duration, intensity of change, etc.) derived from them are then used to characterize the forest dynamics [FRA 14]. These metrics, which are indicative of the past history (silvicultural and natural events, growth and development) of the forest cover, can provide quantitative information on the actual values of the structure variables or the biomass [PFL 12]. In particular, the time since a marked change indicative of stand renewal (such as clear-cutting) can be a proxy to the stand age in homogeneous or managed forests, assuming that they are sustainable and resilient in the face of major environmental or global disturbances. These approaches are undergoing significant development now that long data series of high spatial resolution are becoming available for all land territories, including those from Landsat satellites (several decades), although these satellite observations are too infrequent in time to capture some ecological and functional processes of the ecosystems. Kennedy *et al.* [KEN 14], provide a clear conceptual vision. There is already much hindsight gained in the monitoring of forests from Landsat time series [BAN 14].

7.3. Examples of application

7.3.1. *Estimation and management of forest resource: wood volume, above ground biomass, carbon stocks*

We focus here on the need for quantitative information concerning forest inventory and which is useful in forest management. This involves knowledge, at any given time, of:

– the characteristics of the forest structure, such as the stem diameter, the height, density, and age;

– the location, condition and volume of forest resources;

– its distribution by forest type (species and other), stem diameter or age category, for example.

It also involves periodically updating this knowledge and learning about the gains and losses due to timber harvesting (cutting) and accidental damage. This also includes the need for data to make predictions of biological production and potential harvest in the future – for sustainable resource management - by testing various scenarios of silvicultural systems by means of growth models or yield tables (prospective studies).

The applications concerned are, notably, the drawing up of forest management plans, requiring the gathering of information about each stand of the forest in question, and the planning of the management at regional or national level or beyond, with respect to political considerations in terms of economy, industry (forestry operators and wood industry) and the environment, not requiring an exhaustive and individual observation for all stands.

Optical multispectral remote sensing alone is not able to produce these quantitative data on the variables of structure or biomass. In terms of operational applications, it should be considered as a way of improving traditional methods of *in situ* forest inventory, by reducing the gaps in the temporal or spatial sampling: refining the estimates of variable indicators of forest structure (dominant height, age, category of forest cover, for example) and forest resource (wood volume, above-ground biomass, carbon stocks) and mapping (better geolocalization of forest information, and improving the precision of surface estimates).

But there is also an increasing demand for information relating to the part played by forests in the stocks and fluxes of carbon at a wider scale (continental, global) in terms of:

– forested surface;

– above-ground biomass, carbon stocks;

– variations in stocks over the course of decades under the influence of global changes, such as climate, anthropological activity (deforestation, reforestation, pollution and atmospheric deposition), major weather-related events (fires, storms).

Optical multispectral remote sensing can also supplement knowledge obtained elsewhere on the location of carbon stocks and the changes in regard to this, and improve the accuracy of estimates (see, for example, [HAN 13, BAC 12]).

The approaches used are generally statistical, with no functional basis (no explicit consideration of ecophysiological processes). They are based on observation of states at specific times or the comparison of successive states.

In this section we will present various examples of meaningful contributions of multispectral satellite remote sensing:

– specific contributions of very high spatial resolution (VHR), using stereoscopy or image texture;

– detection of marked anthropogenic changes (harvesting by clear cutting, deforestation), or accidental (fires, storms), or accidental (fires, storms), notably with high resolution (HR);

– multi-source strategies, that is to say, combining multispectral optical data (VHR, HR or medium resolution, MR), not only with *in situ* observations and measurements and pre-existing spatial knowledge (for example, climatic data, topography, land use, etc.), but also with other types of spatial and airborne remote sensing data (LiDAR, hyperpectral and radar).

7.3.1.1. *Contribution of very high spatial resolution stereoscopy using digital photogrammetry*

Few studies have been carried out using satellite images acquired in stereoscopy with sub-meter or approximately sub-meter spatial resolution, such as those of IKONOS (1 m) and PLEIADES (50 cm, 70 cm before resampling) in panchromatic.

St Onge *et al.* [STO 08] and Bélouard *et al.* [BEL 15] studied the possibilities of spatial estimation of the dominant height, above-ground biomass and above-ground standing volume (stems plus branches) of the tree cover, using digital photogrammetry on this type of image: Balsam fir (*Abies balsamea)* mixed with various deciduous and coniferous species (spruce, pine, birch and aspen) observed by IKONOS in boreal forests of Canada in the case of the first, and pure, even-aged maritime pine forest of south-western France observed by PLEIADES in the case of the latter. *In situ* measurements of tree height and stem diameter at 1.3 m above ground (or DBH *Diameter at Breast Height*) provide reference data on the variables of interest: dominant height, above-ground biomass and timber volume, by means of preexisting allometric equations. Additional reference data has been provided on the individual height of the trees of the maritime pine site, by means of photogrammetry on digital aerial photography. For the boreal site, LiDAR aerial measurements, coupled with the acquisition of IKONOS images, have been used to produce biomass estimates for comparison.

The digital model of the surface (DSM), is generated by correlating the images (*image matching*). Bélouard *et al.* [BEL 15] used the open-source MicMac software of the IGN (Institut national de l'information géographique et forestière, which is the French geographical institution). The DSM gives a digital canopy height model (CHM) by subtraction of a digital elevation model (DEM). The procedure includes an optimization phase of the image correlation settings to maximize the altitudinal accuracy of the DSM. A bias of the order of 2 m is observed on the DSM in both studies. The DEMs used in both studies have a xy resolution consistent with that of the DSM. Indeed, on the boreal site, the DEM is from contemporary LiDAR aerial data. On the maritime pine site, where the ground is fairly flat, the DEM used, which is derived from that of the BD Alti® of the IGN, was found to be close to the DEM derived from LiDAR aerial data.

The next step is to model the dominant height of the tree cover and above-ground biomass by unit area of forest stand, based on a variable selection and regressions. The size of the unit areas is designed to be similar to those of the reference plots used to calibrate the models. Determining the dominant height involves finding, by linear regression, the best predictor among the distribution metrics of the height H of the CHM in the unit area. This corresponds to percentiles above 75% for both sites. This is not unexpected, since the dominant height characterizes the tree population with access to light, and it is estimated *in situ* by the height of the largest trees in the stand. For the biomass or volume (y = predicted value), an exponential model is adjusted by nonlinear regression, where the average value of H in the unit area is used as predictor:

$$y = a\ (average(H))^b \qquad [7.3]$$

Bélouard *et al.* [BEL 15] refine this by adding the standard deviation of H, according to the generic model that Bouvier *et al.* [BOU 15] apply to LiDAR measurements to take account of the local heterogeneity of the height and the possible presence of gaps:

$$y = a(average(H))^b\ (standard\ deviation(H))^{-c} \qquad [7.4]$$

The models are then applied to all the unit areas covered by the CHM.

The two experiments differed in the conditions in which the images were acquired (summer, and therefore the deciduous species in full leafing and the sun high in the sky in the case of the IKONOS; and winter, therefore low sun, in the case of PLEIADES), and the stereoscopic configurations (B/H, that is, the relationship between the base and the height of the shot: 0.8 with IKONOS and 0.4 with PLEIADES). Despite this, their results show close agreement on the accuracy of the estimation of the dominant height; the mean square error (RMSE) on the calibration data sample is 2.08 m for the mixed boreal stands observed with IKONOS and 1.75 m for the maritime pine stands observed with PLEIADES. In both cases, the relative error (RMSE, divided by the average observed) is approximately 10%. For the maritime pine, Bélouard *et al.* [BEL 15] validated the quality of prediction using an independent data set: no bias and a RMSE less than 1.75 m. Figure 7.17 provides an example map of the dominant height that can be produced, the land use categories other than the maritime pine being masked using the IGN forest map (BD Forest V2).

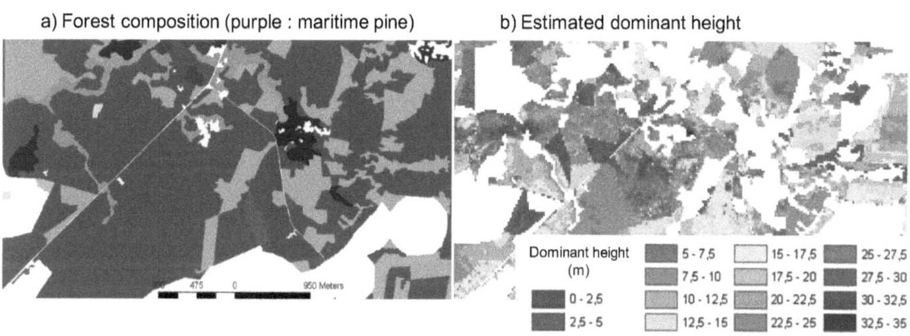

Figure 7.17. *Spatialization of the dominant height of the maritime pine stands of the Landes forest (south-western France) from PLEIADES panchromatic images using digital photogrammetry: example of a sector of the St-Symphorien commune [BEL 15]. a) Composition of the forest provided by the IGN forest map (BD forest V2); b) Estimated dominant height. The dominant height is produced using a 25 m square grid, an area comparable with that of the in situ dendrometric plots used in the calibration. For a color version of this figure, see www.iste.co.uk.baghdadi3.zip*

As for the biomass estimations, St Onge *et al.* [STO 08] have shown that their accuracy, although not as good as that produced from the CHM derived of the LiDAR measurements (3 pulses per m^2), is quite acceptable for a biomass <300 t ha^{-1}: 50 t ha^{-1} (this being ~30 %, relatively speaking) as against 41 with the LiDAR. Above 300 t ha^{-1}, the biomass is underestimated using both methods, but this result is to be confirmed, since it is obtained on a too small sample of in situ plots. For the maritime pine, whose maximum biomass is lower (<194 t/ha, with a standing volume <538 m^3/ha), Bélouard *et al.* [BEL 15] obtain a RMSE of ~20 t/ha (RMSE of standing volume = 53 m^3/ha), that is, a relative error of 25 %. On the same site, with the same ground-based reference database and the same allometric equations, Bouvier *et al.* [BOU 15] obtain, by aerial LiDAR (8 pulses per m^2) a relative error of 12-13 % for the standing biomass and volume.

In conclusion, VHR satellite stereoscopy performances are dependent on the availability of high resolution DEM, such as those obtained from aerial LiDAR measurements. The levels of accuracy achieved for the dominant height and above-ground standing biomass or standing volume are worth noting, even if they are lower than those obtained by aerial LiDAR for biomass or volume. However, for these two variables, the accuracy could be inadequate in the case of complex forests such as those with several tree strata, since the correlation of images does not allow for exploration of the lower levels, in contrast to the LiDAR. One can expect to be able to estimate heights of cover with an accuracy close to that obtained with the CHMs derived from aerial photographs, but the accuracy to be expected for the standing biomass or volume is probably much lower [JAR 12, BOH 12]. However, in situations where there is a pronounced relief, the possibility of occultation of areas on the pair of images - satellite or aerial - reduces the stereoscopy performance. Finally, whatever the source of the CHM (VHR Satellite, aerial photography or LiDAR), reference data is required to calibrate the prediction models of height (particularly to eliminate the possible bias on the DSM) and biomass or volume from the CHM.

7.3.1.2. *Contribution of the texture of very high spatial resolution images*

The potential of textural information from satellite images at sub-metric or metric resolutions has been clearly demonstrated in the estimation of variables of cover structure such as stem diameter, height of trees, their spacing (therefore density) and even the age and biomass. If the age, stem diameter and biomass can be estimated with acceptable accuracy, it is largely due to their allometric relationships with the crown diameter and the height of the trees, which contribute directly to the image texture.

7.3.1.2.1. Estimation of above-ground biomass in tropical forests

We are taking, as examples, three studies [PRO 07, PLO 12, BAS 14], which use the same method on different types tropical forest.

Estimation of biomass relies upon the FOTO method (*Fourier-based Textural ordination*, [PRO 07]) which is described in detail in the chapter by Proisy *et al.* (Chapter 7, Volume 5) on mangroves, and can be summarized as follows. A non-directional frequency spectrum is calculated on each window of the image by 2D Fourier transform, this being repeated for different window sizes. The variability of the spectra is then studied using PCA (principal component analysis), for each window size. The first three principal components are used as texture indices. In these studies, axis 1 orders the textures according to the grain of the canopy, that is, the diameters of the largest to the smallest crowns. Axis 2, on the other hand, relates to the degree of openness of the tree cover ([PLO 12], Figure 1 of this study). The biomass is then modeled as a linear combination of the values of the 3 components. The model is estimated by linear regression for each window size tested, and the best performing of these is retained (criteria: R^2, RMSE, and relative error, i.e. the RMSE divided by the average observed).

In order to calibrate these models and evaluate their performance, the biomass is estimated on reference plots using an allometric model, established elsewhere, having as input the DBH (stem diameter at 1.3 m above ground) measured *in situ*. There has been no validation of the quality of predictions on the independent reference data set. Biomass maps have been produced after verification of their consistency.

The three experiments are the following:

– Proisy *et al.* [PRO 07]: two mangrove sites in French Guyana, each covered by an IKONOS image in 5 bands: panchromatic (1 m) and blue, green, red and near-infrared (4 m). The 26 *in situ* measurement plots cover a range of biomass from 50 to 450 t/ha;

– Ploton *et al.* [PLO 12]: a tropical evergreen rainforest in India, a panchromatic IKONOS image (1 m). The 15 *in situ* measurement plots cover a range of biomass between 100 and 700 t/ha;

– Bastin *et al.* [BAS 14]: a site with diverse forest structures (from open forest to covers where the dominant storey is mono-specific and mature) in the Republic of Congo. It is covered by two partially overlapping panchromatic images of <1 m resolution, acquired in the dry season with differing viewing and solar angles, with two different sensors (differing essentially in spectral sensitivity and resolution): one being GEOEYE and the other QUICKBIRD. The spectra from GEOEYE are corrected by means of an empirical model that connects the spectra of the two images linearly, and calibrated on the common area. The 26 *in situ* measurement plots cover a range of biomass from 25 to 450 t/ha.

The different experiments show that a panchromatic sensor whose resolution is less than or equal to 1 m is able to capture the variability of the canopy grain on the various forest types studied, whatever the stage of development of the trees. This is not the case for greater resolutions (4 m in multispectral with IKONOS), which are especially sensitive to the greater dimension of crowns or gaps. There is no saturation phenomenon of the predicted biomass with the panchromatic over the entire range tested (50-700 t/ha). In the case of closed covers the estimate is without bias and the expected error (RMSE and relative RMSE in %) is quite low, and similar between the 2 studies concerned: 33 t/ha (17 %) for a biomass <500 t/ha [PRO 07], 58 t/ha (13.5 %) for a biomass <700 t/ha [PLO 12]. But in more heterogeneous and more open canopies [BAS 14], the errors are more significant: RMSE = 95 t/ha (35 %) for a biomass <500 t/ha. They increase with observed biomass and are very significant above 200 t/ha. A stratification of the model, based on 3 categories of textural heterogeneity identified using the k-means method, has enabled the diversity of the relationship between texture and forest structure to be taken into account in the area studied. In this way, the errors could effectively be halved: RMSE = 52.5 (19.5 %). These figures have been confirmed by cross-validation.

7.3.1.2.2. Estimation of forest structure variables for single-species and even-aged coniferous stands in temperate forest

We are using the Kayitakire *et al* study on spruce [KAY 06], and the maritime pine study of Beguet *et al.* [BEG 12, BEG 14a, BEG 14b] and Beguet [BEG 14c] as examples. In both cases the stands, single-species and even-aged (high forest issued from plantations), are intensively managed for wood production: final harvesting by clear cutting, reforestation through planting or seeding, periodic thinning.

Both studies are interesting from a thematic generalization point of view, since they use the same method to characterize texture. This is based on co-occurrence matrices or Grey Level Co-occurrence Matrix (GLCM) [HAR 73]. For any given size of image window, a GLCM is a square matrix of dimension n, each element being the frequency of observation of a pair of grey level values (i,j) (grey level = discrete satellite measure on n levels, i and j varying from 1 to n) between two pixels at a distance of d pixels and in a given orientation (in x and y). The matrix is calculated for each pixel for each of the spectral bands under consideration. Derived from this are 14 statistics or attributes that describe the texture in terms of correlation, contrast, variance, homogeneity, etc. In forest studies, the variance of the distribution in the window, indicative of the non-directional local heterogeneity, is often added. It is difficult to interpret the values of these attributes in terms of forest structure. As both studies show, the most interesting aspect is their use as predictors of forest variables after selecting the most efficient, including finding the best spatial configuration of the GLCM: size of window, distance and orientation explored.

In both studies, a linear model is calculated by regression, in order to predict each forest variable separately. In Kayitakire *et al.* [KAY 06], there is only one single predictor in panchromatic. The studies of Beguet *et al.* [BEG 12, BEG 14a, BEG 14b, BEG 14c] take this further. The linear model is multiple, the predictors being selected in both panchromatic and in 4 multispectral bands (blue, green, red and near-infrared), acquired simultaneously at a resolution 2 times lower. Given the large number of predictors to be tested (thousands, in view of the number of texture attributes per spectral band, and the combinations of spatial parameter values), Beguet *et al.* [BEG 12, BEG 14a, BEG 14b, BEG 14c] automate the variable selection procedure, which is done by multiple linear regression, aimed at minimizing the collinearity between predictor variables and balancing their importance in the model. The quality of the models and predictions have been evaluated by cross-validation.

The study of Kayitakire *et al.* [KAY 06] has tested the capabilities of 1 m panchromatic resolution (1 IKONOS image in autumn) to estimate 5 key variables: age, height, stem circumference (= 2π DBH), stand density and basal area. 29 pure spruce stands, aged from 27 to 110 years, were used as a reference. For height and circumference, the accuracy of estimates seems quite high, the relative error (= RMSE/average observed), being of the order of that which is obtained by traditional ground-based inventory: 10 % (RMSE = 2.2 m) for the height and 15 % (RMSE = 13 cm) for the circumference. The accuracy is similar for the age, the relative error being 18% (RMSE = 10 years). By contrast, the quality of the prediction is poor for the density, the variance of errors increasing with density. The relative error is significant: 29 % (RMSE = 276 trees/ha). As for the basal area, the best model found is not truly acceptable: $R^2 = 0.35$, with a non-linearity of the residuals, which are very high, and an effect of saturation of the predicted value when the basal area is reduced.

The first experiment on maritime pine [BEG 12] shows similar results. It was carried out on a QUICKBIRD image (panchromatic at 0.62 m, and multispectral at 2.48 m), acquired in autumn on a very homogeneous site. The results also provide a little information on basal area. Its complex relationship with texture cannot be explained using simple linear models. However, age can be estimated using a multivariate model with low uncertainty: RMSE of 5.25 years, this being a relative error of 20 %; above the age of 35 years there is an underestimation, with a very high variability of errors. This is linked to the fact that, after this age, growth of the maritime pines slows down and the structure becomes less homogeneous. This result is based on a reference set of 184 stands, whose exact age was known (from 3 to 52 years).

The second experiment on maritime pine [BEG 14a, BEG 14b] dealt with the average DBH, average height, average crown diameter, density and spacing[8]. It took place on a site with a large diversity of cover structure, which was sampled with *in situ* measurements on 111 plots of 400 m² (Table 7.1: measured variables and statistics). Three sets of PLEIADES images (panchromatic at 0.50 m and multispectral at 2 m, acquired in 2 seasons and during the same annual growth cycle (June 26 and August 8 2012, February 20 2013), have provided an evaluation of the contribution of a multivariate approach (5 predictors) on the accuracy of prediction and the impact of the acquisition conditions (viewing angles, seasonality of solar

8 Non-linear transformation of density, assuming each tree to be at the top of an equilateral triangle: $= \sqrt{\frac{20000}{\sqrt{3}\,Nha}}$, *Sp* being the spacing (in meters) and *Nha* the density (number of trees per hectare).

angles and foliar development of the understorey and trees). Multispectral bands were generally selected as predictors, in addition to the panchromatic. Thus, even at resolutions twice as low, they provide significant additional forestry information. The most successful models provide quite similar predictions from one date to another, even if they are better in the case of the winter image (Table 7.1). Figure 7.18 shows the results for images where the acquisition conditions seemed the least favorable for accuracy (images of June 26 2012). In the case of the spacing the quality of the predictions is better and more stable, with a constant error regardless of the observed value; the error is from 18 to 20% in relative terms, and always lower than 1 m. The density cannot be satisfactorily predicted directly using a linear model. In the case of height, the winter image significantly improves the expected accuracy, particularly for small trees (<7 m): errors from 18 to 27 % and from 2.5 to 3.6 m, depending on the season. In the case of crown diameter and DBH, which are closely correlated, the relative errors are similar: of the order of 30%, less for the DBH, arguably because of the improved accuracy of field measurements. The absolute errors in the spacing and the crown diameter are quite acceptable, since they are almost double the sensor resolution in panchromatic. The best predictors vary from one image to another (attributes, spatial parameters, and spectral bands). This provides a certain robustness of the method to the conditions of acquisition, but implies a specific calibration for each image.

In conclusion, whether in terms of structure or forest biomass, the VHR image texture provides usable quantitative information. The methods used in the examples have the advantage of automatically identifying the best performing texture indicators (particularly by optimizing the window size), which makes the estimation of variables robust with respect to the diversity of conditions in which the images are acquired: solar angles, viewing angles, and vegetation seasonality in the case of evergreen tree species. This implies a specific calibration for each image and for the different types of forest encountered, and thus the availability of an adequate ground-based reference data set.

In coniferous high forest originating from plantations, one can expect to be able to estimate the stem diameter, the height of the trees and their spacing with a reasonable accuracy, in comparison with the accuracy that can be achieved whether through traditional inventory by in situ sampling, or with aerial LiDAR data or with digital photogrammetry. An operational application, however, requires specification, for a given type of forest, of the sampling effort needed to calibrate the models, and also to validate them with an independent data set in order to ensure a minimum accuracy.

	Month of acquisition	*In situ* observations Average	Min - Max	RMSE	Relative RMSE (%) = RMSE/Average
Cd (m): average diameter of crown	June	3.48	0.7 – 9.5	1.12	32
	August			1.10	31
	February			1.10	31
Sp (m): Spacing	June	4.48	1.3 – 8.8	0.92	20
	August			0.83	18
	February			0.84	18
Ht (m): Average height	June	13.25	1.7 – 26.2	3.6	27
	August			3.1	23
	February			2.5	18
Nha (trees/ha): Density	June	931	150 – 6729	640	68
	August			586	63
	February			657	70
DBH (m): average diameter at 1.3 m above ground	June	0.22	0.02 – 0.56	0.062	28
	August			0.065	29
	February			0.058	26

Table 7.1. *Estimation of structure variables in maritime pine forest (Landes forest in south-western France) using indicators of texture based on panchromatic and multispectral images (PLEIADES 0.50 and 2 m) [BEG 14a, BEG 14c]. The performance of the models is obtained by leave-one-out cross validation on 111 plots of 400m² (value on a plot predicted by the model based on the n-1 other plots). The age of the stands varies from 4 to 68 years. Furthermore, the comparison was carried out between the 3 conditions of acquisition of the images. Only one in situ observation was made for each plot (from the end of June to the beginning of October 2012), the forest structure varying very little between the first (June 2012) and the last dates of acquisition of the images (February 2013)*

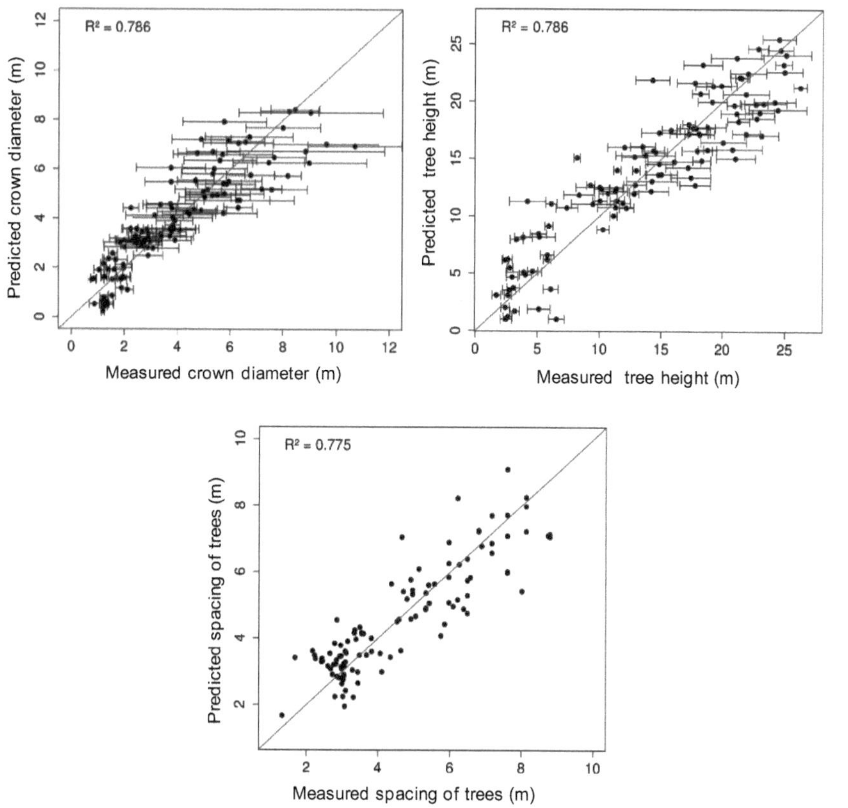

Figure 7.18. *Estimation of structure variables in maritime pine forest (Landes forest in south-western France) using indicators of texture based on panchromatic and multispectral images (PLEIADES 0.50 and 2 m) [BEG 14b]. Comparison between values predicted by the best models and values measured in situ (n = 111 plots). The results are presented for the least favorable conditions of image acquisition in terms of accuracy, (images of 26 June 2012), among those tested. The bars = ± 1 standard deviation*

In estimating the biomass in tropical forests, where there is often very little reference data available, the studies presented suggest that with the method used (FOTO), one can pool several images acquired under different conditions, or even with different sensors, to cover a large area, by means of a simple normalization of Fourier spectra. But this is still to be verified, especially in cases of high intra-regional variability of the three dimensional structure of the tree cover.

7.3.1.3. *Contributions of the detection of significant and sudden forest changes: harvesting by clearcutting, deforestation, fires and storms*

Due to the availability of high spatial resolution data for over 30 years, with regular coverage of forest land, notably with Landsat TM which offers middle infrared bands, much experience has been gained on the detection and mapping of significant and sudden changes such as harvesting by clearcutting, deforestation, fires and storm damage.

An example of an iconic application of the detection is now presented, the dating and mapping of clearcutting at regional level to assist monitoring of forest resource for management purposes. This is an interesting example, since it deals with an operational application.

It was developed in the 1990s and carried out on the Landes maritime pine forest, which covers 1 million ha in south-west France, and has an important economic role in French timber production. The pine stands, both single-species and even-aged, are intensively managed: harvesting by clearcutting at about 50 years' of age, reforestation by planting or artificial seeding, periodic thinning preceded by a cleaning of the understorey by clearing.

As shown in Figure 7.19, clearcutting causes a significant change in the middle infrared reflectance (MIR) of the pine stand (almost double), much greater than that caused by thinning or understorey clearing.

On the basis of supervised classification, pixel by pixel, of biannual pairs of images in the MIR, and on a 7 year series of Landsat TM images (1984-1990), Jolly *et al.* [JOL 96] and Jolly [JOL 93] demonstrated, on an area of ~ 700 000 ha, that the annual mapping of clearcut areas on the entire Landes forest was feasible by using one image per year. An agreement of the order of 95% (differentiation between clearcut and no clearcut) was achieved due to a prior stratification by single-date multispectral classification, and by taking into account the cutting previously detected, in order to limit confusion with reforestation. The approach also required a prior knowledge of the localization of pine maritime areas, forest map data from the forest inventory service in France (IFN, Inventaire forestier national (National Forest Inventory)), and forest reference data to learn the classifications and validate the maps produced. The interannual variability of the cut areas detected were consistent with that of the volumes of harvested wood used by the industry provided by the French annual surveys of the "Commercial logging and Sawing" sector (from the French Ministry of Forests), even if the small clearcut areas (less than 2 ha) were not well detected due to the resolution of the sensor (30 m). In addition to silvicultural clearcutting, the accidental cuts anticipated as a result of catastrophic frost damage (January 1985) and the deforested areas major fires (thousands of ha) were also found.

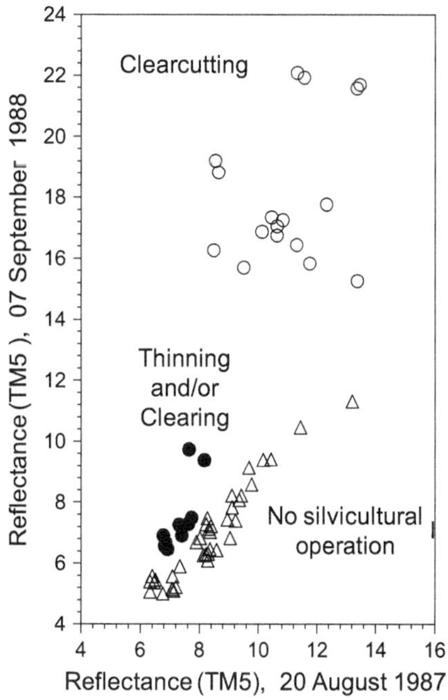

Figure 7.19. *Surface reflectance changes (in %) in the middle infrared induced by clearcutting (mean difference from stands with no silvicultural operation = 9.1), thinning and understorey clearing (1.6), in maritime pine forest (Landes forest, south-west France) [OJ L 93, GUY 96]. Averages observed for stands over 10 years old with Landsat TM at approximately one year intervals*

The studies of Jolly *et al.* [JOL 93] have also shown that a geographical intersection of this annual data on the localization and area of the cuts and the *in situ* measurement plots of the IFN inventories, has enabled an annual update of the statistical distribution of the age of the stands and the distribution of the age at the time of the cutting, for the entire Landes forest. It emphasizes the contribution of the method in the localization of future resources, since in the absence of deforestation, from the date of cutting one can determine, with an uncertainty of one or several years, the date of subsequent reforestation and therefore the age of the new stand. The validity of this hypothesis was demonstrated by Guyon *et al.* [GUY 15] on 30 years of Landsat TM images (1984-2014), but the expected accuracy of the estimated age is reduced after widespread destruction of stands, such as that which resulted from the severe storms of 1999 (Martin[9]) and 2009 (Klaus[10]), which sharply disturbs the dynamics of the reforestation.

Given this potential of annual monitoring using Landsat TM temporal series images in enriching and refining the information produced by its traditional inventory by *in situ* sampling plots, the IFN have created and implemented an operational annual mapping tool for clearcutting on the whole Landes forest in collaboration with wood and forest professionals in Aquitaine. The IFN have adapted the method [STA 05]. The radiometric change is then detected using criteria based on a comparison of the distribution of intra-pair deviations of images to a theoretical model. The identification of probable cuts among the changes detected is carried out using contextual criteria, by visual inspection of the two images. A radiometric normalization is carried out beforehand, between the top of the atmosphere luminance of the two images acquired in different conditions (atmosphere, height of the sun, seasonality of vegetation). This approach has the advantage of overcoming the difficulty of obtaining forest reference data for each year.

The IFN made a validation by comparison with various sources of information, both local (to the plot, forest management data), and global (statistics from French annual surveys of sector, comparisons of successive IFN inventories). These comparisons showed that the method is reliable for the detection of cuts of over 1 ha. The cuts of less than one hectare are not insignificant, which represents an underestimation of approximately 10% of the area exploited in the Landes forest.

A map of the clearcutting, detected annually, has been produced by the IFN for the period between 1990 and 2008, for the entire forest [BEL 08]. In this period, the average annual area of clearcuts was estimated at approximately 2% of maritime pine area in the Landes forest (around 900,000 ha), which is consistent with a lifetime of the stands of 50 years. The clearcutting was at its most extensive between 2000 and 2002, with the exploitation of windfalls due to storm Martin in December 1999, particularly

9 Storm which crossed France in late December 1999.

10 Storm that hit mainly the south west of France in late January 2009.

in the Gironde, the department (in France, a department is a sub-regional administrative division) most seriously affected. The average rate before the storm has been recovered for all departments in 2006. This is illustrated by the statistics in Figure 7.20(a).

The IFN has also effectively demonstrated an enrichment provided by the annual mapping of cutting to its in situ surveys using sampling plots. Thus, the combination of these two sets of data has informed the annual change in the age of the stands at the time of cutting during the twenty years of monitoring. Figure 7.20(b) shows that the timber harvesting affected the younger stands more after 2000 than before, and then again 6 years after the 1999 storm, even though the older stands were always the most harvested. This information, which represents genuine silvicultural scenarios, is valuable for the simulation of future resources by area and volume (prospective studies).

The IFN has also applied this method, with some adaptation, to the production of catastrophe damage mapping for the Landes forest [STA 09] resulting from storm Martin in December 1999. This damage map was established very quickly after the storm, and has provided an initial estimate of the volume of windfall wood, after cross-referencing with *in situ* dendrometric IFN data. The IFN has subsequently improved the methodology, so as to reduce the under-detection of diffuse damage, by means of automating image segmentation before and after the storm, using an object-orientated approach and images of higher spatial resolution (20 m, in the case of the SPOT images). This is how the damage map of storm Klaus was produced, a storm that ravaged the Landes forest in January 2009, with a similar magnitude to that of storm Martin.

Finally, the IFN has extended its "clear cutting" application to other forest regions in France, particularly to more complex situations with marked relief, or mixed species and diverse forest structures. The application used on the Limousin forest (585,000 ha) envisaged an annual regional mapping from 2004 to 2008 using Landsat TM, and a lower frequency mapping at the department scale using remote sensing data with higher spatial resolution (SPOT 4 or 5, 10/20 m resolution). In fact, Landsat TM did not provide for detection of clearcutting for areas of less than 2 ha, which represented 33% of the area of the cuts detected with SPOT, the forest being particularly fragmented. It is the SPOT satellite data that ensured a regional mapping of clearcutting with an appropriate level of detail (0.5 ha of the minimum mapping unit) in the years following. [BEL 08].

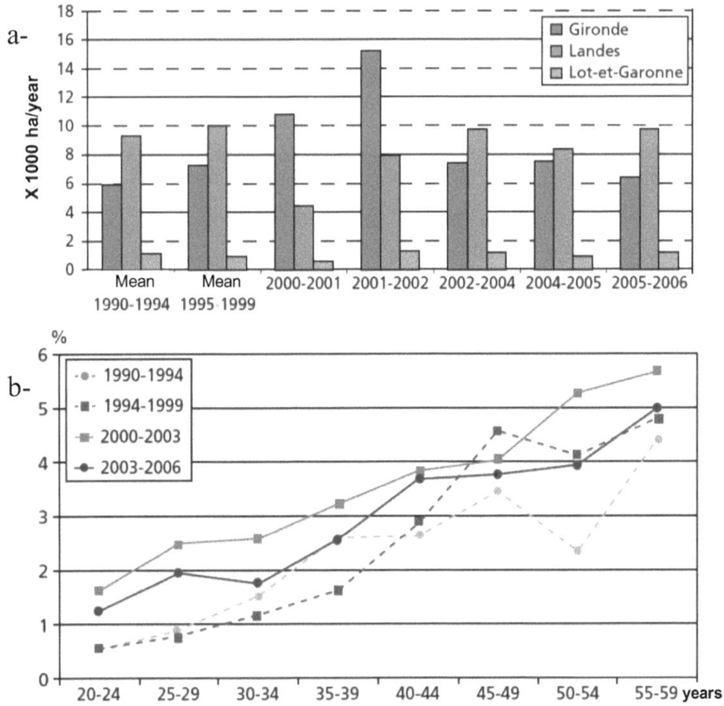

Figure 7.20. *Statistics provided by annual mapping of clearcutting from 1990 to 2008 on the entire Landes maritime pine forest (south-western France) made by the IFN, based on the detection of changes on Landsat TM images [BEL 08]. a) variation in the distribution of the areas harvested annually by clearcutting from 1990 to 2006 according to department. b) surface fraction harvested by clearcutting per age class according to the period under consideration, obtained by cross-referencing with in situ measures on IFN sampling plots. Only the 20 to 60 year old classes are taken into consideration*

In conclusion, this example of application illustrates how the mapping and dating of sudden and extreme forest changes, whether silvicultural or accidental, using multispectral imagery, can enrich the information obtained from forest inventories by sampling on wood harvesting and the future resource on regional levels or above.

The current or forthcoming availability of complete multi-annual series of Landsat TM or Sentinel-2 images on all land areas, by surface reflectance data including quality criteria indicators (clouds and other defects), should facilitate the development of "clearcutting or other substantial forest change" products with high accuracy and low cost, over wide areas. Sentinel-2 (10 m resolution in the visible and near-infrared and 20 m in the mid-infrared) will thus help to refine the accuracy of the maps produced and the statistics derived by cross-referencing with *in situ* inventory data.

Nowadays, data maps are produced, using Landsat data, of the annual changes in forest cover across the globe, showing the gains and losses of forest areas [HAN 13]. However the thematic and geometrical accuracy of this type of product remains to be evaluated in relation to the requirements of the applications envisaged.

These methods of detection of change over two year periods do not separate out the various causes of sudden changes to the standing biomass, which have differing consequences (economic, ecosystem services, carbon footprint, etc.), unless an image shortly before and another shortly afterwards are available in the case of a known disaster. However, the multi-annual profile of the remote sensing signal in the years that follow the change can be used to detect and date the causes of a disturbance separately, if it provides information on the possible dynamics of the forest areas specific to the type of disturbance.

7.3.1.4. *Contribution of multi-source approaches*

VHR, with the exploitation of the texture or stereoscopy, enables access to detailed information about the cover structure or the biomass per stand, and is thus useful for forest management planning. However, this requires a prior calibration of the prediction models (selection of predictors, estimation of parameters) with simultaneous *in situ* dendrometric measurements, that is, reference data, which can be replaced by proxys provided by aerial LiDAR, as well as a stratification of the modeling by type of cover. Furthermore, a combined use of stereoscopy and texture on the same images could also be envisaged, in order to reduce the uncertainties related to the underlying allometric relationships between basic dendrometric variables (such as height and DBH), occurring when each method is separately used to predict biomass.

Using spectral information only - spectral reflectances or vegetation indices – observed at high (Landsat TM for example) or medium resolution (MODIS for example), as predictors of structure or biomass variables, is not advisable, as we have previously seen (Figure 7.9). However, the previous example of the detection of changes due to clearcutting emphasizes the importance of combining remote sensing information with readily available geolocalized information about the structure and biomass, provided *in situ* inventories by sampling. Moreover, the implementation of this application, like those presented on VHR, requires detailed knowledge obtained elsewhere on the localization of the cover types of interest (species or species group, type of forest), which the remote sensing data at hand is not able to provide.

Satellite spectral information can be used to optimize the accuracy of national forest inventories, and to spatialize the estimation of the variables of interest using techniques that perform a spatial interpolation. This can, in particular, make it possible to obtain acceptable bias and accuracy for small areas or rare types of forest formations for which there is insufficient *in situ* data [MCR 07]. These applications have been widely developed in some countries, such as Finland or Sweden. The methods combine satellite images, data produced from *in situ* measurements using sampling plots, and various other numerical geographical database. Various parametric methods (regressions, for example, as we have previously seen with VHR) or non-parametric methods (neural networks, Random Forest, k-nearest neighbors for example) of modeling and selection of variables, can be used to predict a forest variable based on multi-variable satellite observations [BRO 14].

The approaches based on k-nearest neighbors or k-NN are widely used. The advantage lies in being able to predict multiple variables simultaneously, both quantitative and qualitative [MCR 07]. The k-NN method is used in various countries to map a range of variables: age, DBH, height, basal area, volume, biomass, and also the type of forest cover and species group. The value of the forest variable for an area not sampled by the reference measurement plots is predicted by assigning to it the weighted value of the nearest sample plots, the weighting being calculated by means of a statistic metrics of distance (Euclidean distance for example) in the satellite data space. But the realization of the predictions pixel by pixel, or by unit area (a district, for example), based on expansion factors of the plots, and the statistical estimation of the uncertainties, are not trivial, notably because of complex spatial dependences. Tomppo *et al.* [TOM 08] demonstrate some applications to the production of a map of standing volumes for the whole of Finland and Sweden, based on Landsat TM images. The application of Beaudoin *et al.* [BEA 14] on the 4 million km^2 of the Canadian forest area, where the sampling rate of the plots varies greatly according to the region, is interesting. Based on MODIS images and ancillary georeferenced data (climatic, topographic, etc.), it provides a map of above-ground forest biomass for the entire area, whose accuracy at pixel level depends on the rate of sampling of the plots, the level of biomass and the topography, and consequently varies according to the region.

The maps produced - independently of this type of approach or otherwise – can be used for stratification purposes to improve the statistics of the inventories, which is nothing new in itself, since it is already practiced in pre-stratification using maps provided by aerial photography. For example, Nilsson *et al.* NIL 09] demonstrate the improvement in accuracy in the estimation of the area of clearcuts from national inventory data in Sweden, thanks to a post-stratification using a clearcutting map obtained with SPOT and Landsat TM.

The current trend is a move towards multi-source approaches in the broadest sense that is, being able to combine various sources of remote sensing data and *in situ* data in order to optimize the complementary capacities of the different remote sensing techniques (LiDAR, radar, hyperspectral, multispectral). Lidar and radar are better than optical multispectral for the direct estimation of biomass and standing carbon stocks. Hyperspectral is generally better for differentiation of species [KOC 10]. One can quote, for example, the study of Baccini *et al.* [BAC 12], who estimate the Carbon emissions due to deforestation, over a 10 year period, in tropical areas on 3 continents. This makes use of MODIS multispectral surface reflectance data series to spatialize the carbon stock estimated by the GLAS LiDAR satellite, over the thousands of unit areas (of 70 m diameter) which it samples along its track. The carbon stock on these GLAS unit areas is itself derived from a statistical relationship between LiDAR metrics and biomass, calibrated with the aid of on-the-ground sample measurements.

In conclusion, multi-source approaches for quantifying and spatializing forest structure variables, above-ground biomass, timber volume or standing carbon stocks, allow one to maximize the value of the information provided by multispectral satellite remote sensing, reducing the uncertainties of the estimates based on the forest data obtained by sampling on the ground alone, or by other remote sensing techniques.

7.3.2. *Monitoring of the functional response of forests exposed to climate and hazards*

The idea of using remote sensing to describe the forest health is nothing new. It sprung from the realization that, when faced with situations of health crises (extreme drought, pests outbreak, for example), scientists, forest managers and institutional forest organizations are not equipped for a rapid evaluation of the significance of damage, locating it or monitoring its dynamics. It involves detecting, locating, quantifying and tracking, over a period of several years, the damage due to insects, diseases, droughts or dieback with more complex causes (interactions of multiple hazards) on the basis of observation of their aerial symptoms: loss of leaves, color changes in the leaves, death of branches or tree mortality.

Aerial color infrared photography at large scale (~1/5 000-1/10 000, this being of a resolution <30 cm), especially when interpreted visually by stereoscopy, makes the description of the species and the apparent state of health (*crown condition*) fairly simple [DES 06]. Its use has been proven, but it seems difficult to generalize in the long term and over extensive areas. The recourse to satellite imagery VHR (PLEIADES and others) or HR (Landsat TM and others), while guaranteeing observation at the stand level, can prove disappointing for temporal or spatial sampling reasons. The frequency of observation - low revisit frequency, presence of clouds - can be too low to capture damage whose manifestation of symptoms on leaves may be fleeting or spread out with time. Also, HR spatial resolution often proves to be inadequate in cases of slight or diffuse forest decline. In other words, if only a portion of the trees in the pixel are seriously affected at the moment of observation. This corresponds to the majority of the forest decline studied today. Nevertheless, HR can be sufficient in the case of heavy damage, whether defoliation, discoloration of leaves or mortalities. Moreover, without exogenous contextual knowledge, there is no means of identifying the cause of the damage that could be detected on the images. For these reasons, few tools that use the VHR or HR satellite data as input are available for providing information on the health condition of the trees at the stand level.

The recourse to sensors with a higher revisit capacity, currently those of medium and low resolution (250 m – 1,000 m, MODIS and others), can prove profitable in the detection and monitoring of the disruptions to the seasonal development of the green leaf surface, in other words the anomalies of foliar phenology. However, this does not guarantee observation at the forest stand level, but rather of the landscape or the country. In broader terms, given the spatial cover afforded by these satellite sensors and the temporal extent of previous years covered already, one can expect this kind of temporal data series to provide indicators of the long term response of forests (health condition, productivity, phenology for example) faced with changes on a global scale that could threaten the forest.

One can also expect it to provide biophysical data (albedo, vegetation index, LAI, FAPAR, Fcover, phenological indicators, etc.) to supply the input variables, set the parameters, or perform validation of functional models (based on physical and biological processes) for the estimation or prediction of current or future changes in productivity, water and carbon balances in particular.

The various examples presented will illustrate the capabilities and limitations of multispectral optical satellite imagery on these fields.

7.3.2.1. *Detection of heavy defoliation by pest insects*

Numerous studies have concluded that defoliations are detectable and mappable by remote sensing, but little or nothing has been done to validate the anomalies detected by remote sensing by means of *in situ* observations. Two studies illustrate this approach, one for a polyphagous pest in deciduous broadleaf forests, and the other for a monophagous pest in a coniferous evergreen forest. These also illustrate two kinds of satellite data analysis: one by detection of changes on the anniversary period of the defoliations between two Landsat TM images in deciduous forest, the other by analysis of seasonal disturbances of the MODIS time signal in evergreen forest.

7.3.2.1.1. Heavy defoliation by the Gypsy Moth caterpillar (*Lymantria dispar)* in oak forests

This example deals with the damage caused by two outbreaks of this pest in the Appalachian regions of the United States during the years 2000–2001 and 2006-2008. Townsend *et al.* [TOW 12] proposed a method for estimating the percentage of defoliation and evaluated this by repeated application over several years and two sites. The study is based on the weekly ground-based visual assessment, giving the maximum rate of defoliation in fifty plots, according to a sampling designed to fully describe the Landsat pixels. The values of this rate have been validated by means of consumed foliage mass estimates derived from the weekly collection of caterpillar excrement in litter traps.

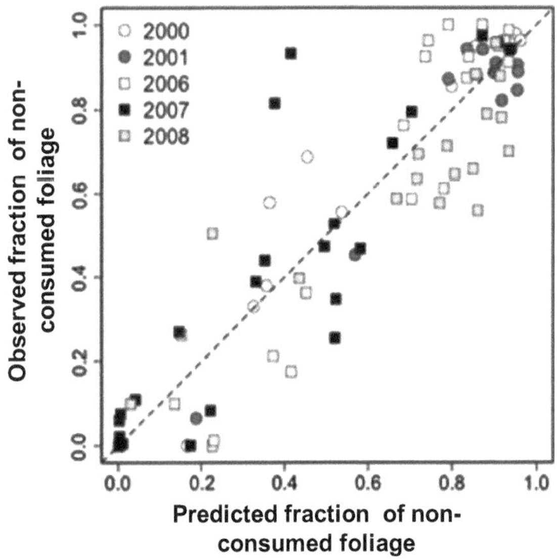

Figure 7.21. *Mapping of the defoliations by the Gypsy Moth caterpillar (Lymantria dispar) in oak deciduous forests in the Appalachians (USA) by detection of changes during the anniversary period of the defoliations between Landsat TM images [TOWN 12]. Cross validation results of the estimation model of the fraction of foliage unconsumed, calibrated over 3 years on one site and two on the other. Circles = Green Ridge site, squares = Savage River site. All the plots from each year-site combination are removed successively, and the model is recalculated with the others each time. The y-axis shows the predictions obtained each time for the year-site removed. For a color version of this figure, see www.iste.co.uk.baghdadi3.zip*

The method consists of detecting changes during the anniversary period of the defoliations between two Landsat TM images, one for the year of the disruption and the other for a year without disruption used as a reference. The indicator of change used is the difference in vegetation index (VI) between the two images:

$$\Delta VI = VI_{reference\ year} - VI_{disturbance\ year} \quad [7.5]$$

The proportion of foliage remaining (FR) is then predicted by modeling of a logistical function:

$$FR = \frac{b}{1+exp\ (c+d\ \Delta VI)} \qquad [7.6]$$

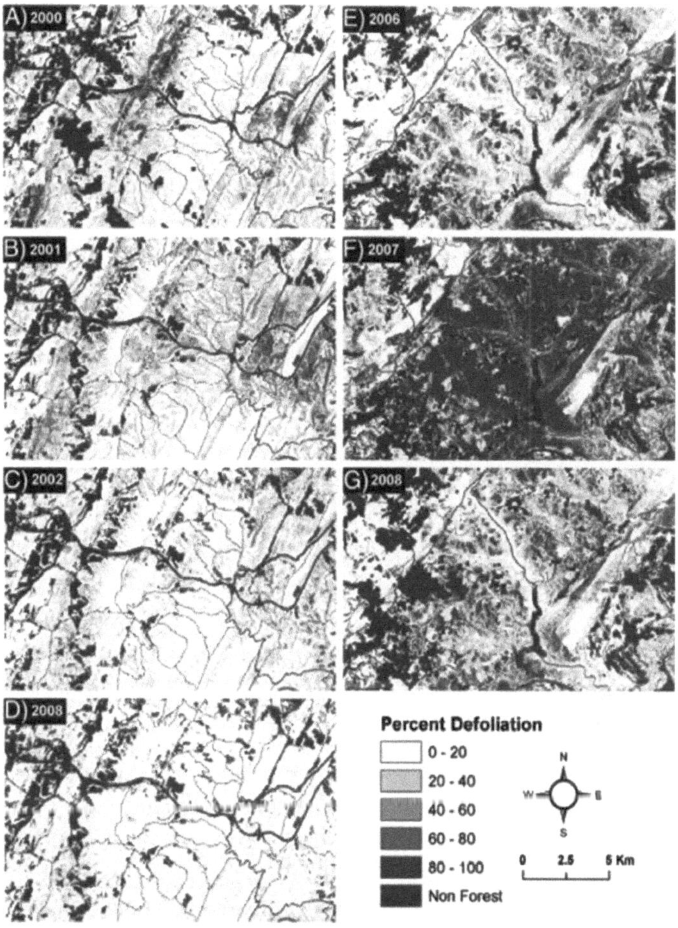

Figure 7.22. *Mapping of the defoliations by the Gypsy Moth caterpillar (Lymantria dispar) in oak deciduous forests in the Appalachians (USA) by detection of changes during the anniversary period of the defoliations between Landsat TM images [TOWN 12]. Map of defoliation levels produced on both sites: Green Ridge in A) 2000, B) 2001, C) 2002, D) 2008 and Savage River in E) 2006, F) 2007, G) 2008. In grey: non-forest, clouds and shadows of clouds. The areas of harvested forest are also masked. For a color version of this figure, see www.iste.co.uk.baghdadi3.zip*

Several vegetation indices calculated on surface reflectance are tested. The indices based on middle (TM5) and near (TM4) infrared: NDII = [TM4 – TM5] / [TM4 + TM5]) and MSI = TM5/TM4 are more effective than the traditional indicators based on visible and near-infrared, such as the NDVI. It is the NDII that has been selected as a predictor. The model, broadly calibrated over three years on one site and two on the other, gives an RMSE by cross-validation of 15%, and an average absolute error of 11% for the predicted defoliation rate, R^2 being 0.805. These validation results, and the maps produced for all given years studied, and the two sites, are provided in Figures 7.21 and 7.22. The risks of errors reside essentially in the difficulty of obtaining images on dates that enable good capture of the maximum defoliation: in mountainous regions, the frequency of completely clear (cloud free) images in the usual defoliation period (the beginning of summer) can be low, and a partial refoliation after the peak of defoliation is possible. Moreover, there is never any certainty that the forest area concerned is fully unscathed by defoliations on the date on which the reference image is acquired.

As the authors underline, to extend this to other sites and other years (that is, to guarantee the reliability of the method), a minimum validation, and if necessary, updating of the model, need to be done on the basis of ground observations. Nevertheless, the results demonstrate strong performances of such an approach in estimating and mapping the levels of defoliation in the case of heavy attacks.

7.3.2.1.2. Heavy defoliation by the pine processionary moth (*Thaumetopoea pityocampa*) in a maritime pine forest

The second example concerns the application proposed by Bories *et al.* [BOR 12] for locating the areas most defoliated by the pine processionary moth (PP) on the landscape scale in the pine maritime Landes forest in south-west France. The PP is the most damaging defoliator caterpillar of this forest, with cyclical outbreaks; the study focuses on the one that peaked in 2009 - 2010.

The approach was founded on MODIS time series at 250 m resolution, by analyzing the seasonal disturbances of the vegetation indices NDVI and EVI (MOD13Q1 product, periodicity 16 days) over several consecutive years from 2000-2011. In order to do this, the seasonal trajectory of the indices was modeled on the 10 years data set, by means of a smoothing using the Savitzky-Golay filter, with the TIMESAT algorithm ([JON 04a]

Figure 7.14). The field observations on the defoliations of winter 2009 – 2010 were carried out on two sites during the following spring, before the emergence of new pine needles. The intensity of defoliation was visually assessed, according to 6 classes from 0 to 1, on all the trees in the plots sampled (Figure 7.23). The levels of defoliation (DR) per plot were derived from this. On the calibration site (18 km²), 269 plots were sampled in order to cover the diversity of the stands and the intensity of the attacks. DR was then integrated on the MODIS grid cells, where all the stands were sampled. On the validation site (16 km × 16 km square), a systematic sampling, using 162 plots in clusters of 2, was carried out. The observed DR level is then interpolated across all MODIS grid cells in the square, using the block kriging method in order to produce a reference map.

Figure 7.23. *Crown condition of maritime pine according to the intensity of defoliation due to the pine processionary moth (Thaumetopoea pityocampa). Source: Département de la Santé des Forêts (Forest Health Department) of the French Ministry of Forests*

Observations of temporal variations in both vegetation indices show a decline in winter 2009 - 2010, which is all the more pronounced given that the defoliation is so intense, as shown in Figure 7.24. Different indicators of this anomaly have been tested by simple linear regression on the calibration site. The best predictor ($R^2 = 0.908$, $p < 0.0001$, RMSE = 0.037) of the defoliation observed is the relative deviation from the minimum EVI in winter to the average minimum EVI of the 4 previous years noted WAM:

$$WAM = \frac{(\sum_{y=2006}^{2009} EVImin) - 4\ EVImin_{2010}}{(\sum_{y=2006}^{2009} EVImin)} \qquad [7.7]$$

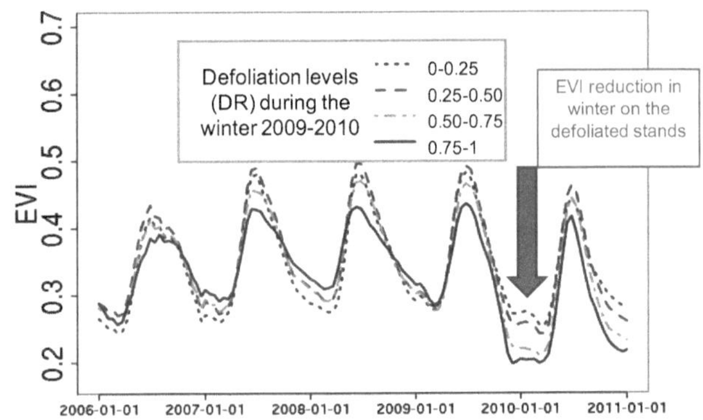

Figure 7.24. *Detection of heavy defoliation by pine processionary moth (*Thaumetopoea pityocampa*) in a maritime pine forest (Landes forest, south-west France) by analysis of the seasonal disturbances of vegetation indices produced from the MODIS sensor (MOD13Q1 product: 16 day periodicity and 250 m resolution) [BOR 12]. Seasonal variations of the EVI vegetation index from 2006 to 2011 on the calibration site (Marcheprime), according to the defoliation level (DR) observed in situ in winter 2009–2010. The EVI curves are averaged by DR class of width = 0.25. For a color version of this figure, see www.iste.co.uk.baghdadi3.zip*

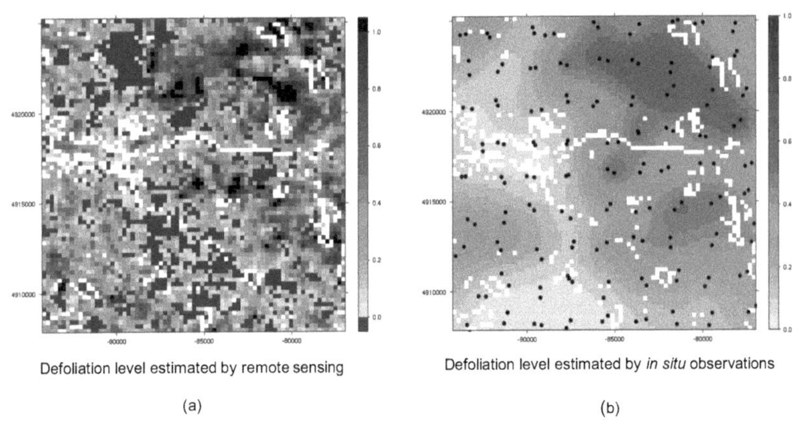

Figure 7.25. *Detection of heavy defoliation by pine processionary moth (*Thaumetopoea pityocampa*) in a maritime pine forest (Landes forest, south-west France) by analysis of the seasonal disturbances of vegetation indices produced by the MODIS sensor (MOD13Q1 product: 16 day periodicity and 250 m resolution) [BOR 12]. Map of the defoliation level attributed to the caterpillar on the 16 km x 16 km validation site (Pontenx) on the MODIS grid: a) Predicted by remote sensing using the winter metric anomaly (WMA) and the calibrated model on the Marcheprime site; b) Estimated using the block kriging method on the in situ observations of sampled plots (black points). In white: pixels where the fraction of forest cover is <0.5. For a color version of this figure, see www.iste.co.uk.baghdadi3.zip*

The DR predictions produced by inversion of the model (DR = (WMA + 0.026) / 0.41)) on the validation site are shown in Figure 7.25, in comparison with a reference map. The accuracy is not very high (RMSE = 0.29 – i.e. ~1/3 of the possible range of variation of DR (0 à 1) – on all pixels, and 0.23 on the pixels including the in situ plots) with a fairly high error dispersion around the line 1:1. These discordances by pixel reflect the impact of factors other than defoliation in the seasonal trajectory of the EVI of each pixel; for example, silvicultural operations and emerging mortalities (isolated or outbreaks) due to a non-herbivore pest (bark beetle) have undoubtedly interacted. But they should also be attributed to the uncertainty around the estimation of reference DR values on the pixels sampled on the ground for calibration and validation, and that of the map produced by kriging by block, probably due to inadequate consideration of spatial variability at short distances in the sampling design used for *in situ* observation.

In spite of this, the spatial structure of defoliation on the landscape scale is well documented, displaying the most defoliated large areas. The authors propose to apply the method to the entire forest (1 million ha) spatially aggregating the predictions using a large cell grid to reduce errors, and thus provide a warning map for areas likely to be the most affected.

A simple measurement of winter anomalies of a vegetation index such as the EVI, by MODIS at 250 m resolution, is able to capture the heaviest defoliations due to the pine processionary moth on a landscape scale. This is possible in spite of the relatively low maximum LAI value (generally < 4, see Figure 7.10) of the tree layer of the maritime pine stands. The fact that, in winter, the LAI of the understorey is small compared to that of the trees, is probably a definite asset to the detection in winter of the defoliations that have taken place during the previous autumn.

In conclusion, at present, the applications that are valid involve heavy defoliations over large areas, and give a spatial accuracy inadequate as an aid to forest management at stand level. There is no generic method available today, and the prospect of operational services for forest management still seems far away.

7.3.2.2. *Detection of effects of drought on forest canopies*

Mapping the response of forest canopies to climatic hazards, extreme drought events in particular, is a frequent issue for managers, especially during episodes of extreme severity known for being likely to induce dieback over an large extent. The short-term reaction, during the days and weeks following the occurrence of a water shortage, often associated with periods of heat wave such as the one that occurred in France in 2003, is a modification of the pigment composition in the cover, and a potential partial loss of leaves, easily detected by satellite reflectance. It can be quantified using indicators derived from the seasonal dynamics of vegetation indices, more (LAI) or less (NDVI and others) elaborated, resulting from satellite reflectance measurements. However, the maps produced are rarely validated.

After the drought episodes of 2003 and 2011 in France, Samalens *et al.* [SAM 12] were able to validate these indicators by means of *in situ* observations of crown conditions carried out on all French forests.

Temporal analysis of vegetation indices (EVI, LAI) on a national scale was carried out using the 1 km resolution temporal series produced from the MODIS and VEGETATION sensors from 2000 to 2011. The EVI index was provided by the MODIS MOD13A2 product with a 16 day periodicity, and the LAI by the VEGETATION CYCLOPES (V3.1 [BAR 07]) product with a 10 day periodicity. The seasonal trajectory of both indices was modeled over the 10 years of time series by means of a smoothing using the Savitzky-Golay filter, with the TIMESAT algorithm [JON 04b]. Subsequently, for the LAI series, the dates of the start and end of the growth season were estimated by thresholding at 20% of the seasonal amplitude of the annual LAI trajectory. The LAI integral between the two dates, designated GS LAI (GS, for *Growth Season)* is derived from it (Figure 7.15). For the EVI series, an indicator of the foliar dynamics anomaly is defined with reference to an average year in the decade. It is calculated for each pixel i for a period p of 16 days for year y as the deviation from the mean (μ) normalized by the standard deviation σ over the decade:

$$Anomaly_{i,p}(y) = (EVI_{i,p}(y) - \mu_{2000}^{2011}(EVI_{i,p})/ \sigma_{2000}^{2011}(EVI_{i,p}) \quad [7.8]$$

An anomaly of value 1 (absolute value) is not insignificant, since it is equal to a standard deviation.

The GS LAI indicator was evaluated as a means of predicting the crown condition during the growth season, which is in itself a forest health indicator used routinely by the European *in situ* surveillance network for forest health (Forest ICP). The "crown condition" indicator of the ICP network is defined as the assumed loss of green leaves in relation to a reference tree (same species) in a local context. This leaf loss is observed during the summer on each tree sampled, and averaged over the plot (n = 540 in France). As shown by Figure 7.26, for the year of extreme drought event, 2003, the GS LAI indicator is a good predictor of the crown condition such as observed by the ICP network ($R^2 = 0.79$, $p<0.001$). This also underlines the high quality of the LAI produced by the CYCLOPES production chain, even though its highest values are underestimated [GAR 08], since it in fact corresponds to an "effective" LAI (leaf distribution supposed to be random, see section 7.2.1.5).

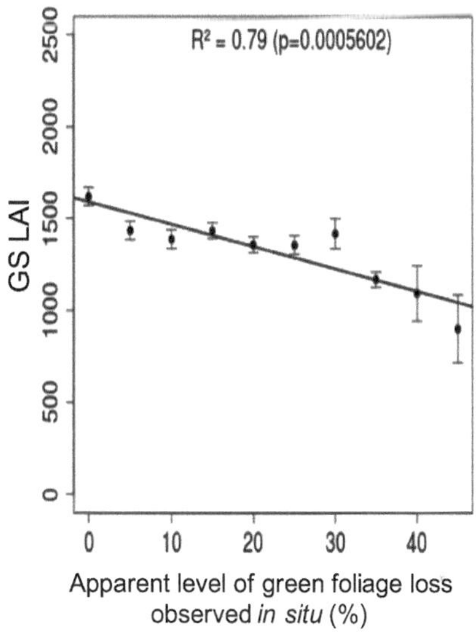

Figure 7.26. *Estimation of changes in crown condition due to the extreme drought event in France during 2003 [SAM 12]. On the abscissa: indicator of the crown condition observed in situ as being the apparent level of green foliage loss. On the ordinate: integral of the LAI over the growing season (denoted GS LAI) estimated on the LAI time series of the CYCLOPES product provided by the VEGETATION sensor (10 day periodicity, 1,000 m resolution). The results obtained for the main deciduous and coniferous species from the 540 French plots of the European Forest Health Monitoring network ICP are averaged by class of 5% of the in situ indicator. The bars correspond to a standard deviation of ± 1*

As to the foliar phenology anomalies, the EVI anomaly indicator for the year 2003 that agrees the most closely with the drought event damages observed *in situ* is that of the end of July 2003, this being the period of extreme water deficit that preceded the heatwave. These reference *in situ* observations consist of a drought damage map in 3 categories per bio-geographical unit of forest, based on the expert observations of the Département de la Santé des Forêts (French Department of Forest Health] between August and October 2003. Figure 7.27 provides a comparison between the two maps, which show great similarity of spatial structure between the negative anomalies and the most serious damage on the reference map. A variance analysis, and a post hoc statistical comparison of means of the anomaly between damage reference categories, have confirmed the validity of the proposed EVI anomaly indicator, for broadleaf and coniferous taken separately.

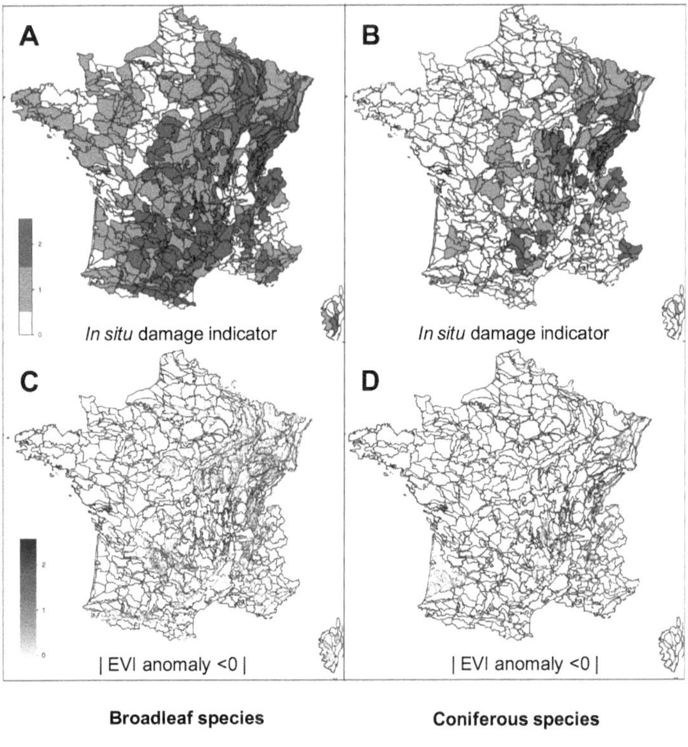

Figure 7.27. *Validation of the map of immediate damage due to the extreme drought event of 2003 in France, produced from the EVI vegetation index anomaly (MODIS 1,000 m, 16 days periodicity) at the end of July 2003 [SAM 12]. In A and B: in situ observations coded into 3 classes of intensity of damage (source: Forest Health Department of the French Ministry of Forests [BEL 04]). In C and D: estimations provided by remote sensing. Only the negative EVI anomalies are represented (expressed as absolute values) on the pixels where the fraction of forest cover is > 50%. The majority land cover category, (>50%) - broadleaf or coniferous - by pixel, is given by the BD Corine Land Cover 2006. The black contours delineate the 680 forest units of the IFN. For a color version of this figure, see www.iste.co.uk.baghdadi3.zip*

The EVI anomaly indicator applied to the spring drought event in year 2011 shows extreme negative anomalies in June, consistent with the surface moisture map measured with SMOS (passive microwave satellite sensor) one month before. The majority of anomalies become positive in September, suggesting a recovery in green leaf surface of the forest vegetation after the return of a rainy summer.

To conclude, the indicators derived from anomaly measurements in the seasonal time course of vegetation indices at medium resolution are able to take account of immediate foliar damage caused by an extreme drought affecting vast forest areas.

7.3.2.3. *Productivity of forests and climatic hazards from the landscape scale to that of continents*

The productivity models, or those which calculate fluxes of water and carbon in ecosystems, describing surface processes (evapotranspiration, carbon assimilation, productivity, fraction of photosynthetically active radiation absorbed) on the scale of wide areas or catchment areas, require spatially distributed data in order to:

– set the state variables for the system under consideration: albedo, LAI, phenology, Fcover;

– evaluate the dynamics of the change in the system throughout a season and over multiple years;

– provide verification of the representations of the processes and contribute to the validation of model outputs: gross primary productivity (GPP) resulting from photosynthesis; net primary productivity (NPP), equal to GPP - Ra, Ra being autotrophic respiration; evapotranspiration; fraction of photosynthetically active radiation absorbed (fAPAR).

The most longstanding optical multispectral remote sensing products mobilized for this type of procedure were drawn from Landsat images to describe land use, and AVHRR sensors to derive the LAI from the NDVI at 1.1 km resolution (after resampling, since if the AVHRR at nadir = 1.1 km, it reaches ~6 km in oblique view at the edge of the image). Thus, Running *et al.* [RUN 89] used this product combination to parameterize and model evapotranspiration (ET) and net photosynthesis (NPP) with the aid of the FOREST-BGC ecosystem process model in an area mainly forested with conifers. This same model was calibrated and applied to oak trees in a Mediterranean region [CHI 02]: the temporal profiles of the leaf area index (LAI) was calculated both by direct measurement and by processing of the NDVI data from AVHRR. In this example, the seasonal LAI profiles taken from NDVI data have enabled improvements in the performance of the model, through greater consideration of the typical effects of water shortage in Mediterranean regions.

This work was at first limited to a scale of one kilometer because of the limited products available (AVHRR) whose revisit frequency was compatible with the need to characterize intra-annual dynamics. The development of MODIS and VEGETATION data has led to a refinement of the scales over which spatial modeling can be parameterized, still in combination with other additional variables and parameters provided through geographical information systems.

The elaboration of multitemporal biophysical products ready for use in the scientific community is also crucial to progress in: LAI, fAPAR, Fcover [BAR 07], phenology, albedo, NPP and others. Some effort to validate these products has been made, but it is difficult because of the gap between their spatial resolutions and the small dimensions of the surfaces where the biophysical variables can be measured on the ground.

As regards productivity, a NPP product MOD17, 1 km resolution) is calculated directly from MODIS data, which some authors have compared to measurements obtained from flux towers or modeling in different climates, and for contrasting types of vegetation, including deciduous broadleaf and coniferous forests. The approach uses the concept of light conversion efficiency ε, which expresses the GPP as the product of the photosynthetically active incident radiation (↓PAR), of the fraction of PAR absorbed by the vegetal cover (fAPAR) and of conversion efficiency ε of the PAR absorbed in biomass:

$$GPP = \downarrow PAR \; fAPAR \; \varepsilon \qquad [7.9]$$

The fAPAR is estimated from reflectance measurements by MODIS.

Evaluations of this NPP product are heterogeneous: there is no systematic bias between measured NPP and MODIS NPP, but the differences vary according to forest type, with a significant overestimation in coniferous forests in dry conditions and a more accurate representation in temperate deciduous broad-leaf or boreal forests [TUR 05]. These authors suggest some possible improvements to the MODIS NPP algorithm, using data obtained from flux towers for water and carbon (*eddy covariance*), in order to calibrate the efficiency of light utilization by forest type.

Regarding the fAPAR, an example of validation of results of modeling of CO_2 exchanges with the atmosphere, and measurements in the European flux measurement site network CARBOEUROPE/FLUXNET, has been provided by Reichstein *et al.* [REI 07] in a study of the effects of the extreme drought event combined with a heatwave during the summer 2003, on the productivity and balance of carbon in European forests. The fAPAR, MODIS and AVHRR products, aggregated into 10km square blocks, highlight a negative anomaly in the fraction of photosynthetically active radiation absorbed in summer 2003, unique in its intensity over the period from 1980 to 2003, more than five standard deviations from previous years in the same season. The deficit of absorption of the PAR, thus of photosynthesis, reflects the decrease in canopy conductance caused by the severe soil water deficit. This deficit is found in the data obtained from the flux towers (13 forest sites), which demonstrate a sharp negative anomaly of the primary productivity (GPP) during summer 2003, with strongly correlated anomalies in carbon and water fluxes, attributed to the soil water deficit rather than the high temperatures. The inter-comparisons carried out by country and type of ecosystem show a strong agreement between the approaches able to describe the spatial distribution of the sensitivity of the ecosystem to the climatic event of 2003. This may be seen in Figures 7.28 and 7.29. Figure 7.28 shows the temporal extension of the spatial anomaly of the fAPAR fraction as observed by the MODIS sensor during summer 2003. Figure 7.29 compares the summer deviations of the fAPAR by country with those of the 4 years before and after, for the MODIS measurements and the simulations obtained from a model based on the soil-plant-atmosphere exchange processes (ORCHIDEE). But the modeled and observed fAPARs do not entirely agree. In fact, the model does not take certain effects into consideration, for example, the contribution of the understorey and the occurrence of large-scale fires. Some possible errors in the determination of the land use class can also be implicated.

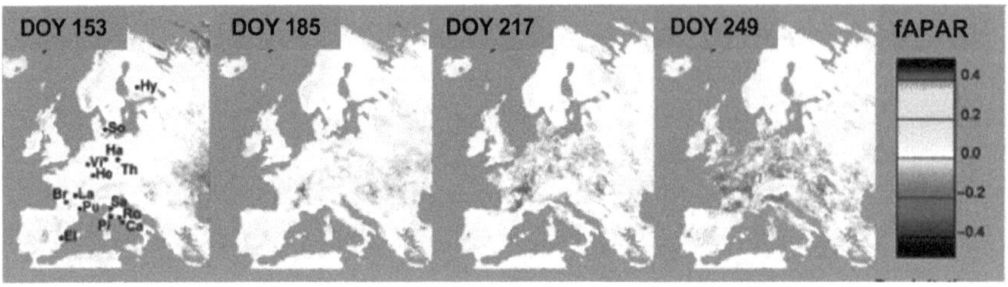

Figure 7.28. *Temporal development (from left to right) of the spatial anomaly of the fraction of photosynthetically active radiation absorbed (fAPAR) during the extreme drought event of summer 2003 in Europe [REI 07]. The fAPAR is that of the MOD15A2 product estimated using MODIS sensor measurements (1,000 m, 16 day periodicity). The location of flux tower sites are shown on the map to the left. DOY: day of year. For a color version of this figure, see www.iste.co.uk.baghdadi3.zip*

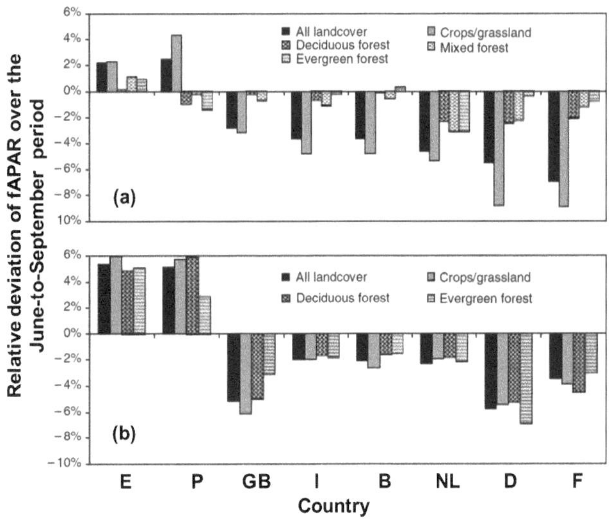

a) The fAPAR of the MOD15A2 product obtained from MODIS measurements. The land use classes are given by the MODIS MOD12Q1 product.

b) The fAPAR obtained from the ORCHIDEE model by functional type of vegetation corresponding to previous land use classes

S: Spain; P: Portugal; GB: Great Britain; I: Italy; B: Belgium; N: Netherlands; G: Germany; F: France

Figure 7.29. *Impact of the extreme drought event of summer 2003 in Europe on the fraction of photosynthetically active radiation absorbed (fAPAR). Relative deviation in fAPAR over the June to September period, 2003 compared with the years 2000, 2001, 2002 and 2004 according to the cover type of the country [REI 07]. It is defined as* $= \frac{fAPAR2003\text{-}average(fAPAR2000, 2001, 2002, 2004)}{average(fAPAR2000, 2001, 2002, 2004)}$

The components of water and carbon balance in forests estimated using remote sensing products are derived from simplified models applicable over large scales of contrasting climate and vegetation. Furthermore, if the temporal series products obtained from medium and low resolution remote sensing are satisfactory at the landscape scales or above, it is because their spatial and temporal smoothing related to pre-treatments (geometric correction, correction of directional effects...), and the large size of the pixel, conceal local heterogeneities. Indeed, the most successful results in terms of applications are obtained on a spatial aggregation or grouping by categories (by area, region, type of forest, for example) of variables predicted. In the future, with the development of Sentinel-2 images (of decametric resolution) one can expect a spatial accuracy compatible with the size of stand units, or management entities (by plot). But the large-scale methods developed will be not directly transferable.

7.4. Prospects

This section draws up a summary of the prospects discussed throughout the chapter.

The prospects that could be offered by the Sentinel-2 satellites, the first of which was launched in 2015, are emphasized. Due to their capability of observing each point of the landscape at high temporal frequency (~5 day revisit) and high spatial resolution (10 m in the visible and the near infrared, 20 m for the middle infrared), one can expect a regular and permanent monitoring of all forests at the stand level. An improvement in performance in relation to the existing one is anticipated, notably in terms of detection and dating of drastic and massive changes, but also gradual and diffuse disturbances. The high level of reflectance sampling throughout the seasons should enable us to obtain useful phenological data elements to better differentiate species and species groups. But in the case of repeated cloudiness, it could be inadequate for a phenological dating or a close seasonal monitoring of the functional response of the ecosystems to climate and hazards.

However, in this respect, the continuity of medium spatial resolution missions with a higher revisit capability (1 day) should be an asset, supported by an improvement in resolution (up to 100 m and 300 m with PROBA-V and Sentinel-3, successors of VEGETATION, for example).

Moreover, the temporal series acquired could be exploited in addition to those obtained by Sentinel-2, to maximize the intra-annual sampling of the reflectance at forest stand level. But this requires progress in the methodological developments in this subject.

There are also some conceivable prospects in terms of development of applications based on the complementarity with other optical remote sensing techniques that have not been mentioned here (LiDAR, hyperspectral) or other wavelength domains (radar) and vector types other than satellite (aircraft, drone).

For example, in France, the IGN produces low frequency mapping of forest formations (stand type, dominant species) obtained from single-date aerial photographs. The cartographic accuracy and thematic richness are high, thanks to the resolution of the images and the use of spectral, textural and morphological information by photo-interpretation. The Sentinel-2 data does not enable this precision to be achieved. It introduces the possibility of making annual updates, but with a less detailed typology. The joint use of hyperspectral satellite data (such as HYPERION) which are more episodic, data should also help to better inform species compositions.

This begs the question of the importance of the exploitation of texture or stereoscopy obtained from very high resolution satellite images in the estimation of forest structure variables and timber volume or above-ground standing biomass. Their performances are indeed lower than those achieved with aerial LiDAR data, or with aerial stereoscopic imagery, acquired from instruments on board a drone or plane. The high cost of implementation of these latest techniques, however, is highlighted as an obstacle to operational applications. Satellite imagery thus retains its importance in frequent updating or covering large areas.

Furthermore, the current trend to estimate and monitor the forest resource is to combine various sources of remote sensing and terrestrial data, thus optimizing the potential of each of these. In the coming years, methodological developments will be reinforced in this respect, due to the increasing availability of satellite or aerial data of interest. It is not only LiDAR (terrestrial or airborne for the moment; high spatial resolution satellite LiDAR in the coming decades) that will have an important place in these innovative approaches, but also radar, particularly with the BIOMASS satellite mission (planned to go into orbit for 2020), dedicated to the estimation of forest carbon stocks and their changes due to global changes.

Finally, the growing accumulation of data acquired at all spatial resolutions (sub-metric to kilometric) for over 30 years, as well as the homogeneity of the pre-treatment of this temporal series of images and the increasing availability of quality biophysical products (surface reflectance, LAI, fAPAR, etc.) derived from them, will favor the intensification of operational applications and the studies of the dynamics, especially long-term, of forest ecosystems.

7.5. Key points

The multispectral reflectance acquired on a single date, and for a single geometric viewing configuration, is quite poor in forest information from a quantitative point of view. On the other hand, the analysis of spatial structure (image texture) or stereoscopic observation (digital photogrammetry) provides quantitative information about forest structure and standing biomass. A comparison between two dates, or a temporal monitoring of the growing season, or over the years, provides much information about the spatio-temporal dynamics of forest cover, especially the sudden and drastic changes (due to clearcutting, fires, storms), the seasonal foliar development (phenology) and its disturbances (drought events, pests), and the multi-annual or long-term dynamics of forest ecosystems. Also, multi-source approaches are often used in operational applications, particularly in the estimation of forest resource, with optical multispectral data enriching the *in situ* observations and inventory measurements. But at present the spatial accuracy and the repeatability of satellite observations are still insufficient to develop methods to aid foresters at the level of management units.

7.6. Bibliography

[ADA 95] ADAMS J.B., SABOL D.E., KAPOS V. *et al.*, "Classification of multispectral images based on fractions of endmembers: Application to land-cover change in the Brazilian Amazon", *Remote Sensing of Environment*, vol. 52, pp. 137–154, 1995.

[AHL 06] AHL D., GOWER S., BURROWS S.N. *et al.*, "Monitoring spring canopy phenology of a deciduous broadleaf forest using MODIS", *Remote Sensing of Environment*, vol. 104, no. 1, pp. 88 – 95, 2006.

[BAD 04] BADECK F., BONDEAU A., BÖTTCHER K. *et al.*, "Responses of spring phenology to climate change", *New Phytologist*, vol. 162, pp. 295–309, 2004.

[BAC 12] BACCINI A., GOETZ S.J., WALKER W.S. *et al.*, "Estimated carbon dioxide emissions from tropical deforestation improved by carbon-density maps", *Nature Climate Change*, vol. 2, pp. 182–185, 2012.

[BAN 14] BANSKOTA A., KAYASTHA N., FALKOWSKI M. *et al.*, "Forest monitoring using landsat time-series data – A review", *Canadian Journal of Remote Sensing*, vol. 40, no. 5, pp. 362–384, 2014.

[BAR 07] BARET F., HAGOLLE O., GEIGER B. *et al.*, "LAI, FAPAR, and FCover CYCLOPES global products derived from Vegetation. Part 1: principles of the algorithm", *Remote Sensing of Environment*, vol. 110, pp. 276–286, 2007.

[BAS 14] BASTIN J.F. BARBIER N., COUTERON P. *et al.*, "Aboveground biomass mapping of African forest mosaics using canopy texture analysis: toward a regional approach", *Ecological Applications*, vol. 24, no. 8, pp. 1984–2001, 2014.

[BEC 06] BECK P.S.A., ATZBERGER C., HOGDA K.A. *et al.*, "Improved monitoring of vegetation dynamics at very high latitudes: a new method using MODIS NDVI", *Remote Sensing of Environment*, vol. 100, pp. 321–334, 2006.

[BEA 14] BEAUDOIN A., BERNIER P.Y., GUINDON L. *et al.*, "Mapping attributes of Canada's forests at moderate resolution through kNN and MODIS imagery", *Can. J. For. Res.*, vol. 44, pp. 521–532, 2014.

[BEG 12] BEGUET B., CHEHATA N., BOUKIR S. *et al.*, "Retrieving forest structure variables from Very High Resolution satellite images using an automatic method", *ISPRS Annals of the Photogrammetry, Remote Sensing and Spatial Information Sciences, I-7, XXII Congress of the ISPRS*, Melbourne, Australia, pp. 1–6, 25 August – 1st September, 2012.

[BEG 14a] BEGUET B., GUYON D., BOUKIR S. *et al.*, "Automated retrieval of forest structure variables based on multi-scale texture analysis of VHR satellite imagery", *ISPRS Journal of Photogrammetry and Remote Sensing*, vol. 96, pp. 164–178, 2014.

[BEG 14b] BEGUET B., CHEHATA N., BOUKIR S. *et al.*, "Quantification et cartographie de la structure forestière à partir de la texture des images Pléiades", *Revue Française de Photogrammétrie et de Télédétection (RFPT)*, vol. 208, pp. 83–88, 2014.

[BEG 14c] BEGUET B., Caractérisation et cartographie de la structure forestière à partir d'images satellitaires à très haute résolution spatiale, thesis, University of Bordeaux, 2014.

[BEL 04] BELROSE V., NAGELEISEN L.M., RENAUD J.P., Les conséquences de la canicule et de la sécheresse sur la santé des forêts: Bilan de la santé des forêts en 2003 , Département de la Santé des Forêts, Ministère de l'Agriculture, de l'Alimentation, de la Pêche et de la Ruralité, December 2004.

[BEL 08] BELOUARD T., FOURCADE C., SALVADO A. *et al.*, "Les images satellitaires pour la gestion forestière", *L'IF Inventaire Forestier National*, vol. 18, 2008.

[BEL 15] BELOUARD T., PY N., MAILLET G. *et al.*, "Pinastéréo – Estimation de la hauteur dominante et de la biomasse forestière dans le massif des Landes de Gascogne à partir d'images stéréoscopiques Pléiades", *Revue française de Photogrammétrie et de Télédétection*, vol. 209, pp. 133–139, 2015.

[BOH 12] BOHLIN J., WALLERMAN J., FRANSSON J.E.S., "Forest variable estimation using photogrammetric matching of digital aerial images in combination with a high-resolution DEM", *Scandinavian Journal of Forest Research*, vol. 27, pp. 692–699, 2012.

[BOR 12] BORIES N. SAMALENS J.C., GUYON D. *et al.*, "Monitoring pine defoliation due to the processionary moth at regional scale from modis time series", *IEEE International Geoscience and Remote Sensing Symposium (IGARSS)*, Münich, Germany, pp. 3383–3386, 22-27 July 2012.

[BOU 15] BOUVIER M., DURRIEU S., FOURNIER R.A. *et al.*, "Generalizing predictive models of forest inventory attributes using an area-based approach with airborne LiDAR data", *Remote Sensing of Environment*, vol. 156, pp. 322–334, 2015.

[BRE 02] BREDA N., SOUDANI K., BERGONZINI J.C., *Mesure de l'indice foliaire en forêt* , GIP-ECOFOR, 2002.

[BRE 03] BREDA N., "Ground-based measurements of leaf area index: a review of methods, instruments and current controversies", *Journal of Experimental Botany*, vol. 54, no. 392, pp. 2403–2417, 2003.

[BRE 08] BREDA N., "Leaf Area Index", in JØRGENSEN S.E., FATII B.D. (eds), *Encyclopedia of Ecology*, Elsevier, 2008.

[BRE 99] BREDA N., "L'indice foliaire des couverts forestiers: mesure, variabilité et rôle fonctionnel", *Revue Forestière Française*, vol. 51, no. 2, pp. 135–150, 1999.

[BRE 02] BRÉON F.M., MAIGNAN F., LEROY M. *et al.*, "Analysis of hot spot directional signatures measured from space", *Journal of Geophysical Research-Atmospheres*, vol. 107, no. D16, p. 428, 2002.

[BRI 16] BRIOTTET X., "Radiometry in the optical domain", in BAGHDADI N., ZRIBI M. (eds), *Optical Remote Sensing of Land Surfaces*, ISTE Press, London and Elsevier, Oxford, 2016.

[BRO 00] BROWN L., CHEN J.M., LEBLANC S.G. *et al.*, "A shortwave infrared modification to the simple ratio for LAI retrieval in boreal forests: An image and model analysis", *Remote Sensing of Environment*, vol. 71, no. 1, pp. 16–25, 2000.

[BRO 14] BROSOFSKE K.D., FROESE R.E., FALKOWSKI M.J. *et al.*, "A review of methods for mapping and prediction of inventory attributes for operational forest management", *Forest Science*, vol. 60, pp. 733–756, 2014.

[BRU 98] BRUNIQUEL-PINEL V., GASTELLU-ETCHEGORRY J.P., "Sensitivity of texture of high resolution images of forest to biophysical and acquisition parameters", *Remote Sensing of Environment*, vol. 65, pp. 61–85, 1998.

[CAR 08] CARDOT H., MAISONGRANDE P., FAIVRE R., "Varying-time random effects models for longitudinal data: unmixing and temporal interpolation of remote sensing data", *Journal of Applied Statistics*, vol. 35, no. 8, pp. 827–846, 2008.

[CHE 07] CHEN J.M., RICH P.M., GOWER S.T. *et al.*, "Leaf area index of boreal forests: theory, techniques and measurements", *Journal of Geophysical Research*, vol. 102, no. D24, pp. 29429–29443, 1997.

[CHI 02] CHIESI M., MASELLI F., BINDI M. *et al.*, "Calibration and application of FOREST-BGC in a Mediterranean area by the use of conventional and remote sensing data", *Ecological Modelling*, vol. 154, pp. 251–262, 2002.

[CHO 08] CHOPPING M., MOISEN G., SU L. *et al.*, "Large area mapping of southwestern forest crown cover, canopy height, and biomass using the NASA Multiangle Imaging Spectro-Radiometer", *Remote Sensing of Environment*, vol. 112, pp. 2051–2063, 2008.

[DEL 05] DELBART N., KERGOAT L., LE TOAN T. *et al.*, "Determination of phenological dates in boreal regions using normalized difference water index", *Remote Sensing of Environment*, vol. 97, no. 1, pp. 26–38, 2005.

[DES 06] DESHAYES M., GUYON D., JEANJEAN H. *et al.*, "The contribution of remote sensing to the assessment of drought effects in forest ecosystems", *Annals For. Sci.*, vol. 63, pp. 579–595, 2006.

[DUF 95] DUFRÊNE E., BRÉDA N., "Estimation of deciduous forest leaf area index using direct and indirect methods", *Oecologia*, vol. 104, pp. 156–162, 1995.

[FIS 06] FISHER J.I., MUSTARD J.F., VADEBONCOEUR M.A., "Green leaf phenology at landsat resolution: scaling from the field to the satellite", *Remote Sensing of Environment*, vol. 100, pp. 265–279, 2006.

[FIS 07] FISHER J.I., RICHARDSON A.D., MUSTARD J.F., "Phenology model from surface meteorology does not capture satellite-based greenup estimations", *Global Change Biology*, vol. 13, no. 3, pp. 707–721, 2007.

[FRA 14] FRAZIER R.J., COOPS N.C., WULDER M.A. *et al.*, "Characterization of aboveground biomass in an unmanaged boreal forest using landsat temporal segmentation metrics", *ISPRS Journal of Photogrammetry and Remote Sensing*, vol. 92, pp. 137–146, 2014.

[GAO 06] GAO F., MASEK J., SCHWALLER M. *et al.*, "On the blending of the landsat and MODIS surface reflectance: predicting daily landsat surface reflectance", *IEEE Transactions on Geosciences and Remote Sensing*, vol. 44, pp. 2207–2218, 2006.

[GAR 08] GARRIGUES S., LACAZE R., BARET F. *et al.*, "Validation and Intercomparison of Global Leaf Area Index Products Derived From Remote Sensing Data", *Journal of Geophysical Research*, vol. 113, p. G02028, 2008.

[GOW 99] GOWER S.T., KUCHARIK C.J., NORMAN J.M. "Direct and indirect estimation of leaf area index, fAPAR, and net primary production of terrestrial ecosystems", *Remote Sens. Environ.*, vol. 70, pp. 29–51, 1999.

[GUY 01] GUYON D., COURRIER G., BERBIGIER P., "Sensitivity analysis of coniferous reflectance with canopy structure and undergrowth characteristics from satellite data and modelling (case study: Landes Maritime Pine Forest)", *8th International Symposium Physical Measurements & Signatures in Remote Sensing*, Aussois, France, pp. 387–392, 8-12 January 2001.

[GUY 11] GUYON D., GUILLOT M., VITASSE Y. *et al.*, "Monitoring elevation variations in leaf phenology of deciduous broadleaf forests from SPOT/VEGETATION time-series", *Remote Sens. Environ.*, vol. 15, no. 2, pp. 615–627, 2011.

[GUY 14] GUYON D., DAYAU S., KRUSZEWSKI A. *et al.*, "Near-surface remote sensing observations for monitoring deciduous broadleaf forest species phenology", *IEEE International Geoscience and Remote Sensing Symposium (IGARSS)*, Québec, Canada, 13–18 July 2014.

[GUY 15] GUYON D., LAVENTURE S., BÉLOUARD T. *et al.*, "Retrieving the stand age from a retrospective detection of multinannual forest changes using landsat data. Application on the heavily managed maritime pine forest in southwestern France from a 30-year landsat time-series (1984-2014)", *IEEE International Geoscience and Remote Sensing Symposium (IGARSS)*, Milan, Italy, 27-31 July 2015.

[GUY 96] GUYON D., RIOM J., JOLLY A. *et al.*, "Evaluation du moyen infrarouge pour la mise en évidence des coupes et éclaircies en forêt landaise (préparation à l'utilisation des données SPOT 4-MIR)", *Conférence internationale "D'une décennie de réalisations... à une décennie de promesses*, Paris, pp. 591–594, 15–18 April 1996.

[GUYO 89] GUYOT G., GUYON D., RIOM J., "Factors affecting the spectral response of forest canopies: a review", *Geocarto International*, vol. 3, pp. 3–18, 1989.

[HAN 13] HANSEN M.C., POTAPOV P.V., MOORE R. *et al.*, "High-Resolution Global Maps of 21st-Century Forest Cover Change", *Science*, vol. 342, no. 6160, pp. 850–853, 2013.

[HAR 73] HARALICK R.M., SHANMUGAN K., DINSTEIN I., "Texture features for image classification", *IEEE Transactions Systems, Man and Cybernetics*, pp. 610–621, 1973.

[HIL 09] HILKER T., WULDER M.A., COOPS N.C. *et al.*, "Generation of dense time series synthetic Landsat data through data blending with MODIS using a spatial and temporal adaptive reflectance fusion model", *Remote Sensing of Environment*, vol. 113, pp. 1988–1999, 2009.

[HE 12] HE L., CHE, J.M., PISEK J. *et al.*, "Global clumping index map derived from the MODIS BRDF product", *Remote Sensing of Environment*, vol. 119, pp. 118–130, 2012.

[HUA 10] HUANG C., COWARD S.N., MASEK J.G. *et al.*, "An automated approach for reconstructing recent forest disturbance history using dense Landsat time series stacks", *Remote Sensing of Environment*, vol. 114, pp. 183–198, 2010.

[JAC 09] JACQUEMOUD S., VERHOEF W., BARET F. *et al.* "PROSPECT + SAIL models: A review of use for vegetation characterization", *Remote Sensing of Environment*, vol. 113, pp. S56–S66, 2009.

[JAR 12] JÄRNSTEDT J., PEKKARINEN A., TUOMINEN S. *et al.*, "Forest variable estimation using a high-resolution digital surface model", *ISPRS Journal of Photogrammetry and Remote Sensing*, vol. 74, pp. 78–84, 2012.

[JOL 93] JOLLY A., Estimation par télédétection satellitaire de la récolte annuelle en bois dans la futaie pure de pin maritime du massif des Landes de Gascogne, Apports pour la prévision de la ressource forestière, PhD Thesis, University Paul Sabatier Toulouse, 1993.

[JOL 96] JOLLY A., GUYON D., RIOM J., "Utilisation des données du moyen infrarouge de Landsat TM pour la mise en évidence des coupes rases sur le Massif forestier landais", *International Journal of Remote Sensing*, vol. 17, no. 18, pp. 3615–3645, 1996.

[JON 04a] JONCKHEERE I., FLECK S., NACKAERTS K. *et al.* "Review of methods for *in situ* leaf area index determination - Part I. Theories, sensors and hemispherical photography", *Agricultural and Forest Meteorology*, vol. 121, nos. 1–2, pp. 19–35, 2004.

[JON 04b] JÖNSSON P., EKLUNDH L., "TIMESAT - a program for analyzing time-series of satellite sensor data", *Computers and Geosciences*, vol. 30, pp. 833–845, 2004.

[JON 10] JÖNSSON A.M., EKLUNDH L., HELLSTROM M. *et al.* "Annual changes in MODIS vegetation indices of Swedish coniferous forests in relation to snow dynamics and tree phenology", *Remote Sensing of Environment*, vol. 114, no. 11, pp. 2719–2730, 2010.

[KAY 06] KAYITAKIRE F., HAMEL C., DEFOURNY P. "Retrieving forest structure variables based on image texture analysis and IKONOS-2 imagery", *Remote Sensing of Environment*, vol. 102, pp. 390–401, 2006.

[KEN 07] KENNEDY R.E., COHEN W.B., SCHROEDER T.A., "Trajectory-based change detection for automated characterization of forest disturbance dynamics", *Remote Sensing of Environment*, vol. 110, pp. 370–386, 2007.

[KEN 10] KENNEDY R.E., YANG Z., COHEN W.B., "Detecting trends in forest disturbance and recovery using yearly landsat time series: 1. LandTrendr - Temporal segmentation algorithms", *Remote Sensing of Environment*, vol. 114, pp. 2897–2910, 2010.

[KEN 14] KENNEDY R.E., ANDRÉFOUËT S., COHEN W.B. *et al.*, "Bringing an ecological view of change to landsat-based remote sensing", *Front Ecol. Environ.*, vol. 12, no. 6, pp. 339–346, 2014.

[KOC 10] KOCH B., "Status and future of laser scanning, synthetic aperture radar and hyperspectral remote sensing data for forest biomass assessment", *ISPRS Journal of Photogrammetry and Remote Sensing*, vol. 65, pp. 581–590, 2010.

[LEB 99] LEBLANC S.G., BICHERON P., CHEN J.M. *et al.*, "Investigation of directional reflectance in boreal forests with an improved four-scale model and airborne POLDER data", *IEEE Transactions on Geoscience and Remote Sensing*, vol. 37, no. 3, pp. 1396–1414, 1999.

[LEM 08] LE MAIRE G., FRANÇOIS C., SOUDANI K. *et al.*, "Calibration and validation of hyperspectral indices for the estimation of broadleaved forest leaf chlorophyll content, leaf mass per area, leaf area index and leaf canopy biomass", *Remote Sensing of Environment*, vol. 112, no. 10, pp. 3846–3864, 2008.

[MCR 07] MCROBERTS R.E., TOMPPO E., "Remote Sensing Support for national forest inventories", *Remote Sensing of Environment*, vol. 110, pp. 412–419, 2007.

[MON 09] MONTPIED P., GRANIER A., DREYER E., "Seasonal time-course of gradients of photosynthetic capacity and mesophyll conductance to CO_2 across a beech (*Fagus sylvatica* L.) canopy", *Journal of Experimental Botany*, vol. 60, no. 8, pp. 2407–2418, 2009.

[NIL 08] NILSON T., SUVISTE S., LUKK T. *et al.*, "Seasonal reflectance course of some forest types in Estonia from a series of Landsat TM and SPOT images and via simulation", *International Journal of Remote Sensing*, vol. 29, nos. 17–18, pp. 5073–5091, 2008.

[NIL 09] NILSSON M., HOLM S., WALLERMAN J. *et al.*, "Estimating annual cuttings using multi-temporal satellite data and field data from the Swedish NFI", *International Journal of Remote Sensing*, vol. 30, no. 19, pp. 5109–5116, 2009.

[NIL 12] NILSON T., RAUTIAINEN M., PISEK J. *et al.*, "Seasonal reflectance courses of forests, new advances and contributions to forestry research", in ANDREW A., OTENG-AMOAKO (ed.), available from: http://www.intechopen.com/books/new-advances-and-contributions-to-forestry-research/seasonal-reflectance-course-of-forests, 2012.

[PAR 09] PARDE J., BOUCHON J., *Dendrométrie*, 2nd edition, AgroParisTech 2009.

[PFL 12] PFLUGMACHER D., COHEN W.B., KENNEDY R.E. "Using landsat-derived disturbance history (1972–2010) to predict current forest structure", *Remote Sensing of Environment*, vol. 122, pp. 146–165, 2012

[PIS 12] PISEK J., RAUTIAINEN M., HEISKANEN J. *et al.*, "Retrieval of seasonal dynamics of forest understory reflectance in a Northern European boreal forest from MODIS BRDF data", *Remote Sensing of Environment*, vol. 117, pp. 464–468, 2012.

[PLO 12] PLOTON P., PELISSIER R., PROISY C., "Assessing aboveground tropical forest biomass using Google Earth canopy images", *Ecological Applications*, vol. 22, pp. 993–1003, 2012.

[PRO 07] PROISY C., COUTERON P., FROMARD F., "Predicting and mapping mangrove biomass from canopy grain analysis using Fourier-based textural ordination of IKONOS images", *Remote Sensing of Environment*, vol. 109, pp. 379–392, 2007.

[RAU 11] RAUTIAINEN M., MOTTUS M., HEISKANEN J. *et al.*, "Seasonal reflectance dynamics of common understory types in a northern European boreal forest", *Remote Sensing of Environment*, vol. 115, no. 12, pp. 3020–3028, 2011.

[REI 07] REICHSTEIN M., CIAIS P., PAPALE D. *et al.*, "Reduction of ecosystem productivity and respiration during the European summer 2003 climate anomaly: a joint flux tower, remote sensing and modelling analysis", *Global Change Biology*, vol. 13, no. 3, pp. 634–651, 2007.

[RIC 07] RICHARDSON A.D., JENKINS J.P., BRASWELL B.H. *et al.*, "Use of digital webcam images to track spring green-up in a deciduous broadleaf forest", *Oecologia*, vol. 152, no. 2, pp. 323–334, 2007.

[RUN 89] RUNNING S.W., NEMANI R.R., PETERSON D. *et al.*, "Mapping regional forest evapotranspiration and photosynthesis by coupling satellite data with ecosystem simulation", *Ecology*, vol. 70, pp. 1090–1101, 1989.

[SAM 12] SAMALENS J.C., GUYON D., BORIES N. *et al.*, "Satellite-based forest health monitoring using coarse resolution data: focus on the 2003 and 2011 droughts in France", *IEEE International Geoscience and Remote Sensing Symposium (IGARSS)*, Münich, Germany, pp. 3367–3370, 22-27 July 2012.

[SOM 11] SOMERS B., ASNER G.P., TITS L. *et al.*, "Endmember variability in spectral mixture analysis: A review", *Remote Sensing of Environment*, 115, pp. 1603−1616, 2011.

[SOU 08] SOUDANI K., LE MAIRE G., DUFRENE E. *et al.*, "Evaluation of the onset of green-up in temperate deciduous broadleaf forests derived from Moderate Resolution Imaging Spectroradiometer (MODIS) data", *Remote Sensing of Environment*, vol. 112, pp. 2643–2655, 2008.

[SOU 12] SOUDANI K., HMIMINA G., DELPIERRE N. *et al.* "Ground-based Network of NDVI measurements for tracking temporal dynamics of canopy structure and vegetation phenology in different biomes", *Remote Sensing of Environment*, 123, pp. 234-245, 2012.

[SPA 90] SPANNER M.A., PIERCE L.L., PETERSON D.L. *et al.*, "Remote sensing of temperate coniferous forest leaf area index. The influence of canopy closure, understory vegetation and background reflectance", *International Journal of Remote Sensing*, vol. 11, no. 1, pp. 95–111, 1990.

[STA 09] STACH N., DESHAYES M. "Estimation des dégâts de tempête: l'œil, l'avion et le satellite", BIROT Y., LANDMANN G., BONHEME I. (eds), *La forêt face aux tempêtes,* Edition Quae, 2009.

[STA 05] STACH N., DESHAYES M., DURRIEU S., "Mapping clear-cutting in French forests by satellite remote sensing", *Proceedings of ForestSat, Borås, Swedish National Board of Forestry*, May 31–June 3, 2005.

[STE 08] STENBERG P., MÕTTUS M., RAUTIAINEN M., "Modeling the spectral signature of forests: application of remote sensing models to coniferous canopies", in LIANG S. (ed.), *Advances in Land remote Sensing: System, Modeling, Inversion and Application*, Springer-Verlag, New York, 2008.

[STO 08] ST-ONGE B., HU Y., VEGA C., "Mapping the height and above-ground biomass of a mixed forest using LiDAR and stereo Ikonos images", *International Journal of Remote Sensing*, vol. 29, no. 5, pp. 1277–1294, 2008.

[TOM 08] TOMPPO E., OLSSON H., GORAN S. *et al.*, "Combining national forest inventory field plots and remote sensing data for forest databases", *Remote Sensing of Environment*, vol. 112, no. 5, pp. 1982–1999, 2008.

[TOW 12] TOWNSHEND J.R., MASEK J.G., HUANG C. *et al.*, "Global characterization and monitoring of forest cover using landsat data:opportunities and challenges", *International Journal of Digital Earth*, vol. 5, no. 5, pp. 373–397, 2012.

[TOWN 12] TOWNSEND P.A., SINGH A., FOSTER J.R. *et al.*, "A general landsat model to predict canopy defoliation in broadleaf deciduous forests", *Remote Sensing of Environment*, vol. 119, pp. 255–265, 2012.

[TUR 05] TURNER D.P., RITTS W.D., COHEN W.B. *et al.*, "Site-level evaluation of satellite-based global terrestrial gross primary production and net primary production monitoring", *Global Change Biology*, vol. 11, no. 4, pp. 666–684, 2005.

[VER 10] VERBESSELT J., HYNDMAN R., NEWNHAM G. *et al.*, "Detecting trend and seasonal changes in satellite image time series", *Remote Sensing of Environment*, vol. 114, pp. 106–115, 2010.

[WEI 04] WEISS M., BARET F., SMITH G.J. *et al.*, "Review of methods for in situ leaf area index (LAI) determination. Part II. Estimation of LAI, errors and sampling", *Agricultural and Forest Meteorology*, vol. 121, nos. 1–2, pp. 37–53, 2004.

[WID 07] WIDLOWSKI J.L., TABERNER M., PINTY B. *et al.*, "The third RAdiation transfer Model Intercomparison (RAMI) exercise: Documenting progress in canopy reflectance models", *Journal of Geophysical Research*, vol. 112, p. D09111, 2007.

[WIL 91] WILLIAMS D., "A comparison of spectral reflectance properties at the needle, branch, and canopy level for selected conifer species", *Remote Sensing of Environment*, vol. 35, pp. 79–93, 1991.

[YAU 08] YAUSCHEW-RAGUENES N., GUYON D., KRUSZEWSKI A. *et al.*, "Discrimination of the respective contributions of understory and tree canopy to the seasonal dynamics of reflectance of the maritime pine forest in the Southwest of France", *IEEE International Geoscience and Remote Sensing Symposium (IGARSS'08)*, Boston, Massachusetts, USA, vol. 3, pp. III-808–III-811, 6–11 July 2008.

[YAU 12] YAUSCHEW-RAGUENES N., Dynamique saisonnière de la végétation forestière (arbres et sous-bois) dans le massif des Landes - Application de la télédétection optique au suivi des hétérogénéités à l'échelle régionale, PhD thesis, University of Bordeaux, 2012.

[ZHA 03] ZHANG X., FRIEDL M.A., SCHAAF C.B. *et al.*, "Monitoring vegetation phenology using MODIS", *Remote Sensing of Environment*, vol. 84, pp. 471–475, 2003.

[ZHU 14] ZHU X., LIU D., "Accurate mapping of forest types using dense seasonal landsat time-series", *ISPRS Journal of Photogrammetry and Remote Sensing*, vol. 96, pp. 1–11, 2014.

8

Characterization of Forests with LiDAR Technology

8.1. Introduction

Forest surfaces on a global scale represent one-third of the world land area [HAN 10]. Human civilizations have always interacted closely with these ecosystems, notably through the broad range of environmental services that they provide: protection of water resources, soil protection, mitigation of the excesses of local climate, reduction of impacts of gas emissions, and conservation of natural habitat and biodiversity [ALE 97]. Apart from these environmental services, forests are also of interest to the social life of these civilizations, especially through economic and recreational functioning, as well as their cultural dimension. This interest in forest ecosystems results, in particular, in a desire to monitor or even control them. In the 20th Century, the advent of aerial and satellite imagery revolutionized the traditional image of the forestry officer, patiently counting and measuring his trees on the ground.

Remote sensing technologies provide quick and repeated information about extensive areas, which cannot be matched by the scale of perception of an individual on the ground. The first applications of remote sensing to forestry were based largely on passive optical systems such as aerial or satellite imagery (*Landsat Thematic Mapper*, [GOW 97]) or, to a lesser

Chapter written by Adrien MICHEZ, Sébastien BAUWENS, Stéphanie BONNET and Philippe LEJEUNE.

extent, on active radar sensors, including ERS, ASAR/ENVISAT, ALOS and RADARSAT [KAS 97].

The lightning development of LiDAR technology during the past 15 years marks a real turning point in the potential of three-dimensional description of forest cover. This three-dimensional description has been valued mainly for the purpose of quantitative characterization of the forest, through the construction of models linking the LiDAR data with conventional dendrometric variables (stand height, basal area, number of stems, biomass, etc.). Within the context of biomass estimation, LiDAR data have enabled successful models of ecosystems with high levels of biomass to be constructed without the saturation phenomenon typically observed with traditional remote sensing data (optical and RADAR data) [KOC 10, ZOL 13].

8.2. The LiDAR technology

The applications of LiDAR technology in forestry can be gathered into two main approaches based on the management of the transmitted laser pulse: full waveform LiDAR and discrete return LiDAR. From a given laser impulse (Figure 8.1), the discrete return LiDAR sensors perform a synthesis of the return signal corresponding to the various surfaces encountered by the beam, this synthesis resulting in a series of points. Full waveform sensors are still less used in forestry [VAN 10]. They store the entire return signal without the loss of information relating to the point discretization stage that is inherent to the discrete return sensors. Full waveform LiDAR sensors detect and record the amount of energy reflected over time, represented in the form of a continuous curve (Figure 8.1).

In addition to the signal management method (discrete vs. full wave return), LiDAR applications are also differentiated by the size of the laser footprint on the ground. One can distinguish small footprints, of less than 1 m and generally corresponding to discrete return sensors, and large footprints, between 10 and 70 m, implemented in full wave sensors [MEA 99]. Given the less frequent use of the second category, the content of this chapter is mainly devoted to small footprint discrete return LiDAR. In order to simplify the reading of this chapter, if authors do not specifically mention the type of LiDAR survey, they refer to small footprint discrete return

LiDAR. For details relating to LiDAR technology itself, the reader should consult [CHA 16] and [MAL 16].

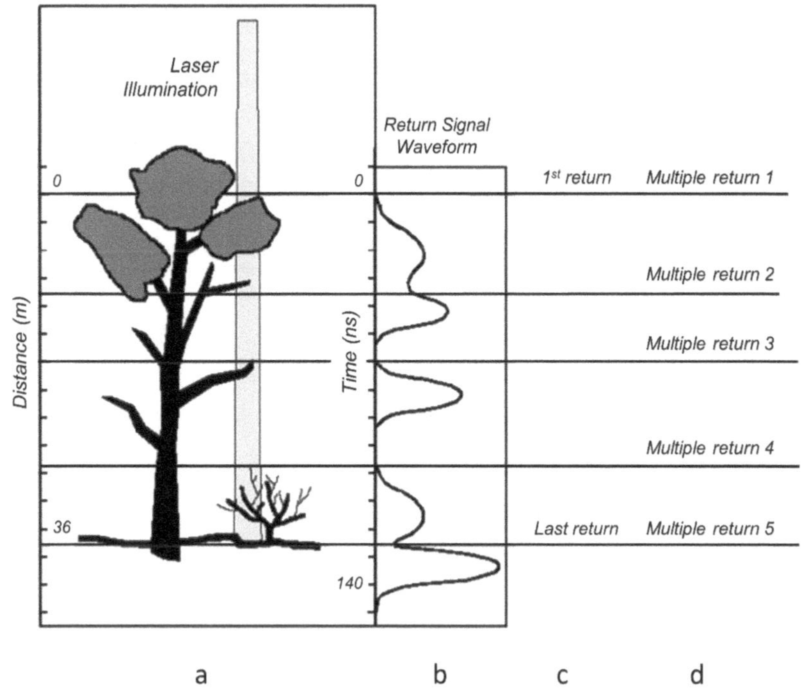

Figure 8.1. *a) Illustration (adapted from [LEF 02]) of various signal management methods for a hypothetical LiDAR impulse on a simplified tree; b) curve representing the return signal as recorded by a full waveform return LiDAR sensor. The succession of peaks and troughs of this curve enable the elements encountered by the laser beam to be identified; c) and d) the discrete return LiDAR sensors perform a simplification of the signal in the form of points corresponding to the main elements encountered by the beam. The systems known as "multi-echo" record not only the first and the last echo, but also the intermediate echoes*

8.3. LiDAR technology in forestry: platforms and applications

The first uses of airborne LiDAR sensors for environmental studies occurred in the 1960s, and involved non-forestry themes such as bathymetry [HIC 69] or the study of sea ice [HIB 75]. These early results mentioned the detection of forests and vegetation, but no special attention was devoted to them. The first forestry applications of LiDAR appeared in the 1980s [ALD 85, ARP 82, MAC 86, NEL 84].

This work was based on profiles captured by an airborne discrete return LiDAR Since these first studies, LiDAR technology was implemented on a wide variety of platforms (Figure 8.2): ground platform (in the case of the terrestrial LiDAR scanner), mobile terrestrial platform, aerial platform (with or without pilot), or on board a satellite. These different possibilities are presented in the following paragraphs and illustrated by four applications (Figure 8.2). These applications are representative of the use of LiDAR technology in forestry for the diffirent study objectives (mapping, resource inventory, ecological monitoring) in relation to platform and investigation scale (local < regional < global).

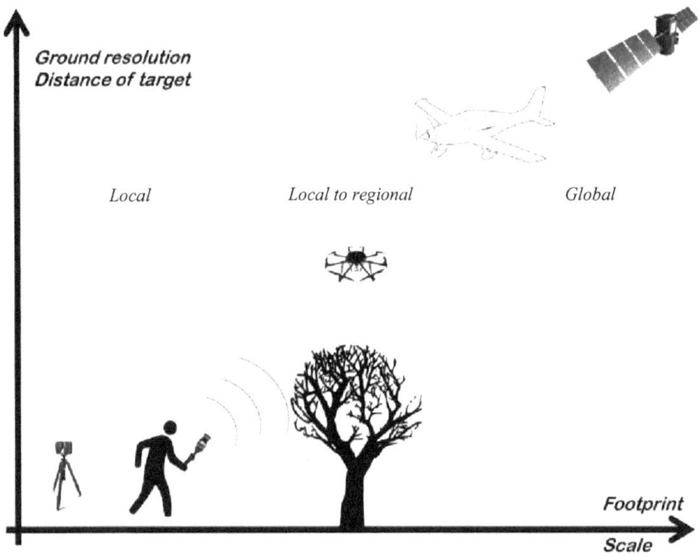

Figure 8.2. *The various platforms able to use LiDAR technology based on spatial resolution (ordinate axis) and scale of application (abscissa): from left to right, fixed and mobile terrestrial LiDAR scanners, aerial (drone and traditional aircraft) and satellite scanners*

8.3.1. *Satellite LiDAR : example of an ICESat project*

8.3.1.1. *State of the art*

The ICES at satellite (*Ice, Cloud and land Elevation Satellite*) was launched by the National Aeronautics and Space Administration of the United States (NASA) in 2003, and is currently the most important operational mission of a spatial LiDAR sensor. This satellite was equipped with a full waveform LiDAR and large footprint sensor (GLAS – *Geoscience Laser Altimeter System*). The GLAS sensor was operating discontinuously from 2003 to 2009. The main mission of the satellite involved mapping the thickness of the polar icecap, studying the altitude of the Earth's surface and observing the quantity of aerosols. Nevertheless, the sensor collected almost 250 million LiDAR beams relating to forest areas. These beams were acquired on the basis of rapid and occasional impulses (5 ns) covering the entire surface of the Earth discontinuously. The footprint of the GLAS LiDAR beam has a width of almost 70 m, the beams being spaced apart by a distance of around 170 m [LEF 10].

The data obtained from the ICESat project aroused enthusiasm amongst the forestry scientific community. It was valued in a number of studies mainly concerned with estimating the height of the forest cover, and the biomass contained within these ecosystems.

Lefsky *et al.* were the first to produce estimates of biomass with the aid of models based mainly on the maximum height, extracted from GLAS data, within temperate (United States) or tropical (Brazil) forests. The height and biomass estimation models constructed as part of this first study achieved a performance (R^2) ranging from 59 to 68%, with a biomass model whose R^2 was 73% in the case of Brazilian tropical forests. Using similar methods in temperate forests, the introduction of topographical indices into some of these models has significantly improved predictive performance [DUN 10, LEG 07, ROS 08].

The spatially discontinuous GLAS data were combined with other sources of data (ground-based and remote sensing) in order to obtain models enabling the modeling of height and forest biomass on a regional scale. Boudreau *et al.* [BOU 08], for example, combined the GLAS data with data obtained from airborne LiDAR sensors, satellite data (Landsat and SRTM – *Shuttle Radar Topography Mission*) together with thematic data aimed at establishing a biomass map on the scale of Quebec (1.3×10^6 km²). Similar methods were also carried out on more restricted areas by Simard *et al.* [SIM 08] and Nelson *et al.* [NEL 09] in Colombia (1.3×10^3 km²) and in Siberia (8.1×10^5 km²). The method developed by Simard *et al.* [SIMS 08] shows originality in basing its models exclusively on a combination of SRTM, GLAS data and reference biomass field measurements without the use of optical satellite data.

The "methodological leap" of making a world map of forests using ICESat data was realized by Lefsky [LEF 10] and Simard *et al.* [SIM 11], who produced a map of the height of the world's forests at a resolution of 500 m. The results obtained by these two studies summarized in application no. 1 (section 8.3.1.2). More recently, these methods have been extended by Baccini *et al.* [BAC 12] and Saatchi *et al.* [SAA 11], who produced forest biomass maps on a pantropical scale. The models used were built on ICESat data and a network of biomass data collected in the field; 4079 plots; in the case of Saatchi *et al.* [SAA 11], and 283 for Baccini *et al.* [BAC 12]) mainly combined with MODIS (500 m) and SRTM (90 m) data. The accuracy of these global approaches on a regional scale is, however, called into question by the literature and still needs improvement [GUI 15, MAR 14, MIT 14].

The encouraging results obtained with the ICESat satellite have convinced NASA to get to work on an improved version of this: an ICESat 2 satellite should be launched in October 2017, for a period of 15 years. However, the purpose of quantification of biomass will again not be central to the mission, causing certain limitations to the applications in this field [GOE 11].

8.3.1.2. *Application 1: forest height on a global scale*

The work of Lefsky [LEF 10] and Simard *et al.* [SIM 11] led to the publication of a first comprehensive mapping of forest height using data obtained from the GLAS LiDAR sensor on board the American ICESat satellite. Over the course of seven years (2003-2009), the GLAS LiDAR mission (fullwave form sensor) collected almost 250 million LiDAR beams in forest areas. Lefsky's study [LEF 10] is based on a forest height model combining spatially discontinuous LiDAR data with various spatially continuous worldwide databases (see Figure 8.3). Besides LiDAR data from the GLAS sensor, this study also relies MODIS data (resolution 500 m). Simard *et al.* [SIM 11] undertook a similar project, combining the same MODIS data with other sources of global and spatially continuous data including the *Worldclim* database (meteorological data), SRTM, protected areas maps, etc.

These studies have led to the first height maps forest environments on a global scale (500 m resolution), according to a method that is unified both in terms of input data and the way in which it is processed (Figure 8.3).

Lefsky [LEF 10] summarizes the extreme forest height values observed, while emphasizing the fact that the estimated height corresponds to the height of the 90th percentile of the heights of trees in the forest patch mapped. As expected, the tallest forest stands are located on the western coast of North America. They consist of conifers such as Sequoia (*Sequoiadendron giganteum*) and Douglas Fir (*Pseudotsuga menziesii*). The observed average heights often exceed 40 m, with some individuals reaching 70 m or higher. The boreal forests of northern Asia, with average heights consistently below 20 m, make up the forest formations of lowest height.

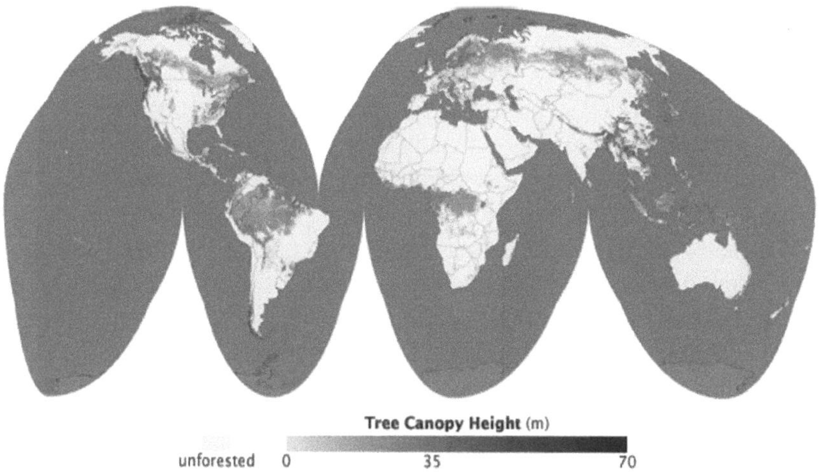

Figure 8.3. *Heights of world forests – the first global map of the heights of the world's forests (0–70 m), derived from the work of Lefsky [LEF 10]. Source: http://www.nasa.gov/topics/earth/features/forest-height-map.html. For a color version of this figure, see www.iste.co.uk/baghdadi/3.zip*

8.3.2. *Airborne LiDAR*

8.3.2.1. *State of the art*

The first forest applications of airborne LiDAR focusing on forest structure description and quantification were developed by Arp *et al.* [ARP 82] in Latin America (Venezuela), by NASA in Pennsylvania [NEL 84] and by the forest services of Canada [ALD 85, MAC 86]. These first studies were limited to a local level, centered on an individual flight line along which vertical profiles were extracted. The emergence of kinematic GPS positioning systems coupled with inertial stations led to the development of scanning sensors (scanning LiDAR), the referencing and merging of different frames of acquisition [WEH 99], as well as the acquisition of spatially continuous LiDAR surveys. In addition to the constantly decreasing costs, the improvement in the referencing of LiDAR acquisitions, coupled with various technological advances (accuracy of GPS, quality of inertial stations, increase in transmission frequency, storage capacities and in-flight processing) has led to the expansion and commercial success of LiDAR technology over the last twenty years [LIM 03, MAS 14].

Since the earliest studies, researchers have tried to enhance the information obtained about the canopy surface and the ground, but also about all the elements of the various vegetation layers in between the ground and the top of the canopy. The purpose was generally to test the link between the variables derived from LiDAR data and the height, biomass or marketable timber volume [ARP 82, MAC 86, NEL 88]. These pioneering studies formed the basis of many others which, during the 1990s, confirmed LiDAR as a source of input data for numerous characterization and quantification models on a tree or plot scale. In addition to the general LiDAR derived mapping products (Figure 8.4), which are digital terrain models (DTMs), digital surface models (DSMs) and canopy height models (CHMs), most of these studies synthesize LiDAR data in the form of different point cloud parameters (Figure 8.5). These parameters are extracted at different scales (individual tree < plots (several trees) < entire area studied). These applications involve, for example, elements related to the vertical distribution of points (such as the standard deviation of heights) or the value of the intensity of the returns (as an average). The models produced by these studies have been mainly related to the height of the canopy (as the individual height and the maximum height), the biomass, or any other variable describing the supply of stock of standing timber (such as the basal area and the number of trees per hectare) [LIM 03]. These modeling approaches reached an operational stage thanks to National Forest Inventory data.

Thanks to the data collected in the field at plot level, in national or regional forest inventories, dendrometric models can be calibrated and validated for large areas. Application no. 2 (section 8.3.2.2) describes a regional mapping project of forest structure variables related to the height of the stands (dominant height). This project makes use of an extensive "field" database obtained from a regional forest inventory. It provides a fine example of a forestry application recognizing the importance of LiDAR data as part of a decision support instrument functioning on the scale of a region of 17, 000 km^2 (see [DED 15] for further details).

In addition to the dendrometric information, LiDAR data makes it possible to evaluate forest ecological parameters, such as the species composition [DON 07]. In this context, the intensity value of the return signal can be interesting if a normalization is performed between the various lines of flight [KOR 10]. Due to the information that it contains in relation to the vertical structure of the forest cover, LiDAR data present valuable input data for the detailed modeling of forest species habitat [GOE 10, NEL 05, VIE 08]. It is also used in the evaluation of the ecological integrity of forest ecosystems [JOH 10, MIC 13]. Application no. 3 (section 8.3.2.3) summarizes a case study carried out on the characterization of the ecological integrity of riparian forests on the basis of an airborne LiDAR dataset.

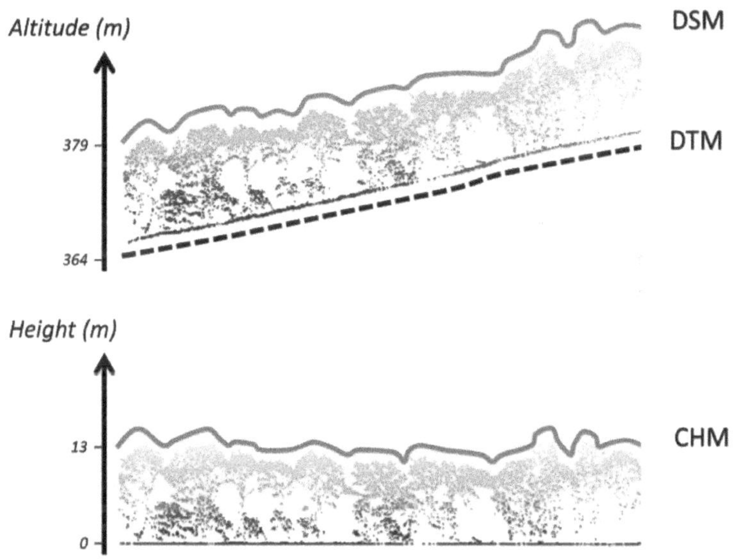

Figure 8.4. *Examples of LiDAR products used in forestry: digital terrain model (DTM), digital surface model (DSM) and canopy height model (CHM). The CHM is the nomalization of the point cloud by means of the DTM. The illustration is based on a discrete LiDAR dataacquired from an airborne. For a color version of this figure, see www.iste.co.uk/baghdadi/3.zip*

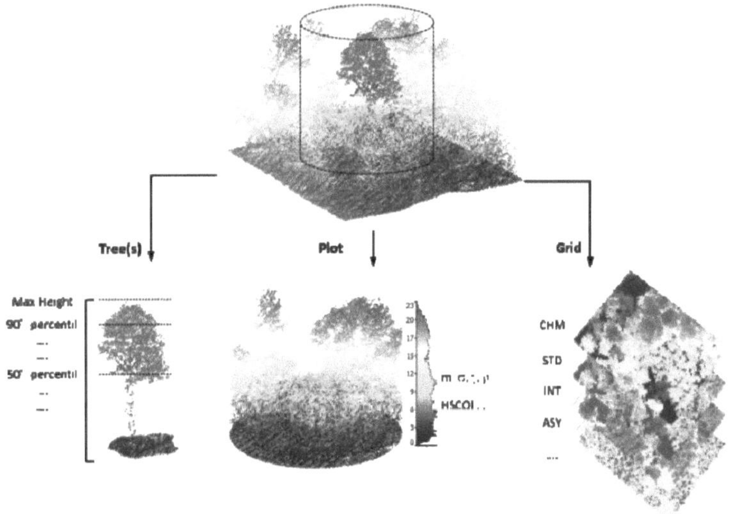

Figure 8.5. *The extraction of variables derived from a LiDAR point cloud can be perfomed at three levels: single tree, tree plot (e.g. circular one) and continuous grid (raster data). These variables, also known as metrics, are mainly statistical parameters calculated on the basis of the distribution of the height of the points: maximum height, height percentile (90th, 50th…), mean (m), standard deviation (STD or σ), kurtosis (γ), skewness (ASY or ρ). The signal intensity (INT), or the number of returns, can also be used in addition to the height. More complex parameters, such as Height-Scaled Crown Openness Index (HSCOI) [LEE 07], can also be implemented. The HSCOI is based on the concept of voxels, and represents a measure of the porosity of the canopy. In the case of a tree or a plot, these variables are calculated for the entire point cloud under consideration. In the case of a raster, these same variables are computed at the level of the pixel. For a color version of this figure, see www.iste.co.uk/baghdadi/3.zip*

8.3.2.2. *Application 2: regional characterization of forests, a case of dominant height in Wallonia, Belgium*

Sampling field inventory still the main source of information for the characterization and monitoring of forest resources.

They generally involve taking dendrometric measurements (diameter at breast height, total height, etc.) of trees within a set number of plots of several acres. These measurements provide variables to describe standing volume stocks, the stand structure, the health condition... The repeating of some measurements over time also makes it possible to calculate indicators of evolution of the timber resource and estimate its production.

The capacity of LiDAR to capture the three-dimensional structure of vegetation has opened up the way for extensive research into the description and quantification of forest environments. In particular, the exploitation of airborne LiDAR data has been an answer to the significant costs of data field acquisition by an individual. The accuracy of dendrometric variables derived from LiDAR data (mainly height) can be considered as equal to, if not better in the case of height, those obtained from field inventory methods [NAE 04, MAL 09].

The Wallonia public services (southern Belgium) have a complete coverage of the area (≈17,000 km^2) with airborne LiDAR data of a density of the order of 1 ground classified point/m^2. On the basis of the raw LiDAR point cloud, two models have been produced with a spatial resolution of 1 m: a DTM describing the elevation of the ground in each pixel, and a DSM describing the altitude of elements present on the ground surface. Subtraction of these two layers enables the production of a CHM. The CHM obtained from LiDAR data represents the maximum height reached by the vegetation in each pixel (Figure 8.4), offering a height measurement of the canopy.

The dominant height is an indicator widely used by forest managers, mainly in forest stands with uniform structure. It is defined as the average height of the 100 largest trees per hectare [RON 93]. Combined with the stand age, it provides a very good indication of the productivity (site index). It can also be involved in the estimation of standing timber volume.

On the basis of the CHM produced on the scale of Wallonia, the dominant trees were identified by the detection of local maxima [POP 04], which assumes that a local maximum of the CHM corresponds to the apex of a tree (Figure 8.6).

An estimation model of the dominant height was produced using the height values of these local maxima, linking them to field height estimations. These total height measurements were conducted on a sub-sample of the plots of the Walloon Regional Forest inventory. The dominant "field" height is consequently based on the average of the 100 highest local maxima per hectare, to which a constant is added. This may be considered as a correction of the underestimation of the dominant height. This underestimation is explained by the fact that the laser impulse rarely reaches the exact position of the apex of the dominant trees, but a part of the crown located on average 1 m lower in the case of spruce, and 1.3 m in the case of Douglas firs. Further analysis of the models failed to show any effects of the slope of the ground or the stand age.

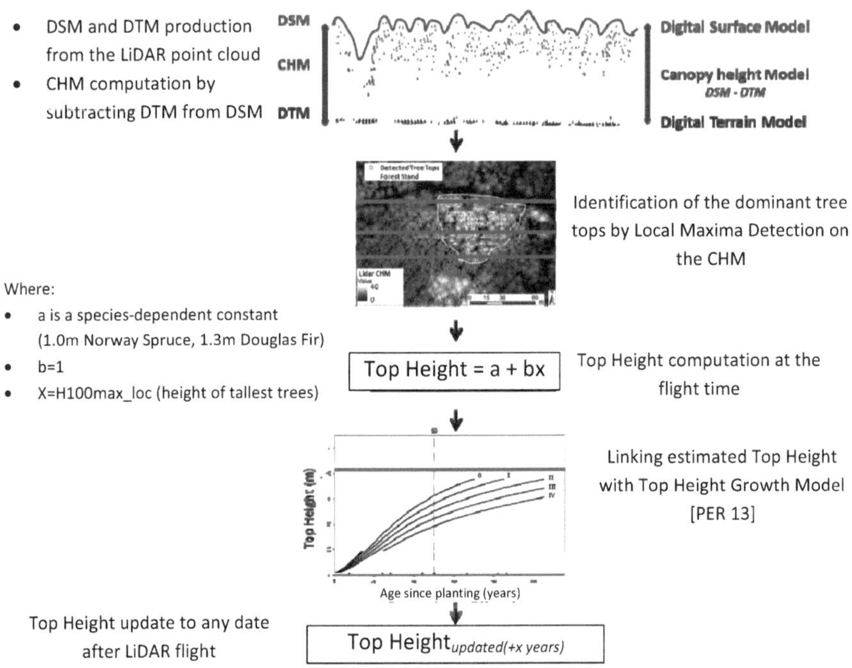

Figure 8.6. *Modeling of the dominant height of Walloon coniferous forests (≈17,000 km²). The dominant height at the time of the LiDAR acquisition is extracted from the point cloud by means of the CHM. The dominant height is then updated by using height growth model. See Dedry et al. [DED 15] for further details*

In order to facilitate the use of this model, it was integrated into the open source environment QGIS (GIS software) throughout a plug-in which dominant height estimation was combined with a growth prediction model to update dominant height [PER 13]. On the basis of information linked to the nature and age of the stand, the plug-in also provides an estimation of productivity in the form of a *Site Index*, corresponding to the dominant height reached at 50 years old.

This example describes how a low density LiDAR dataset, acquired on a regional scale, can be used to characterize the productivity of even-aged coniferous stands. The importance of airborne LiDAR in this application resides in its capacity to deliver useful indicators on a scale adapted to stand management. This kind of application is becoming an accepted addition to the inventories obtained by traditional sampling carried out in the field; in some countries, LiDAR is already used in operational terms for forest inventories [WHI 13, NAE 07, MAL 11, WOO 11].

8.3.2.3. *Application 3: ecological integrity of riparian forest in Wallonia*

As well as locating and describing forest vegetation as a timber resource (height, volume, biomass...), LiDAR data can also provide information relating to the functioning of the ecosystem, such as the ecological integrity.

The concept of ecological integrity can be described as the degree of integrity of a given environment in relation to its capacity to assume all the ecological functionalities associated with it. This application describes how, based solely on indicators derived from LiDAR datasets with a high point density (>40 points/m^2), the ecological integrity of riparian forests may be characterized on a scale that corresponds to a river system of almost 30 km (Table 8.1). The indicators are constructed from spatial analysis of various parameters directly extracted from the LiDAR point cloud: (e.g. CHM and DTM) in order to characterize riparian forests as well as various variables relating to the morphology of the watercourse, such as the water surface and the width of the valley bottom. As an example, Figure 8.7 represents the evolution of the density of the forest cover dominated by exotic conifer stands in the Houille Valley (Namur Province, Belgium). It highlights the presence of coniferous stands at a distance less than the statutory distance of 12 m (*Natura 2000* protection status) on virtually all the study area. Furthermore, the most important rise of density in coniferous stand is recorded within an area of great biological interest.

Riparian forest parameter	Indicator of ecological function
Longitudinal continuity	Dispersion corridor for plants, habitat and migration area for birds and mammals
Overhanging riparian forest	Effect of shade, regulation of temperature, habitat creation and nutrient supply
Density of coniferous stands	Reduction in the stability of riverbanks and biodiversity
Floodplain height above water level	Frequency of flooding, proximity of the water, directly related to the specificity and diversity of riparian area
Riparian forest height (average, variation coefficient)	Location of mature stands (average height); spatial diversity (high variation coefficient) in relation to species diversity

Table 8.1. *Indicators of ecological integrity of riverside forests derived from a LiDAR point cloud (adapted from Michez et al. [MIC 13])*

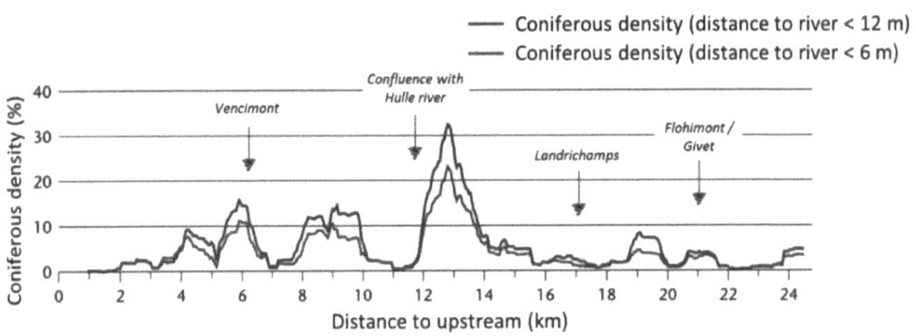

Figure 8.7. *Density of the forest cover dominated by exotic conifer stands (mobile average over 1 km) in the Houille Valley (50°N 4°58E, Namur Province, Belgium) in February 2011 – adapted from Michez et al. [MIC 13]. Abscissa: density of conifers. Ordinate: distance in relation to the upstream point of the study (Gedinne). In italics, particular points along the watercourse studied. For a color version of this figure, see www.iste.co.uk/baghdadi/3.zip*

LiDAR data also make it possible to obtain valuable information about the "topographical proximity" of riparian forests. Such information, in relation to the watercourse and thereby the riparian character, enables us to distinguish the riparian forest from the neighboring stands. The floodplain height above water level (Figure 8.8) was developed in order to express, for a specified part of the riparian forest, its proximity with respect to the base of the riverbank. As well as being an indicator of the typical characteristics of riparian forests, this parameter also provides information on the physical condition of the riverbank and the modified character arising from this. The two cases shown in Figure 8.8 represent contrasting scenarios along the study site. The first scenario (Figure 8.8(a)) corresponds to a section with a higher mean height above water level, following an artificial reshaping of the watercourse and its banks. The second scenario (Figure 8.8(b)) corresponds to a more natural situation, with a lower mean height above water level, and thus a riparian forest in much closer contact with the watercourse.

Figure 8.8. *Two contrasting scenarios of height above water level: Ha > Hb*

8.3.3. *Terrestrial LiDAR*

8.3.3.1. *State of the art*

The use of the terrestrial LiDAR scanner (TLS) in forests follows the introduction onto the market of portable LiDAR scanners with a range exceeding 10 meters [DAS 11]. The difference of TLS compare to other LiDAR systems are its static configuration (system attached to a tripod), the absence of an inertial unit and its capacity to measure the surrounding areas in 3D with millimeter accuracy. There are two categories of TLS, which differ in their measurement methods of the distance traveled by the LiDAR laser: the time of flight (*tof*) TLS and the phase-shift TLS. These two systems differ mainly in the time required to acquire a given resolution (higher for phase-shift TLS), signal range (higher for the *tof* TLS), as well as the type and accuracy of the information collected. Both record the position where the laser is reflected, but some tof TLS may record several returns of one emitted signal or even the full waveform of it, whereas phase-shift TLS record only one return. Moreover, phase-shift TLS suffer from the mixed pixels (ghost points) and range/intensity cross talk effects which may limit the accuracy of the points at the edge of objects. These two systems are explained in more details in Petrie and Toth [PET 09].

TLS technology has quickly established itself as an additional remote sensing tool besides aerial or satellite remote sensing systems the use of TLS in forests can provide a relatively accurate three-dimensional description of the surrounding ecosystem from the ground (Figure 8.9). The forestry sector is starting to be more and more interested by this technique since it makes it possible to acquire non-destructive three-dimensional data on trees or groups of trees with unprecedented precision. The parameters collected or derived from data obtained from TLS can be dendrometric (position, diameter, height, crown radius, volume and biomass of trees) or ecological (leaf area index, distribution of gaps, foliar distribution, distribution of micro-habitats), or even ecophysiological (interaction between the light and the canopy). Some of these parameters are difficult or even impossible to collect by traditional inventory or measurement methods.

The use of TLS as a tool for forest inventory is detailed in the following section. For more information on other uses of this technology in forest environments, we recommend the synthesis work of Dassot *et al.* [DAS 11] and Van Leeuwen and Nieuwenhuis [VAN 10].

8.3.3.2. *TLS as a tool for forest inventory*

The estimation and spatialization of parameters such as timber volume or biomass using aerial or spatial remote sensing data first require a calibration and validation phase. The calibration stage consists of developing models relating attributes acquired or generated by remote sensing to forest parameters deduced from measurements carried out in the field (basal area, volume and biomass).

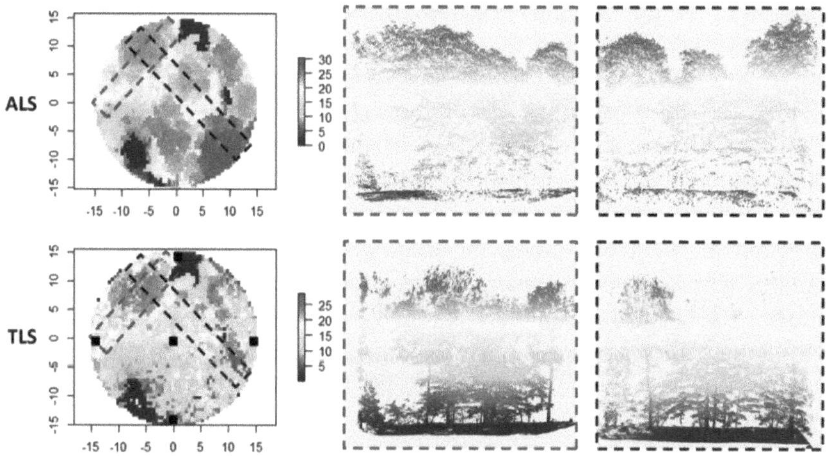

Figure 8.9. *Canopy height models (CHMs) (in m) of an uneven deciduous forest extracted from airborne LiDAR scanning (ALS) and terrestrial LiDAR scanning (TLS). Above, illustrations are derived from CHM and point cloud profiles of high density ALS data (≈45 points/m²). Below, the illustrations are based on data from five TLS scans of a FARO Focus 3D 120 (black squares). For a color version of this figure, see www.iste.co.uk/baghdadi/3.zip*

The samples used in the calibration of models are taken from plots in which various measurements have been carried out on the trees. The most commonly recorded measurements in a conventional field inventory are species, diameter at breast height (DBH) and height. These measurements are then used, by means of allometric equations [8.1], [8.2], to estimate parameters such as timber volume or above-ground biomass. The biomass of a standing tree is then estimated indirectly by means of an equation. This equation is defined either (1) for one species, by relating biomass to one or two parameter(s) measured in the field, such as the DBH (equation [8.1] illustrates the type of equation commonly used in temperate forests [ZIA 05]), or (2) for multiple species integrating other parameters such as tree height and basic wood density (an example of equation [8.2] which is generally used in tropical forests [CHA 14, CHA 05]).

$$AGB = a\,DBH^{b} \quad [8.1]$$

$$AGB = a\,(DBH\,\rho H)^{b} \quad [8.2]$$

with AGB, which represents the above-ground biomass (kg), DBH is the diameter at breast height (cm), ρ is the basic wood density (g/cm³), *H* is the total height of the tree (m), *a* and *b* are statistically adjusted parameters.

TLS offer the possibility of conducting most of these measurements in the field (DBH, height and location of trees) in a way that is objective, accurate and repeatable. It also opens up the possibility of improving the timber volume estimates by a direct estimate of the volume through adjustment of cylinders along the trunk [RAU 13] instead of the indirect estimates with allometric equation. The improvements may also arise from a number of additional parameters that can be measured on each tree making of the plot (diameter at various heights, taper, volume of a section of trunk...). These parameters could ultimately serve as additional input for allometric equations.

The three-dimensional scan of a plot using TLS is carried out according to one of the following three acquisition methods: single-scan, multi-scan and multi-single-scan.

Single-scan involves acquiring one scan of the plot by placing the scanner at its center. The acquisition time is thus minimal (a few minutes). On the other hand, with only one point of view, it does not provide a complete observation of the plot, since some trees or parts of trees are out of the scanner's field of view: this is the occlusion effect. Multi-scans reduce the occlusion by carrying out scans at various places within the plot. However, this approach requires artificial references in the plot (reflective targets or simple geometrical elements: cylinders or spheres) in order to combine the scans. Setting up these reference elements is time-consuming because they must be placed so as to be visible from several positions of the scanner. Finally, the third method, multi-single-scan, involves making several scans of the plot and combining these scans without the use of artificial references, but by using the position of the trees detected on each individual scan [LIA 13]. This method assumes that the base of the trunks have circular cross-sections, with an assumption that is not encountered in all kinds of forests (for example: temperate and tropical deciduous forests).

The point cloud obtained from (x) scans of the plots is then processed automatically or semi-automatically in order to extract the desired dendrometric parameters. An example of the required processing stages for the extraction of these parameters is shown in Figure 8.10. It should be noted that the volume and the biomass of a plot remain difficult to estimate automatically, since the detection of trees, the mitigation of occlusions and the species determination still need improvement, especially in tropical forests. The case study (application 4) shown below illustrates one possible solution to the mitigation of the occlusion effect by using mobile LiDAR scanning.

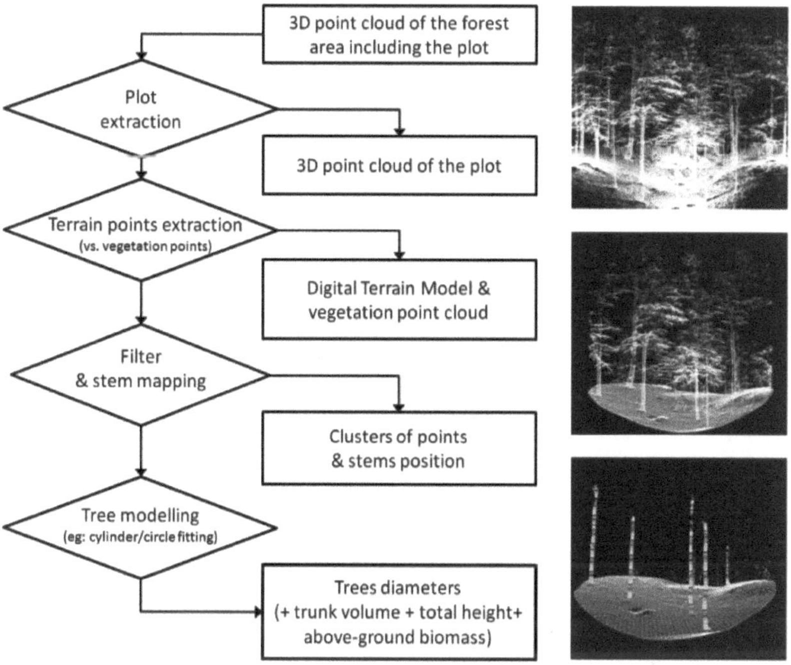

Figure 8.10. *TLS processing stages for the extraction of dendrometric parameters. (1) Extraction of plot within the point cloud, (2) separation of "ground" points from "vegetation" points (generation of DTM), (3) filtering vegetation points (cluster of points and stems positions), (4) tree modeling (cylinder/circle fitting) of clusters identified as one tree for the extractions of diameters as the DBH (+ total height + trunk volume) and, by means of allometric equations, the estimation of above-ground tree volume or biomass*

8.3.3.3. *Application 4: ground-based LiDAR scanner in forest inventories, comparison of acquisition methods*

In recent years, the use of a mobile terrestrial platform has gradually emerged as an alternative to the TLS for forest plots scanning [LIA 14a, LIA 14b, RYD 15]. In fact, the mobile character of the scanner has the advantage of minimizing the occlusion phenomenon and covering a wider area in a given time. The study presented in this section aims to evaluate the potential of the portable MLS as an alternative to the traditional acquisition methods of single-scan and multi-scan. The three acquisition methods tested are thus the single-scan (FARO1 method), multi-scans with five scanner positions, including a central one (FARO5 method) and the mobile scan (ZEB1 method). The names of the methods refer to the scanners used: the FARO Focus 3D 120 for the TLS and the ZEB1 of Geoslam for the MLS. The scans were carried out on 10 circular plots (15 m radius), which differ in terms of forest type (deciduous, coniferous or mixed), stand structure (coppice, even-aged or uneven-aged forest) and topography (flat or sloping ground). The processing was carried out using the Computree open-source sofware (plug-in ONF/ENSAM).

The occlusion effect, which results in partial three-dimensional information of the scanned trees, is more or less an issue according to the acquisition method. In order to evaluate the importance of this occlusion in three scanning methods, a slice 10 cm thickness at a height of 1.3 m above ground level was carried out on the point cloud for each trees of the plots (DBH >10 cm). The trees were then classified regarding the proportion of the scanned section (Figure 8.12).

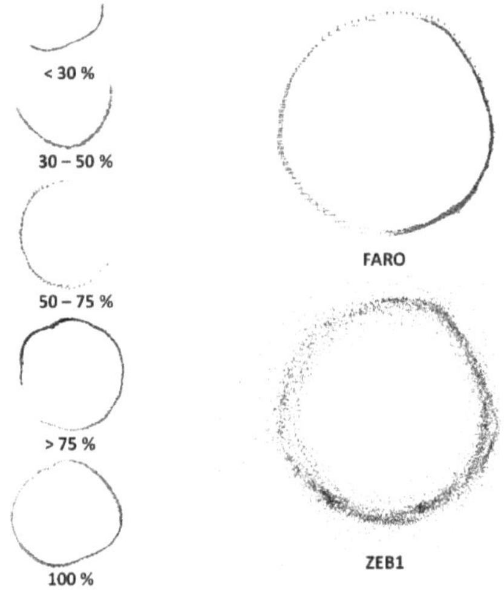

Figure 8.11. *Cross-sections of the scanned trees at a height of 1.3 m above ground-level. On the left, percentage of the cross-section closure. On the right, the precision of positioning of the scanning points for the two technologies used in the study (terrestrial LiDAR scanner: FARO Focus 3D 120; portable mobile LiDAR scanner: ZEB1) is presented*

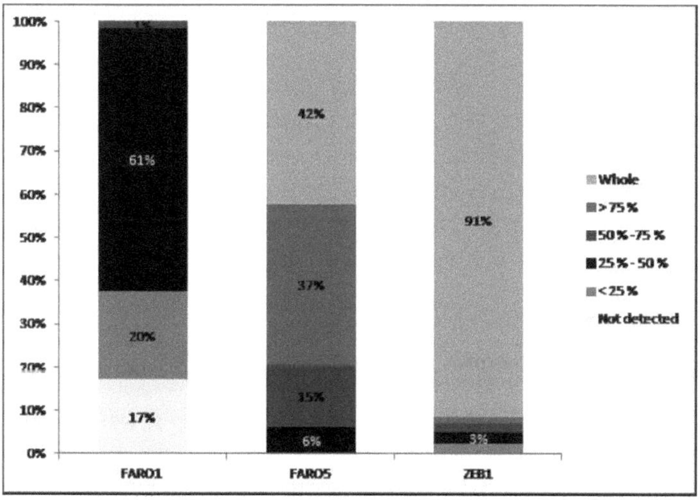

Figure 8.12. *Percentage of the cross-section closure between the three different scanning methods according to the visual interpretation of the point cloud slices at 1.3 m height (thickness of 10 cm) of the eight plots, according to the three acquisition methods: single-scan (FARO1), multi-scans (FARO5) and mobile LiDAR scanner (ZEB1). For a color version of this figure, see www.iste.co.uk/ baghdadi/3.zip*

The estimation of diameter at 1.3 m is, on average, good for multi-scan and MLS acquisitions (bias lower than –0.2 cm and RMSE lower than 1.5 cm in the case of this study) and acceptable for single-scan acquisitions (bias of –1.2 and RMSE of 3.8 cm) for trees detected automatically. However, for the single-scan method, a significant number of the trees are missed (17% in this study; see Figure 8.12). The estimation of forest parameters thus remains limited for this acquisition method. Despite its lower scanning accuracy (Figure 8.11), MLS seems to be a good alternative for the scanning of forest plots, since it leads to diameter estimations similar to those of the multi-scans method, while being more efficient in the field (average acquisition time of 25 min as opposed to 1h 15 min for the multi-scans method). Moreover, the higher coverage of the plots by MLS (better completeness of sections at a height of 1.3 m, Figure 8.12) should provide a better adjustment of the cylinders for trunk volume estimation. Portable MLS should therefore rapidly become a future technology for three-dimensional data acquisition of the forest from the ground.

8.4. Future of LiDAR technology in forestry?

The contribution of LiDAR technology towards the characterization of forest ecosystems is now very widely documented in the literature. The use of LiDAR data have made it possible to obtain estimates of tropical forest biomass with a greater accuracy than ever before [KOC 10]. In the field of forestry, several prospects are worth highlighting, as much for their technological evolution as their method of implementation.

8.4.1. *Technological evolution*

The constant miniaturization of LiDAR sensors is especially encouraging. Several business solutions are now available, involving LiDAR sensors of less than 1 kg in weight and with an operating range of a 100 m (for example, the Velodyne HDL32E and VLP16 LiDAR). Some slightly heavier sensors, but whose mass does not prohibit mounting on an Unmanned Aerial Vehicle (<4 kg) are also being marketed (for example, the Riegl VUX-1). This has a more suitable operating range for forestry applications (flight altitude of 300 m), notably in the case of resource inventories. However, few studies that highlight these newly commercialized LiDAR solutions are currently available [PFE 14, TUL 14]. On a trial basis, some projects have been developed with LiDAR scanners embedded within a UAV (multirotor type), particularly for scanning trees: Lin *et al.* [LIN 11] as well as Wallace *et al.* [WAL 12]. These projects are still very experimental, however, and function over restricted areas (several trees, or perhaps a stand) and further research is required to integrate this technology on a more operational basis.

Furthermore, the recent coming of "flash LiDAR", also known as *Photon Counting Lidar,* could further improve the implementation of LiDAR in a UAV. Functioning in a similar way to a camera (instant recording of photons returning from a LiDAR beam split into different directions by a lens), these sensors have a lower weight and energy consumption than that of traditional LiDAR scanners. This type of system will equip the future satellite ICESAT 2 [MOU 14]. It could potentially be installed on UAV platforms, or on mobile terrestrial LiDAR scanners. Several general pilot studies have already been undertaken [DUO 12, ZHO 12].

8.4.2. *Multispectral LiDAR and combination of ALS LiDAR data with optical data*

The late 2000s saw the emergence of several projects combining the simultaneous acquisition of airborne LiDAR data with other data types, especially from spectrometers. Among these, the results obtained by the Carnegie Airborne Observatory of Stanford University (United States) [ASN 12, ASN 07] proved to be very promising in terms of detailed characterization of complex ecosystems, such as the species composition in tropical rain forests [SOM 15] or the study of ecological traits of forest species on a regional scale [ASN 14].

Multispectral LiDAR systems also constitute an interesting prospect. Given that bathymetric LiDAR sensors have already been using two spectral bands (one green LiDAR sensor and one near-infrared LiDAR sensor) for many years, the multispectral character is not an innovation in itself. The acquisition of a "colored" point cloud on the basis of the information recorded by a combination of LiDAR sensors with different emission frequencies provides a convincing three-dimensional and colored view of the forests observed. The whole colored point cloud could thus provide spectral information about all vertical layers of forest ecosystems (dominant and dominated vegetation), comparable with that obtained using passive remote sensing methods on the canopy, but not limited by the light conditions. This development prospect would facilitate numerous applications, especially regarding a more detailed characterization of the forest vegetation structure (wood vs. leaves, species composition etc.), even for the vegetation beneath the forest cover. Pioneering studies have modeled and tested experimental prototypes with encouraging results. [MOR 09, WAL 14]. The first systems are now available on the market (Optech Titan) and cover three spectral bands, one in the visible region (532 nm) and two in the infrared (1064 and 1550 nm).

8.4.3. *Towards a global mapping of world forests using LiDAR data?*

Whether aerial or satellite, the importance of LiDAR data for the characterization of forest biomass is widely documented in the literature. Comprehensive studies of forest biomass (notably [BAC 12, SAA 11]) have thus far been confined to the spatial extrapolation of LiDAR data obtained from the ICESat satellite. Since ICESat satellites are not operational anymore, any improvement in these results will involve an intensifying of the information on the vertical structure of forests. The ICESat 2 satellite (http://icesat.gsfc.nasa.gov/icesat2/) will make it possible to obtain a new spatially discontinuous coverage of the world's forests by 2017. However, this project, still hypothetical and not specifically dedicated to the study of forests, would only provide a limited improvement on these first models [GOE 10, GOE 11]. The GEDI mission (http://science.nasa.gov/missions/gedi/) of NASA could also offer data in addition to the ICESat 2 satellite, with its main objective being the mapping of biomass by 2018.

Another emerging issue in the scientific community is the coverage of tropical forests using airborne LiDAR data. This application, especially promoted and evaluated by Mascaro *et al.* [MAS 14], would make it possible, in four years, to obtain a global coverage of tropical forests using high resolution airborne LiDAR data, with a budget that represents barely 5% of the United Nations REDD program budget (*Reducing Emissions from Deforestation and Forest Degradation*).

8.5. Key points

LiDAR is an active remote sensing technology that makes possible the characterization of the forest vertical structure on scales ranging from an individual tree to the world's forests.

The first technological developments took place in the early 1980s, but its use on an operational basis has only been going on since the 2000s, particularly in support of national forest inventory projects (in Canada, Finland, etc.).

In forestry, LiDAR technology is generally used for the estimation of dendrometric variables (heights, basal areas, timber volume, biomass...) but also for more qualitative characterizations (species composition, ecological integrity of forest ecosystems, health condition...).

LiDAR makes it possible to estimate the biomass in tropical forests, where there are significant standing stocks, without the saturation phenomenon typically observed using traditional remote sensing data.

The costs of acquisition represent the main limitation to the development of the large-scale use of LiDAR in forestry.

8.6. Bibliography

[ALD 85] ALDRED A., BONNER G., "Application of airborne lasers to forest surveys", *Petawawa National Forestry Institute*, Canadian Forestry Service, Information Report PI-X-51, 1985.

[ALE 97] ALEXANDER S., EHRLICH P.R., GOULDER L. *et al.*, *Ecosystem Services: Benefits Supplied to Human Societies by Natural Ecosystems*, Ecological Society of America, vol. 2, 1997.

[ARP 82] ARP H., GRIESBACH J.C., BURNS J.P., "Mapping in tropical forests a new approach using the laser APR (Airborne Profile Recorder)", *Photogrammetric Engineering and Remote Sensing,* vol. 48, pp. 91–100, 1982.

[ASN 07] ASNER G.P., KNAPP D.E., KENNEDY-BOWDOIN T. *et al.*, "Carnegie Airborne Observatory: in-flight fusion of hyperspectral imaging and waveform light detection and ranging for three-dimensional studies of ecosystems", *Journal of Applied Remote Sensing*, vol. 1, pp. 013536–013536–21, 2007.

[ASN 12] ASNER G.P., KNAPP D.E., BOARDMAN J. *et al.*, "Carnegie airborne observatory-2: increasing science data dimensionality via high-fidelity multi-sensor fusion", *Remote Sensing Environment*, vol. 124, pp. 454–465, 2012.

[ASN 14] ASNER G.P., ANDERSON C.B., MARTIN R.E. *et al.*, "Landscape-scale changes in forest structure and functional traits along an Andes-to-Amazon elevation gradient", *Biogeosciences*, vol. 11, pp. 843–856, 2014.

[BAC 12] BACCINI A., GOETZ S.J., WALKER W.S. *et al.*, "Estimated carbon dioxide emissions from tropical deforestation improved by carbon-density maps", *Nature Climate Change*, vol. 2, pp. 182–185, 2012.

[BOU 08] BOUDREAU, J., NELSON, R.F., MARGOLIS, H.A. *et al.*, "Regional aboveground forest biomass using airborne and spaceborne LiDAR in Québec", *Remote Sensing Environment*, vol. 112, pp. 3876–3890, 2008.

[CHA 05] CHAVE J., ANDALO C., BROWN S. *et al.*, "Tree allometry and improved estimation of carbon stocks and balance in tropical forests", *Oecologia*, vol. 145, pp. 87–99, 2005.

[CHA 14] CHAVE J., RÉJOU-MÉCHAIN M., BÚRQUEZ A. *et al.*, "Improved allometric models to estimate the aboveground biomass of tropical trees", *Global Change Biology*, vol. 20, no. 10, pp. 3177–3190, 2014.

[CHA 16] CHAZETTE P., TOTEMS L., HESPEL L., "Principle and physics of the LiDAR measurement", in BAGHDADI N., ZRIBI M. (eds), *Optical Remote Sensing of Land Surfaces*, ISTE Press, London and Elsevier, Oxford, 2016.

[DAS 11] DASSOT M., CONSTANT T., FOURNIER M., "The use of terrestrial LiDAR technology in forest science: application fields, benefits and challenges", *Annals of Forensic Science*, vol. 68, pp. 959–974, 2011.

[DED 15] DEDRY L., DETHIER O., PÉRIN J., *et al.*, "FORESTIMATOR : un plug-in QGIS d'estimation de la hauteur dominante et du Site Index de peuplements résineux au départ de données LiDAR aérien : application à la Wallonie (Belgique)", *Revue Française de Photogrammétrie et de Télédétection*, vol. 211–212, Spécial Forêt, pp. 119–127, 2015.

[DON 07] DONOGHUE D.N.M., WATT P.J., COX N.J., *et al.*, "Remote sensing of species mixtures in conifer plantations using LiDAR height and intensity data", *Remote Sensing Environment, ForestSAT 2005 Conference, 'Operational Tools in Forestry Using Remote Sensing Techniques'*, vol. 110, pp. 509–522, 2007.

[DUN 10] DUNCANSON L.I., NIEMANN K.O., WULDER M.A., "Estimating forest canopy height and terrain relief from GLAS waveform metrics", *Remote Sensing Environment*, vol. 114, pp. 138–154, 2010.

[DUO 12] DUONG H.V., LEFSKY M.A., RAMOND T. *et al.*, "The electronically steerable flash LiDAR: a full waveform scanning system for topographic and ecosystem structure applications", *IEEE Transactions of Geoscience and Remote Sensing*, vol. 50, pp. 4809–4820, 2012.

[GOE 10] GOETZ S.J., STEINBERG D., BETTS M.G. *et al.*, "LiDAR remote sensing variables predict breeding habitat of a Neotropical migrant bird", *Ecology*, vol. 91, pp. 1569–1576, 2010.

[GOE 11] GOETZ S.J., "Editorial Board", *Remote Sensing Environment*, DESDynI VEG-3D Special Issue, vol. 115, 2011.

[GOW 97] GOWARD S.N., WILLIAMS D.L., "Landsat and Earth systems science: development of terrestrial monitoring", *Photogrammetric Engineering and Remote Sensing*, vol. 63, pp. 887–900, 1997.

[HAN 10] HANSEN M.C., STEHMAN S.V., POTAPOV P.V., "Quantification of global gross forest cover loss", *Proceedings of the National Academy of Sciences*, vol. 107, pp. 8650–8655, 2010.

[HIB 75] HIBLER III W., "Characterization of cold-regions terrain using airborne laser profilometer", *Journal of Glaciology*, vol. 15, pp. 329–347, 1975.

[HIC 69] HICKMAN G.D., HOGG J.E., "Application of an airborne pulsed laser for near shore bathymetric measurements", *Remote Sensing Environment*, vol. 1, pp. 47–58. 1969.

[HYY 96] HYYPPÄ J., HALLIKAINEN M., "Applicability of airborne profiling radar to forest inventory", *Remote Sensing Environment*, vol. 57, pp. 39–57, 1996.

[JOH 10] JOHANSEN K., ARROYO L.A., ARMSTON J. *et al.*,, "Mapping riparian condition indicators in a sub-tropical savanna environment from discrete return LiDAR data using object-based image analysis", *Ecological Indications*, vol. 10, pp. 796–807, 2010.

[KAS 97] KASISCHKE E.S., MELACK J.M., DOBSON M.C., "The use of imaging radars for ecological applications a review", *Remote Sensing of Environment*, vol. 59, no. 2, pp. 141–156, 1997.

[KOC 10] KOCH B., "Status and future of laser scanning, synthetic aperture radar and hyperspectral remote sensing data for forest biomass assessment", *ISPRS Journal of Photogrammetric Remote Sensing*, vol. 65, pp. 581–590, 2010.

[KOR 10] KORPELA I., ØRKA H.O., MALTAMO M. *et al.*, "Tree species classification using airborne LiDAR–Effects of stand and tree parameters, downsizing of training set, intensity normalization, and sensor type", *Silva Fennica*, vol. 44, pp. 319–339, 2010.

[LEE 07] LEE A.C., LUCAS R.M., "A LiDAR-derived canopy density model for tree stem and crown mapping in Australian forests", *Remote Sensing Environment*, vol. 111, pp. 493–518, 2007.

[LEF 02] LEFSKY M.A., COHEN W.B., PARKER G.G. *et al.*, "LiDAR remote sensing for ecosystem studies", *BioScience*, vol. 52, pp. 19–30, 2002.

[LEF 05] LEFSKY M.A., HARDING D.J., KELLER M. *et al.*, "Estimates of forest canopy height and aboveground biomass using ICESat", *Geophysical Research Letters*, vol. 32, 2005.

[LEF 07] LEFSKY M.A., KELLER M., PANG Y. *et al.*, "Revised method for forest canopy height estimation from Geoscience Laser Altimeter System waveforms", *Journal of Applied Remote Sensing*, vol. 1, pp. 013537–013537–18, 2007.

[LEF 10] LEFSKY M.A., "A global forest canopy height map from the moderate resolution imaging spectroradiometer and the geoscience laser altimeter system: a global forest canopy height map", *Geophysical Research Letters*, vol. 37, 2010.

[LIA 13] LIANG X., HYYPPÄ J., "Automatic stem mapping by merging several terrestrial laser scans at the feature and decision levels", *Sensors*, vol. 13, pp. 1614–1634, 2013.

[LIA 14a] LIANG X., HYYPPA J., KUKKO A. *et al.*, "The use of a mobile laser scanning system for mapping large forest plots", *IEEE Geoscience and Remote Sensing Letters*, vol. 11, pp. 1504–1508, 2014.

[LIA 14b] LIANG X., KUKKO A., KAARTINEN H. *et al.*, "Possibilities of a personal laser scanning system for forest mapping and ecosystem services", *Sensors*, vol. 14, pp. 1228–1248, 2014.

[LIM 03] LIM K., TREITZ P., WULDER M. *et al.*, "LiDAR remote sensing of forest structure", *Progress of Physical Geography*, vol. 27, pp. 88–106, 2003.

[LIN 11] LIN Y., HYYPPA J., JAAKKOLA A., "Mini-UAV-Borne LIDAR for fine-scale mapping", *IEEE Geoscience and Remote Sensing Letters*, vol. 8, pp. 426–430, 2011.

[MAC 86] MACLEAN G.A., KRABILL W.B., "Gross-merchantable timber volume estimation using an airborne LiDAR System", *Canadian Journal of Remote Sensing*, vol. 12, pp. 7–18,1986.

[MAL 09] MALTAMO M., PACKALÉN P., SUVANTO A. *et al.*, "Combining ALS and NFI training data for forest management planning: a case study in Kuortane, Western Finland", *European Journal of Forest Research*, vol. 128, no. 3, pp. 305–317, 2009.

[MAL 11] MALTAMO M., PACKALÉN P., KALLIO E. *et al.*, "Airborne laser scanning based stand level management inventory in Finland", *Proceedings of SilviLaser, 11th International Conference on LiDAR Applications for Assessing Forest Ecosystems*, University of Tasmania, Australia, 16–20 October, pp. 1–10, 2011.

[MAL 16] MALLET C., CHEHATA N., BAILLY J.-S., "Airborne LiDAR data processing", in BAGHDADI N., ZRIBI M. (eds), *Optical Remote Sensing of Land Surfaces*, ISTE Press, London and Elsevier, Oxford, 2016.

[MAR 14] MARVIN D.C., ASNER G.P., KNAPP D.E. *et al.*, "Amazonian landscapes and the bias in field studies of forest structure and biomass", *Proceedings of the National Academy of Sciences*, vol. 111, no. 48, pp. E5224–E5232, 2014.

[MAS 14] MASCARO J., ASNER G.P., DAVIES S. *et al.*, "These are the days of lasers in the jungle", *Carbon Balance Management*, vol. 9, no. 7, 2014.

[MEA 99] MEANS J.E., ACKER S.A., HARDING D.J. *et al.*, "Use of large-footprint scanning airborne LiDAR to estimate forest stand characteristics in the Western Cascades of Oregon", *Remote Sensing Environment*, vol. 67, pp. 298–308, 1999.

[MIC 13] MICHEZ A., PIÉGAY H., TOROMANOFF F. *et al.*, "LiDAR derived ecological integrity indicators for riparian zones: Application to the Houille river in Southern Belgium/Northern France", *Ecological Indices*, vol. 34, pp. 627–640, 2013.

[MIT 14] MITCHARD E.T., FELDPAUSCH T.R., BRIENEN R.J. *et al.*, "Markedly divergent estimates of Amazon forest carbon density from ground plots and satellites", *Global Ecology and Biogeography*, vol. 23, no. 8, pp. 935–946, 2014.

[MOR 09] MORSDORF F., NICHOL C., MALTHUS T. *et al.*, "Assessing forest structural and physiological information content of multi-spectral LiDAR waveforms by radiative transfer modelling", *Remote Sensing Environment*, vol. 113, pp. 2152–2163, 2009.

[MOU 14] MOUSSAVI M.S., ABDALATI W., SCAMBOS T. *et al.*, "Applicability of an automatic surface detection approach to micro-pulse photon-counting LiDAR altimetry data: implications for canopy height retrieval from future ICESat-2 data", *International Journal of Remote Sensing*, vol. 35, pp. 5263–5279, 2014.

[NAE 04] NÆSSET E., "Practical large-scale forest stand inventory using a small-footprint airborne scanning laser", *Scandinavian Journal of Forest Research*, vol. 19, no. 2, pp. 164–179, 2004.

[NAE 07] NÆSSET E., "Airborne laser scanning as a method in operational forest inventory: status of accuracy assessments accomplished in Scandinavia", *Scandinavian Journal of Forest Research*, vol. 22, no. 5, pp. 433–442, 2007.

[NEL 84] NELSON R., KRABILL W., MACLEAN G., "Determining forest canopy characteristics using airborne laser data", *Remote Sensing Environment*, vol. 15, pp. 201–212, 1984.

[NEL 88] NELSON R., KRABILL W., TONELLI J., "Estimating forest biomass and volume using airborne laser data", *Remote Sensing Environment*, vol. 24, pp. 247–267, 1988.

[NEL 05] NELSON R., KELLER C., RATNASWAMY M., "Locating and estimating the extent of Delmarva fox squirrel habitat using an airborne LiDAR profiler", *Remote Sensing Environment*, vol. 96, pp. 292–301. 2005.

[NEL 09] NELSON R., RANSON K.J., SUN G. *et al.*, "Estimating Siberian timber volume using MODIS and ICESat/GLAS", *Remote Sensing Environment*, vol. 113, pp. 691–701, 2009.

[PER 13] PERIN J., HÉBERT J., BROSTAUX Y. *et al.*, "Modeling the top-height growth and site index of Norway spruce in Southern Belgium", *Forest Ecology and Management*, vol. 298, pp. 62–70, 2013.

[PET 09] PETRIE G., TOTH C.K., "Terrestrial laser scanners", in SHAN J., TOTH C.K. (eds), *Topographic Laser Ranging and Scanning: Principles and Processing*, CRC Press, 2009.

[PFE 14] PFENNINGBAUER M., RIEGL U., RIEGER P. *et al.*, "UAS based laser scanning for forest inventory and precision farming", *Proceedings of the International Workshop on Remote Sensing and GIS for Monitoring of Habitat Quality*, Vienna, Austria, pp. 24–25, 2014.

[POP 04] POPESCU S.C., WYNNE R.H. "Seeing the trees in the forest: using LiDAR and multispectral data fusion with local filtering and variable window size for estimating tree height", *Photogrammetric Engineering and Remote Sensing*, vol. 70, no. 5, pp. 589–604, 2004.

[RAU 13] RAUMONEN P., KAASALAINEN M., ÅKERBLOM M. *et al.*, "Fast automatic precision tree models from terrestrial laser scanner data", *Remote Sensing*, vol. 5, no. 2, pp. 491–520, 2013.

[RON 93] RONDEUX J., *La mesure des arbres et des peuplements forestiers*, Les Presses Agronomiques de Gembloux, 1993.

[ROS 08] ROSETTE J.A.B., NORTH P.R.J., SUÁREZ J.C., "Vegetation height estimates for a mixed temperate forest using satellite laser altimetry", *International Journal of Remote Sensing*, vol. 29, pp. 1475–1493, 2008.

[RYD 15] RYDING J., WILLIAMS E., SMITH M.J. *et al.*, "Assessing handheld mobile laser scanners for forest surveys", *Remote Sensing*, vol. 7, pp. 1095–1111, 2015.

[SAA 11] SAATCHI S.S., HARRIS N.L., BROWN S. *et al.*, "Benchmark map of forest carbon stocks in tropical regions across three continents", *Proceedings of the National Academy of Science*, vol. 108, pp. 9899–9904, 2011.

[SIM 08] SIMARD M., RIVERA-MONROY V.H., MANCERA-PINEDA J.E. *et al.*, "A systematic method for 3D mapping of mangrove forests based on Shuttle Radar Topography Mission elevation data, ICEsat/GLAS waveforms and field data: Application to Ciénaga Grande de Santa Marta, Colombia", *Remote Sensing Environment*, vol. 112, pp. 2131–2144, 2008.

[SIM 11] SIMARD M., PINTO N., FISHER J.B. *et al.*, "Mapping forest canopy height globally with spaceborne LiDAR", *Journal of Geophysics. Research*, vol. 116, G04021 2011.

[SOM 15] SOMERS B., ASNER G.P., MARTIN R.E. *et al.*, "Mesoscale assessment of changes in tropical tree species richness across a bioclimatic gradient in Panama using airborne imaging spectroscopy", *Remote Sensing Environment*, vol. 167, pp. 111–120, 2015.

[TIC 01] TICKLE P.K., WITTE C., LEE A. *et al.*, "Use of airborne scanning LiDAR and large scale photography within a strategic forest inventory and monitoring framework", *Geoscience and Remote Sensing Symposium*, vol. 01, no. 3, pp. 1000–1003, 2001.

[TOM 99] TOMPPO E., GOULDING C., KATILA M., "Adapting Finnish multi-source forest inventory techniques to the New Zealand preharvest inventory", *Scandavian Journal of Forest and Research*, vol. 14, pp. 182–192, 1999.

[TUL 14] TULLDAHL H.M., LARSSON H., "LiDAR on small UAV for 3D mapping", *Proceedings of SPIE 9250, Electro-Optical Remote Sensing, Photonic Technologies, and Applications VIII; and Military Applications in Hyperspectral Imaging and High Spatial Resolution Sensing II*, p. 925009, 2014.

[VAN 10] VAN LEEUWEN M., NIEUWENHUIS M., "Retrieval of forest structural parameters using LiDAR remote sensing", *European Journal of Forest Research*, vol. 129, pp. 749–770, 2010.

[VIE 08] VIERLING K.T., VIERLING L.A., GOULD W.A. *et al.*, "Lidar: shedding new light on habitat characterization and modeling", *Frontiers in Ecological Environment*, vol. 6, pp. 90–98, 2008.

[WAL 12] WALLACE L., LUCIEER A., WATSON C. *et al.*, "Development of a UAV-LiDAR system with application to forest inventory", *Remote Sensing*, vol. 4, pp. 1519–1543, 2012.

[WAL 14] WALLACE A.M., MCCARTHY A., NICHOL C.J. *et al.*, "Design and evaluation of multispectral LiDAR for the recovery of arboreal parameters", *IEEE Transactions of Geoscience and Remote Sensing*, vol. 52, pp. 4942–4954, 2014.

[WEH 99] WEHR A., LOHR U., "Airborne laser scanning – an introduction and overview", *ISPRS Journal of Photogrammetric. Remote Sensing*, vol. 54, pp. 68–82, 1999.

[WHI 13] WHITE J.C., WULDER M.A., VARHOLA A. *et al.*, , "A best practices guide for generating forest inventory attributes from airborne laser scanning data using an area-based approach", *The Forestry Chronicle*, vol. 89, no. 6, pp. 722–723, 2013.

[WOO 11] WOODS M., PITT D., PENNER M. *et al.*, "Operational implementation of a LiDAR inventory in Boreal Ontario", *The Forestry Chronicle*, vol. 87, no. 4, pp. 512–528, 2011.

[ZHO 12] ZHOU G., YANG J., LI X. *et al.*, "Advances of flash LiDAR development onboard UAV", *International Archives of the Photogammetry, Remote Sensing and Spatial Information Sciences*, vol. XXXIX-B3, pp. 193–198, 2012.

[ZIA 05] ZIANIS D., MUUKKONEN P., MÄKIPÄÄ R. *et al.*, "Biomass and stem volume equations for tree species in Europe", *Finnish Society of Forest Science, Finnish Forest Research Institute*, Silva Fennica Monographs, vol. 4, p. 63, 2005.

[ZOL 13] ZOLKOS S.G., GOETZ S.J., DUBAYAH R. "A meta-analysis of terrestrial aboveground biomass estimation using LiDAR remote sensing", *Remote Sensing of Environment*, vol. 128, pp. 289–298, 2013.

9

Forest Biomass From Radar Remote Sensing

9.1. Forest biomass at the global scale

9.1.1. *Forest biomass, an essential climate variable*

9.1.1.1. *Of vital importance, but still poorly known at the global scale*

Forests play a primordial role for life on Earth. Beyond their contribution as a major source of raw materials and renewable energy, they also hold an inestimable treasure of biodiversity. They ensure the protection of arable land, are a continuous source of water and contribute to improved air quality. Whether for food or pharmacopoeia, forests are the principal source of subsistence for almost 2 billion people.

On a global scale, the need for information regarding forest resources is why the Global Forest Resources Assessments have been produced at 5–10 year intervals since 1946 by the FAO.

Beyond their importance in terms of resources, the role of forests in climate regulation throughout the carbon cycle is also essential. Indeed, forests play a crucial role in carbon absorption associated with the growth of vegetation and the emissions caused during the degradation of forests, both

---

Chapter written by Ludovic VILLARD, Thuy LE TOAN, Dinh HO TONG MINH, Stéphane MERMOZ and Alexandre BOUVET.

natural and anthropogenic. By absorbing more than a quarter of CO_2 emitted by the combustion of fossil fuels (Global Carbon Project, 2015), the terrestrial biosphere is an important carbon sink. However, of the three components which are involved in the carbon cycle, namely the atmosphere, the ocean and the terrestrial biosphere, it is the last element that is the least understood, both qualitatively and quantitatively. There are major uncertainties regarding the spatial distribution and the dynamics of stocks and fluxes of carbon associated with forests.

A basic parameter for characterizing the spatial distribution of carbon in the biosphere, the so-called biomass – defined here as the amount (mass) of dry organic matter of plant origin – is about twice the mass of carbon in a forest. Biomass is in fact the basic unit for measuring carbon, and is the variable used in order to quantify the contribution of forests to the carbon cycle. Over 80% of the above-ground terrestrial biomass is contained in forests, it is therefore necessary to measure and follow this variable in time in order to reduce uncertainties in our knowledge of the climate system (Global Climate Observing System, or GCOS). Forest biomass is therefore identified as an essential climate variable by the GCOS [GCO 06, SES 08]. Also worth mentioning, carbon sequestration, e.g. into forest biomass, is the only climate change mitigation mechanism recognized by the Kyoto Protocol, other than the reduction of human-induced greenhouse gas emissions.

We must also remember that forests represent almost a third of the Earth's total land surface (see figures and definition of wooded land as forest [FAO 04]) and more than 70% of the total biomass of vegetation. While the distribution of temperate, boreal and tropical forests are essentially of the same order of magnitude in terms of area (respectively, 1040, 1370 and 1750 million hectares [04 FAO]), biomass per hectare is much higher for tropical forests – they alone contain nearly half of the world's forest carbon stocks.

9.1.1.2. *The link between forests and carbon: the "biomass" variable*

Within the terrestrial carbon cycle, forests are not only a carbon sink (under the atmospheric CO2 conversion into biomass through photosynthesis) but is also a source of carbon through the decomposition of organic matter, as well as CO_2 emissions from fires, whether natural or caused for deforestation purposes. The carbon balance exchange between the atmosphere and ecosystems can be expressed in two ways using two simple equations: one [9.1] refers to the functioning of the ecosystem, and the other to mass conservation [9.2] [LET 11]. The net flux of carbon (C) can be written:

$$\Delta C = GPP - R_p - R_a - D \quad [9.1]$$

$$\Delta C = \Delta AGB + \Delta BGB + \Delta L + \Delta S \quad [9.2]$$

In the first equation [9.1], ΔC is the difference between the Gross Primary Production (GPP), the primary production of the ecosystem, and the loss of carbon because of heterotrophic and autotrophic respiration (respectively, R_p and R_a) as well as deforestation, D (because of forest fires).

In the second equation [9.2], ΔC is expressed as the sum of the variations of carbon stocks in a forest ecosystem, in trees as well as in the soil. These stocks are divided mainly among Above Ground Biomass, or AGB, Below Ground Biomass, or BGB, plant litter (L) and soil (S).

AGB is not only an explicit term in equation [9.2], but is also strongly linked to below-ground biomass and the production of plant litter. Through plant litter, it is linked to changes in soil carbon. In equation [9.1], AGB is the result of carbon stocks linked to GPP, and the carbon loss *D*.

It is also worth noting that even though a forest carbon stock is present in the soil, both as dead biomass and fresh below-ground biomass, we are particularly interested in fresh above-ground biomass, since the latter allows us on the one hand to make indirect estimations of other components and most importantly because it is the most accessible component in terms of remote sensing and *in situ* validation, as well as being the most sensitive interface to natural and anthropogenic disturbances.

Models of the functioning of ecosystems have been developed, from the local to the global scale in order to estimate the net flux of carbon using equation [9.1]. However, by using equation [9.2], access to the variation in carbon stocks contained in the forests with the help of inventories allows for a more practical approach in terms of measuring the amount of carbon in a forest, a region, a nation, or at the global level. It is the preferred method of international initiatives such as the Kyoto Protocol or REDD (Reduction of Emissions from Deforestation and Forest Degradation).

9.1.1.3. *Climate regulation participation: carbon fluxes*

After anthropogenic emissions linked to the burning of fossil fuels, land-use changes are the second most significant contributor to the increase in atmospheric CO_2, mainly because of deforestation, degradation and natural forest fires. Conversely, the role of forests as carbon sinks through their growth, regrowth and the plantation of new forests, results in a slightly positive net balance [PAN 11, LEQ 13].

The carbon balance estimates of terrestrial ecosystems, however, remain very imprecise [PAN 11, LEQ 13]. In the report published by the Intergovernmental Panel on Climate Change (IPCC) in 2013 [IPCC 13], the estimated share of emissions because of deforestation, both from the cleared surface and the biomass lost as a result, is extremely uncertain (65%). The carbon uptake by forests has not been calculated, but merely deduced from the carbon fluxes emitted, minus the amount which is retained in the atmosphere and which is absorbed by the ocean, as shown in Figure 9.1.

Figure 9.1. *Carbon flux in billions of tons of carbon per year (PgC.y^{-1}). On the left are the anthropogenic sources of emissions because of the burning of fossil fuels and deforestation, while on the right are the carbon sinks; namely the atmosphere, forests and the oceans, with the percentage of their carbon absorption (47%, 27% and 26%, respectively). Circled in red is the flux associated with the uptake of carbon by forests, deduced from net global fluxes (statistics: www.globalcarbonproject.org/). For a color version of this figure, see www.iste.co.uk/baghdadi/3.zip*

Forests constitute the majority (~90%) of the net balance of vegetation on continental surfaces, and it is tropical forests that demonstrate the most variation, either because of deforestation or the absorption of atmospheric CO_2 because of growth (of wooded areas). Pan *et al.* [PAN 11] estimate that the annual increase in biomass of tropical forests resulted in an equivalent carbon flux of around 2300 TgC/year between the years 2000 and 2007, while the total flux of forests is 2900 TgC/year (boreal forest share is 120 TgC/year, temperate forests 454 TgC/year).

9.1.2. *Expectations for a better understanding of biomass globally: users and needs*

The afore-detailed level of uncertainty about the role of vegetation in the carbon cycle raises of course many questions: on the one hand it is a block which prevents a better understanding of the Earth's system, but it is also a source of leverage which may be used to justify the strengthening of research efforts.

9.1.2.1. *Assimilation in land surface models*

Essential in the simulation of climate scenarios, land surface models allow us to simulate the exchanges between the atmosphere and ecosystems, in particular the water and carbon cycles.

As a core component of land surface models, dynamic vegetation models quantify the carbon exchanges between ecosystem and the atmosphere. In addition, these models also enable the simulation of future trends to different climate, land use and ecosystem management scenarios. In order to do this, they simulate the temporal dynamics of the matter (water, carbon) and energy budgets of the ecosystems, particularly via the allocation of photosynthetic product to the different plant compartments, as well as mortality and stage of senescence. Consequently, these models make it possible to calculate the long-term carbon storage in an ecosystem.

However, the estimations by these models of the carbon budgets relative to forest ecosystems are currently faced with strong uncertainties linked to the lack of knowledge of the conditions other than the climate, such as the soil nutrient content, the floristic composition, forest management, or the age of forests stands. Most of the efforts in terms of data assimilation for the carbon cycle nevertheless relate to short-term processes, controlling the diurnal cycle and seasonal and inter-annual variability, particularly through (optical) remote sensing observations of leaf area or phenology. For the long-term accumulation of carbon, it is crucial to be able to assimilate the biomass data already in place in order to improve the various functioning parameters such as carbon allocation, growth, mortality and fire occurrence.

More specifically, it implies:

– obtaining the correct spatial distribution of forest biomass and using this information to validate, constrain and improve biosphere models;

– an improved characterization and consideration of biomass growth with age, whether modified or not by forestry practices within the models;

– integrating into these models the impact of anthropogenic and/or fire disturbances and improving the natural mortality parameterization.

While the biomass data assimilation methods are not yet generalized, a number of studies and test cases have shown the impact of biomass data in estimating the productivity of ecosystems.

Among the models tested, we are able to cite ORCHIDEE, developed by the French laboratory LSCE[1] [KRI 05]. The biomass simulated by ORCHIDEE was compared with the biomass measured in Amazonian forest sites [DEL 10] and it turned out that the large apparent differences were caused partly by the overestimation of the mortality of forests. The introduction of an adequate mortality rate has allowed for the reduction of these divergences. Another example concerns the net ecosystem productivity (NEP) of forests in France, simulated by ORCHIDEE for two cases:

– by supposing that all of the forest stands have an average age of 40–50 years (Figure 9.2) which is equivalent to simulations performed without information on the age or biomass of forests;

– by using a reconstruction of the age of forests thanks to biomass data.

As illustrated by the NEP maps (Figure 9.2), this comparison clearly shows weaker levels of NEP in the case where biomass information is introduced, in regions such as the center and the north-east of France, where the forests are older.

1 Laboratoire français des sciences du climat et de l'environnement (French laboratory for climate sciences and the environment).

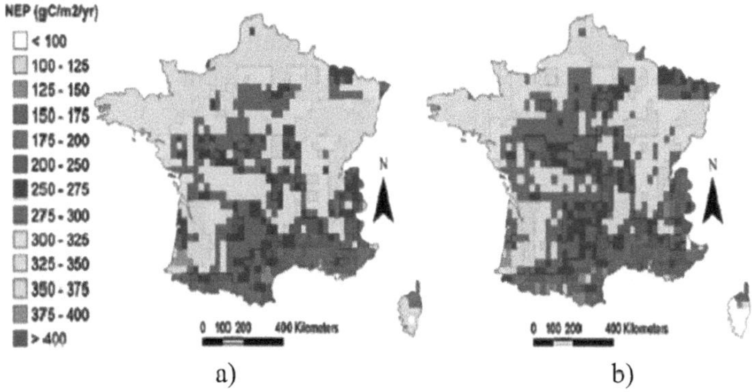

Figure 9.2. *Net Ecosystem Productivity (NEP) simulated by ORCHIDEE-FM (Forest Management module) with a) and without b) information about the age of forest stands reconstructed using biomass data [BEL 11]. For a color version of this figure, see www.iste.co.uk/baghdadi/3.zip*

On the European continent scale, by using age estimations obtained from biomass data and from the Forest Management module of ORCHIDEE, a forest NEP close to 175 ± 52 g C m^{-2} $year^{-1}$ was calculated [BEL 11]. This estimation of the NEP seems to be consistent with independent estimations based on an interpolation of sites measured over the long term and flux measurements by eddy covariance, or from national forest inventories.

9.1.2.2. *Support of international treaties*

As previously mentioned, deforestation is mainly an anthropogenic phenomenon. Whether for the timber trade, conversion to arable land for agriculture or forestry, or the extension of inhabited areas, forests are currently more threatened by socio-economic pressures than by problems related to climate or other naturally occurring destructive agents, such as insects or disease. In view of the fact that over 70 countries in the world are not able to contain deforestation, international initiatives have been set up, such as the REDD+ program (Reduced Emissions from Deforestation and Forest Degradation) with the objective of containing this threat through the use of economic incentives supporting developing countries so they may better preserve and manage their forests. This initiative, which focused on controlling the emission of greenhouse gases, follows the Kyoto Protocol (the international agreement signed in 1997 linked to the United Nations Framework Convention on Climate Change), in which incentivizing mechanisms were created to encourage developing countries, which are generally emitters, to contribute to the reduction of deforestation and to replanting via a system of carbon credits. Recent studies [HAN 13], based on a series of temporal images from optical satellites (Landsat), and which have provided interactive maps online at a resolution of 30 m, confirm the previous results on a global scale (with losses of around 2.3 million km^2 between 2000 and 2012, and around 800,000 km^2 in new forests). They also serve to bring to light the large disparity of deforestation between regions: Indonesia, for example, has the largest increase in deforestation, with a loss during this period (2000–2012) of more than 6 million hectares of virgin forest to make room for plantations (notably for palm oil).

To estimate the carbon flux (ΔC_{em}) caused by deforestation it is necessary to take into account the variations in surface area ΔA_i of different types of forests having a biomass B_i and efficiency F_i which quantifies the fraction of biomass converted into carbon dioxide. In the same way, the absorption of carbon because of growth of trees ΔC_{ca} is expressed via the biomass increment ΔB_i:

$$\Delta C_{em} = \sum_i \Delta A_i . B_i . F_i \qquad [9.3]$$

$$\Delta C_{ca} = \sum_i \Delta B_i A_i$$

The net flux of carbon is the algebraic sum of ΔC_{em} and ΔC_{ca}.

To calculate the net flux, we therefore need to be able to estimate both surface deforestation and the fraction of biomass converted to CO_2, as well as biomass variations in the preserved areas. To evaluate deforested areas, which for the most part are located in tropical regions, optical remote sensing, limited by cloud cover, cannot provide us with the frequency of measurements necessary for the fight against deforestation.

9.1.3. *Towards forest biomass mapping globally: issues and solutions*

9.1.3.1. *In situ measurements of the forest biomass*

In this chapter, the term biomass refers to plant biomass density per unit area. This is the mass of the plant matter in a dry state. The measurement units are g/m^2 or, more commonly used by foresters, ton per hectare.

In a forest, biomass is found in both above-ground and below-ground plant parts, as well as in the litter. Only the above-ground biomass is measurable. This is the biomass of tree elements including the trunk, branches, bark, foliage and fruit. It is this biomass that is involved in carbon exchanges (expressed for example by equations [9.2] and [9.3]).

Starting from a pragmatic basis, forest biomass can only be measured through the felling of the trees that compose it. As these destructive methods cannot be conducted on a large scale, non-destructive measurements are therefore used. The methods rely mainly on trunk volume measurements, performed to quantify forest resources using the diameter (DBH, or diameter at breast height), the height (H), and a shape factor taking into account the shape of the trunk. To obtain the trunk biomass, the volume is multiplied by the density of the dry wood. The total biomass of a tree is then derived from the biomass of the trunk, by means of a factor to account for the proportion of mass of the trunk relative to the total mass of the tree. Wood density and different factors are unique to each species [CHA 04]. The biomass per unit area (hectare), or the sum of tree biomass values which form the surface area, is often estimated using the diagram in Figure 9.3.

While this method is reasonably convenient for mono-specific forest stands (such as plantations which are usually composed of a single species) or limited in the number of species (such as the case for most temperate and boreal forests), it is, however, much more complex to conduct in tropical forests, given the difficulties in the identification of often unknown species and the much higher number of species (often greater than 300 per hectare, compared with less than 50 species throughout the whole Europe), and also for reasons of accessibility.

In addition, if these methods lead to accuracies of 5–7% compared with the destructive measurements of biomass (with a confidence interval of 90% at the hectare scale) for plantations, temperate or boreal forests, the errors could be far superior for tropical forests, certain studies reporting errors of up to 100% in carbon stocks [HEN 10]. Several studies have nonetheless sought to identify critical points in the propagation of errors, particularly with the use of allometric equations being used outside of their domain, typically for excessive DBH values [CHA 04] or regions that are too remote [BUV 06], pan-tropical models [CHA 05, BRO 97], however, being recommended as long as local allometries are not determined from the felling of at least 100 sample trees [CHA 04].

Given the difficulty in measuring tropical forests, very few national data inventories exist, unlike for temperate or boreal forests. Despite great efforts to expand the existing network of forest study sites, such as the CTFS (Center for Tropical Forest Science[2]), AFRITRON (African Tropical Rainforest Observation Network[3]) or RAINFOR (Amazon Forest Inventory Network[4]), a number of areas are not being measured. The Democratic Republic of Congo, for example, a country where conflicts are recurrent, has very few local allometric relations, while it has the largest forest cover in Africa, with 1.54 million km^2 of forest in 2010 [FAO 10].

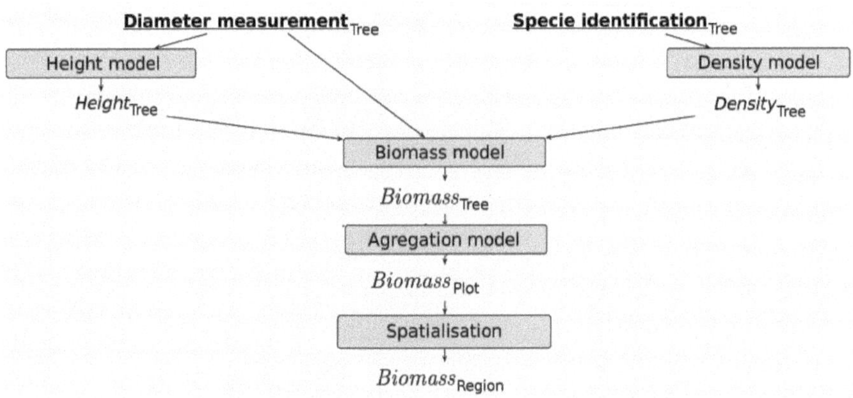

Figure 9.3. *Multi-scale estimation of forest biomass*

These data are, however, valuable for the validation of estimates from remote sensing measurements as well as other inverse methods based on measurements from remote sensing. While many of these methods are valid only in the vicinity of the reference data, the possibility of spatial measurements *in situ* is often very interesting, if only for the spatial collocation between measures of various kinds. A multi-scale spatialization of *in situ* biomass estimations can also be envisaged, by considering the aerial optical measurements, LiDAR or radar, that are detailed below.

2 www.forestgeo.si.edu.

3 www.afritron.org.

4 www.rainfor.org.

9.1.3.2. *The biomass of forests by passive optical remote sensing systems*

Optical remote sensing measurements (from passive sensors) have a limited interest for bioamass estimation, given that the multi or hyperspectral reflectances that are used (in the visible range or beyond) come mainly from the top canopy scatterers, which correlation with the total above ground biomass is variable.

While this information may be useful for species identification, phenological vegetation monitoring and the detection of areas of degradation or deforestation (see Chapter 7), the correlation with biomass is much more difficult, based on – in addition to spectral signatures and vegetation indices [DON 03] or BRDF [DEW 02] – texture analysis of forest stands combined with allometric [BAR 11, MUI 01] or morphological [CEA 03, PEK 02] relationships. Let us note, however, that works based on the algorithm FOTO [COU 02] applied to very high resolution optical data, through the texture of the image, allow the estimation of the biomass of dense tropical forests with relative errors of less than 20% [PRO 07, BAS 14]. These results can only be obtained, however, in small, unclouded areas.

Nonetheless, while these methods may be used locally, systematic acquisition that would assure pertinent sampling is difficult in tropical regions. An example of this is in certain areas of Gabon, where no image from the Landsat series (30 years of archives, currently revisited every 16 days) or from the SPOT-Take5 series (experimental method over a period of almost 6 months revisited every 5 days, https://www.spot-take5.org) is able to be used.

9.1.3.3. *Forest biomass by LiDAR remote sensing*

LiDAR is an active system that measures the propagation time of photons using laser pulses, and thereby the distance travelled [CHA 16]. Depending on the density of photons per square meter, penetration even through dense vegetation is possible, and through the analysis of the echo distribution it is possible to identify the varying levels, including those from the ground as well as from the top of the canopy, in order to estimate height (see Chapter 8), and from these, in turn, the biomass. Depending on the resolution of the LiDAR, biomass–height relationships developed at an individual level are not always relevant, and different metrics are used (Lorey's height[5]), in order to give more emphasis to bigger trees in the biomass average.

While significant progress, with the emergence of technologies for photon counting or complete waveforms, attests to a real maturity in recent years, the LiDAR measurements from space (ICESat/GLAS from 2003 to 2009 and ICESat-2 for which the launch is planned for 2017) do not provide us with continuous coverage but rather each time produce samples and transects from the satellite. For forests, the extension of the aperture or beam width (to allow more coverage) is limited by errors related to topography.

The flight heights of airborne configurations make it possible to obtain continuous coverage, the details of the estimates of height and biomass are therefore of the same order of magnitude as *in situ* measurements (see for example the work of [ENG 13]).

Just as is the case with data from passive optical remote sensing, LiDAR data can be very relevant for spatializing regions of reference estimated from field measurements.

9.1.3.4. *Forest biomass from synthetic aperture radar (SAR) remote sensing*

Unlike with passive optical systems or LiDAR, cloud cover or lighting conditions do not result in observation limitations for radar frequencies below the X-band (a wavelength of about 3 cm except in cases of extreme weather events, see also the atmospheric transmittance curves contained in [ULA 13] for water absorption lines at higher frequencies).

Another major advantage of radar backscatter, the coherence of the backscatter radiation allows for synthetic aperture processing, and for the production of images with a much better spatial resolution (typically 1 to 50 m) than passive microwave systems (radiometers).

Benefitting from these two advantages, SAR allows for continuous global spatial coverage and systematic acquisitions, fundamental for constructing relevant temporal series. A multiannual independently acquired sampling of

5 Lorey's height is defined by : $H_{Lorey} = \frac{\sum_{i=1}^{n} DBH_i^2 \times H_i}{\sum_{i=1}^{n} DBH_i^2}$, n being the number of selected trees (minimum DBH criterion) within a given perimeter.

observation conditions is obviously fundamental for the phenological monitoring of crops (see Chapter 4), but also the remote sensing of forests to track deforestation, and to increase the accuracy of observations to detect seasonal variations in biomass or more generally in response to environmental conditions.

While these features of SAR imagers make them operational systems, the growth of remote sensing also comes from the microwave's sensitivity to macroscopic scattering from plant cover, mainly because of their water content and their geometric structure. This sensitivity to the geometry of scatterers (which we will detail in the next section), and in particular their volume, is the reason for the use of active microwaves to determine the biomass of natural environments such as forests.

9.2. Quantifying forest biomass with radar remote sensing: physics and measurement techniques

9.2.1. *Interaction mechanisms between microwaves and vegetation scatterers, relationship with biomass*

The theory of wave scattering with matter [ISH 78, TSA 85] distinguishes several domains, according to the various dimensions of a scatterer with respect to the wavelength: the homogenization domain associated with Rayleigh scattering, the resonance domain corresponding to Mie scattering, and finally, the specular domain whereby the specific dimensions of the scatterer are respectively smaller, comparable to or larger than the wavelength.

Considering the question of the optimal radar signal sensitivity to forest biomass, it becomes clear to use the resonant domain, for which an optimal sensitivity to a significant part of the vegetation in terms of biomass can be expected. As a result, we will focus on dielectric elements for which at least one dimension is of the same order as the wavelength, so that volume variations and thus biomass variations (proportional to the volume in first approximation) can be detected through variations of the radar measurements. Indeed, even if the sensitivity of the radar signal is not sufficient to get a univocal relation with the biomass, it is obviously a necessary condition that any inverse method must tackle.

For each wavelength possibly used by spaceborne SAR from 3 cm to 70 cm (or from X to P-band frequencies), it is worth studying the various relevant types of scatterers in the forest. In the case of X-band, the relevant scatterers are those of smaller dimensions, such as leaves, needles, small stems, etc. Of course, they are extremely numerous, but not very correlated with the total biomass of the forest. Alternatively at P-band (wavelength about 70 cm), key scatterers are the bigger elements of the tree (trunk and primary or secondary branches), wooden elements which mainly drive the total biomass. Radar backscattering depends on the overall geometrical and dielectric features of these various scatterers, whose geometrical volume and wood density give the whole forest biomass. As a general rule, increasing backscattering values goes with increasing biomass whether at L- or P-bands, until a saturation value, which occurs earlier at L-band [LET 92].

9.2.1.1. *Modeling woody elements*

Woody elements of the tree can be considered as dielectric cylinders, composed of timber and filled with water (sap) in the alveoli. To model the electromagnetic (EM) interactions with woody elements, the most effective description (in terms of finding a trade-off between complexity of the description and impact of simplifications) is to use a homogeneous regular cylinder. The relevance of this description as well as those for the other shapes of scatterers will be discussed further in the section dedicated to the full simulator, which aims at modeling the overall forest scenes.

To explain the relationship with biomass, let us consider the case of a regular dielectric cylinder of a uniform circular cross-section (Figure 9.4), characterized by three parameters: the height and radius of its geometry and a complex value for the dielectric permittivity. Compared with the general case of electromagnetic environments, plant materials are rather simple: they actually have no magnetic properties, so the relative magnetic permeability is 1, and constitute a medium that is called "linear", whose dielectric constant is independent of the intensity of the electric field.

Conventionally in electromagnetism, the scattering problem from a target is addressed using the propagation equation, which is specified through the boundary conditions defined according to the interface separating the empty space from a dielectric or conductive medium. The solving complexity of these equations including boundary conditions is strongly related to the interface geometry. In the case of scattering from a dielectric cylinder, the resolution of the wave propagation equation leads, in a cylindrical coordinate system (ρ, θ, z), to decomposable solutions according to Hankel functions (of first and second kinds [ISH 78]). While there is no exact analytical formulation for cylinders of any type of dimension, an approximation, commonly used for elongation values of vegetation scatterers (ratio of height to radius typically more than 100), consists of neglecting the non-uniform effects at the edges, which are anyway not entirely found in a natural configuration with juxtaposed cylinders. The approximation of a finite long cylinder assumes that the internal field is the same as that of an infinitely long cylinder and can be solved using the same incident field.

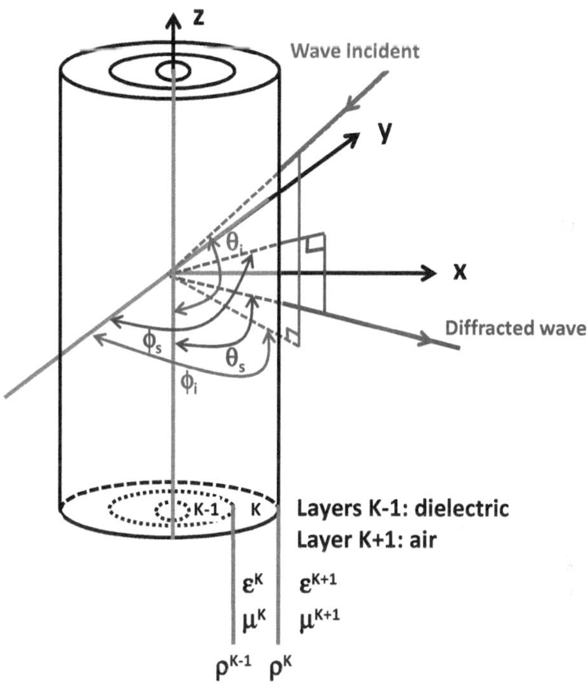

Figure 9.4. *Geometry of the scattering modeling problem of a stratified regular cylinder with the direction (θ_s, ϕ_s) of the incident wave from the direction (θ_i, ϕ_i). The dielectric and magnetic parameters (ε^k, μ^k) are indexed based on their membership in the layer K, as well as radius ρ^k (where superscript K does not mean raised to the power K but belongs to layer K)*

The internal field is found using the continuity of transversal electric and magnetic fields at the surface of the cylinder. Once the internal field is known, the radiation from this field can be calculated with the finite length assumption, using an integral equation based on the Huygens principle

[TSA 92] or volume [KAR 86]. These two integral equations are equivalent, except for the directions of sight close to the cylinder axis, but of secondary importance compared with the assumption regarding the internal field of an infinite cylinder.

From the calculation of the scattered field E^s and knowing the field on transmit E^i (incident), the effective radar cross-section (RCS) of a target at distance R is defined by:

$$\sigma = 4\pi R^2 \cdot \frac{|E^S|^2}{|E^i|^2} \qquad [9.4]$$

Expressed in m² (or in dBm² as a ratio compared with a power reference), it represents the surface which would scatter a power in an isotropic manner equivalent to the target, and which constitutes a key variable for radar users in the power budget equation [RAN 95, FER 16]. This formulation can be generalized for the various possible combinations of polarizations, depending on the polarization basis at emission and reception. In the standard case of a plane wave (far-field approximation) and with a linear field decomposition according to horizontal (denoted H) and vertical (denoted V) polarizations, the polarimetric RCS is expressed as

$$\sigma_{pq} = 4\pi |S_{pq}|^2 \qquad [9.5]$$

where p and q correspond respectively to the state of polarization at the emission and the reception (H or V on a straight-line basis), and where the term S_{pq} corresponds to one of the four complex coefficients of the scattering matrix S (also called the Sinclair matrix), relating the scattered field to the incident field by:

$$\vec{E}^S = \frac{e^{jkR}}{R} S\, \vec{E}^i = \frac{e^{jkR}}{R} \cdot \begin{bmatrix} S_{hh} & S_{hv} \\ S_{vh} & S_{vv} \end{bmatrix} \vec{E}^i \qquad [9.6]$$

where k is the radar wave number, equal to $2\pi/\lambda$.

Our analysis of the interaction between an incidental wave (formulated by an incident field) and a cylinder will be here after quantified through this concept of polarimetric RCS σ_{pq} , taking into account the different sizes and orientations according to the Euler angles in three dimensions, and particularly for the specific case of backscattering, where incident and scattering directions are the same (corresponding to the monostatic radar configuration, a special case of the bistatic general configuration still barely used in remote sensing given its operational complexity [VIL 09]). Figure 9.5 shows the variation of the RCS of a cylinder as a function of radius (r) to the HH and VV polarizations. Note the importance of the dielectric permittivities, values derived from different water content volume (20, 50 and 80%), resulting in a translation of the horizontal axis (radius), thereby justifying the use of scaled parameters in the study of the resonance phenomenon, such as:

$$k_e.r = \frac{2\pi.r}{\lambda_e} = \frac{2\pi.r}{\lambda/\sqrt{\varepsilon_r}} \qquad [9.7]$$

where k_e and λ_e denote the effective wavenumber and wavelength of the medium (corrected index propagation = $\sqrt{\varepsilon_r}$). For example, increasing the water content from 30% to 80%, it results that the real part of the dielectric constant at 430 MHz (wavelength λ of 70 cm) is multiplied by approximately 4, the effective wavelength divided by 2 so that the equivalent radius to keep the same scaled parameter $k_e{\cdot}r$ has to be twice as small as the original one.

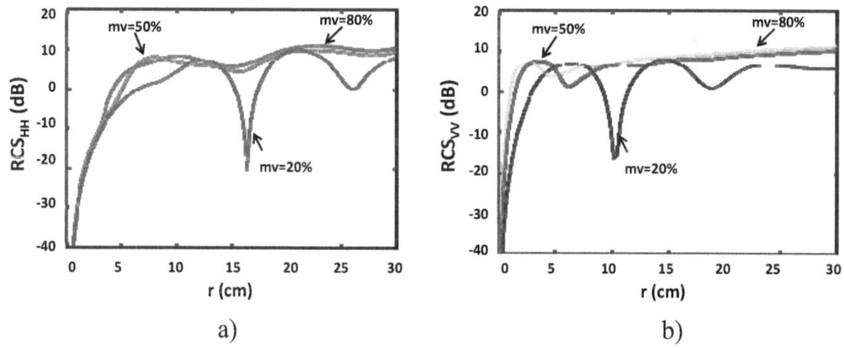

Figure 9.5. *Radar backscatter (expressed as the Radar Cross-Section (RCS) in dB) at 430 MHz, for the polarizations HH a) and VV b) and at a normal incidence (θ_i=θ_s=90°) of a straight regular cylinder and as a function of its radius r, illustrating the resonance phenomenon (at a wavelength of 70 cm) for different values of water content (mv). For a color version of this figure, see www.iste.co.uk/baghdadi/3.zip*

While these results at the scale of a single cylinder represent the basis of the sensitivity of the radar backscatter to woody elements of a full forest, the effects of having a set of scatterers, distributed within a given volume, must also be taken into account. In this context, Figure 9.6(a) shows the backscattering in a free space (still without attenuation) by a set of cylinders uniformly distributed and oriented in spherical geometry (probability density uniform in azimuth) and following sin(φ_{ins}) in elevation, where φ_{ins} represents the orientation of the branches in the vertical angle, also called the insertion angle. Because of their properties of symmetry, angular distributions of the branches can be discriminated from polarimetric measurements [VIL 09], and in the simple case of uniform angular distribution, the backscattered signal is unaffected by polarization change. Polarization dependence occurs once the cylinders are oriented in specific directions, under the effect of mechanical stresses (as for the trunks), but also by heliotropism in the case of the branches (to maximize the capture of light or on the contrary to minimize evapotranspiration).

Figure 9.6(b) also shows the forward scattering (viewing direction parallel to the direction of incidence), equally important in the study of the scattering of a set of scatterers because it is the origin of the attenuation of waves propagating in the medium. The role of attenuation (formalized by the transmissivity matrices, reflecting the propagation losses in the environment) is essential in the sensitivity of the radar backscattering coefficient in the biomass of a forest.

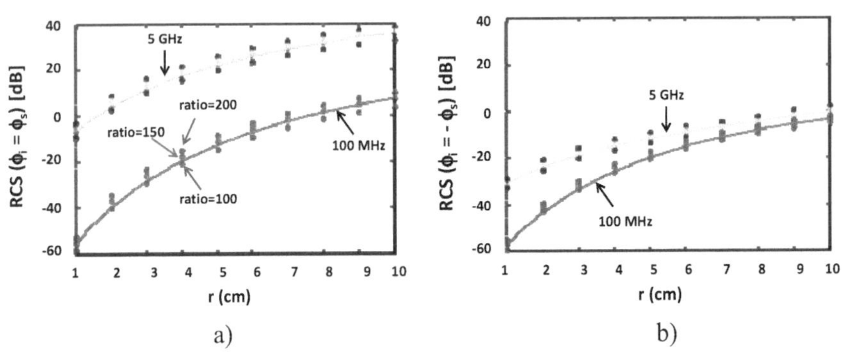

Figure 9.6. *Backscatter signal (radar cross-section RCS) a) and forward scatter b) of a set of identical cylinders uniformly oriented and separated according to their radius, at 100 MHz and 5 GHz, for elongations (height to radius ratio) of 100, 150 and 200 (respectively associated with three increasing values per radius). For a color version of this figure, see www.iste.co.uk/baghdadi/3.zip*

To complete the study of the backscatter sensitivity to the dimensions of scatterers, it is also interesting to note, with the results of Figure 9.6, the relative lack of sensitivity to the cylinder height compared with the cylinder radius, which is consistent with the dependence on the dielectric volume proportional to the height but to the square of the effective radius.

At this stage of the chapter, we can highlight the fact that it is precisely the sensitivity of backscatter radar to the dielectric volume and therefore to fresh biomass (as opposed to dry biomass deprived of water) of the modeled woody element, which is at the heart of the key relationship between radar backscatter and forest biomass.

In order to best model the electromagnetic interactions (EM) with inhomogeneous scatterers from a dielectric point of view, the most complete description – for which an analytical expression of the scattering coefficients can be calculated – is to use a straight layered and rough generalized dielectric cylinder, namely a cylinder that is geometrically characterized by its elliptical cross-sections, its surface roughness parameters and by the layers between two given radii; within which the dielectric permittivity is the same. Given the very large amount of scatterers which need to be considered in order to model a realistic scene (typically on the scale of 1 ha of forest), numerical methods such the moments method or Finite Difference Time Domain (FDTD) usually based on grids at $\sim\lambda/10$ would lead to inappropriate computing time, hence the interest in the development of analytical formulations. As another general point in modeling, but also essential here, the relevance of a more complex description has to be appreciated given our knowledge of the associated extra input parameters. As a concrete example to illustrate this trade-off, the very complex variability of the dielectric constant, not only according to radial variations or along the height of trunks or branches, but also between species and along with time, make an accurate local parameterization from *in situ* measurements impossible.

9.2.1.2. *EM models integrating a realistic description of coverage*

Given their high spatial variability (especially in terms of geometric and dielectric structures), forests are very complex media. Nevertheless, regarding the key interaction features for microwaves, a discrete description with individual dielectric elements modeled by canonical geometric shapes (ellipsoids, cylinders) can be retained. It is also very appropriate insofar as volume fractions of scatterers are relatively moderate (typically less than 5 %), making possible the use of Born approximation (DWBA – Distorted Wave Born Approximation [ISH 78]). Regarding wave propagation, its application leads to the Foldy-Lax approximation, also known as Effective Field Approximation (EFA), being derived from the Forward Scattering Theorem, resulting itself from the use of the stationary phase theorem (see [TSA 00] for the associated detailed calculations). Moreover, given the azimuthal symmetry of a forest scene (reflection symmetry whatever the plane of incidence) and usually confirmed by all natural environments, the wave propagation rule via the environment can be simplified using the following formulation:

$$\frac{d\langle E_q \rangle}{dr} = j\left(k - \frac{\kappa_q}{2}\right)\langle E_q \rangle \quad q = [V, H] \qquad [9.8]$$

where κ_q denotes the extinction coefficient for the state of polarization q, r the distance along the line of sight (as in Figure 9.12), and the brackets the average field resulting from the interactions with the scatterers, which make up the environment. This fundamental equation representing wave propagation in the environment also results in diagonal transmissivity matrices, expressed for a multi-layered environment l by:

$$[t(\vec{r})] = \begin{bmatrix} \Pi_l \, e^{-\kappa_H^l(\theta)d_l/2} & 0 \\ 0 & \Pi_l \, e^{-\kappa_V^l(\theta)d_l/2} \end{bmatrix} \qquad [9.9]$$

where d_l denotes the distance traveled by the wave in the layer l, and Π_l represents the product following all of the homogeneous layers traveled. Similarly, the received scattered field results from the application of DWBA, considering not the averaged value, but the coherent sum of the contributions of the involved scatterers. The phase of the scattered field, an essential feature for interferometric simulations, is therefore preserved (unlike incoherent models, typically based on the resolution of the radiative transfer equation). Illustrated in Figure 9.7, simple contributions (from vegetation or ground scatterers) and double bounces (resulting from the coupling between the vegetation and the underlying ground, with respect to the transmitter or the receiver [VIL 15a]) represent the major mechanisms of radar scattering.

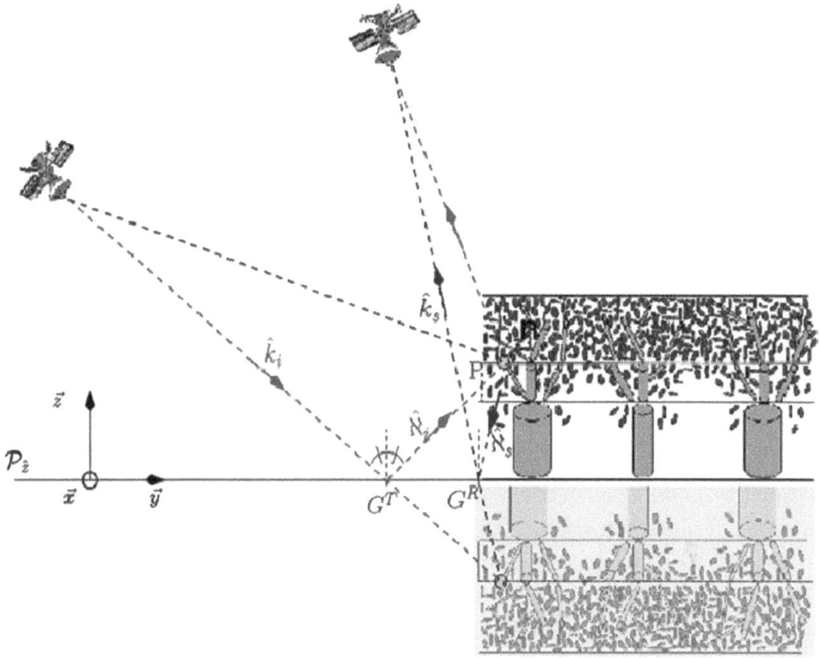

Figure 9.7. *Illustration of the scattering mechanisms resulting from the coupling between ground scatterers and vegetation, also called double bounce mechanisms. These mechanisms are modeled after image theory [VIL 15a] from points of specular reflection at ground level, here denoted G^T and G^R respectively regarding the transmitter (T) and receiver (R). In the case of monostatic radar (negligible gap position between T and R), G^T and G^R are then combined. For a color version of this figure, see www.iste.co.uk/baghdadi/3.zip*

Concerning the scattering matrices of the canonical shapes, analytical expressions derived from Rayleigh-Gans approximations are considered for ellipsoids, whereas the afore-mentioned infinite length approximation is used for the cylinders (and formulated by means of surface integrals [TSA 85]). Models of dielectric constants coming from the mixing law (between vegetation components, free and bound water) highlight the fundamental importance of the water content. Various improvements are possible from these canonical forms, such as layered cylinders (shown in Figure 9.4 and typically encountered with the trunks of trees), surface roughness [LIN 95], or curved axis or scalable sections. Although potentially very interesting, these improvements also raise the issue of the input parameterization, and the capability to fill the associated inputs based on comprehensive *in situ* data.

Concerning the ground contribution, its description by a dielectric (see [PEP 95] for the permittivity model as a function of frequency and water content) and rough surface (with exponential statistical profiles generally preferred to Gaussian one) is used to parameterize EM models such as the Integral Equation Method (IEM, see [FUN 94]) in order to calculate the scattered field. As for the vegetation scatterers, more sophisticated analytical models have been proposed in the literature (see [ELF 04]), especially with the improvements brought by the Advanced IEM, and with hybrid approaches involving partially numerical methods like the above-mentioned FDTD.

This framework provides the basis for a simulator like MIPERS (Multistatic Polarimetric Interferometric Electromagnetic model for Remote Sensing [VIL 09]) to generate observable and/or measurable remote sensing SAR data from realistic scenes of natural vegetation (Figure 9.8). Besides the possibilities of integrating ellipsoids or layered curved cylinders, the MIPERS model was developed for an optimal versatility in the way realistic scenes can be generated, in which spatial heterogeneities could be taken into account by describing multiple regions, with various vegetation structure can be tailored according to plants or trees architecture, involving specific homogeneous sub-layers that can be overlaid vertically and arranged horizontally. Within each layer, characterized by its extinction coefficient, different modes of statistical generation can be tuned according to the most suitable pdf, determining geometrical dimensions and orientation of scatterers, in a independent manner or following specific rules driven by growth models, whether for the whole tree or a single branch (as for the class of curved juxtaposed scatterers). Concerning the ground contribution, a multi-faceted subdivision enables us to specify slope angles, roughness and soil permittivity locally, plus the phase made of a deterministic component because of the travelling wave path and a random one to reproduce the speckle effect.

The multi-zone mode (modeling of vegetation patches or homogeneous ground) is particularly relevant to simulate the specific effects of forest edges (particularly visible in high resolution (<2 m) SAR images) as well as to better understand the distorting effects (layover, shadowing) the importance of the double bounce mechanism behind the observed reinforced intensity from the layover [VIL 07]. Furthermore, this feature allows us to take into account covered internal areas, in particular to model the presence of artificial targets for applications related to the detection of areas of degradation or deforestation. These artificial targets can therefore be considered as classic scatterers with field incident resulting from DWBA hypothesis associated with the ray tracing method by linked lists, the target's contribution being derived from the hybridization numerical method (FDTD [BER 94]).

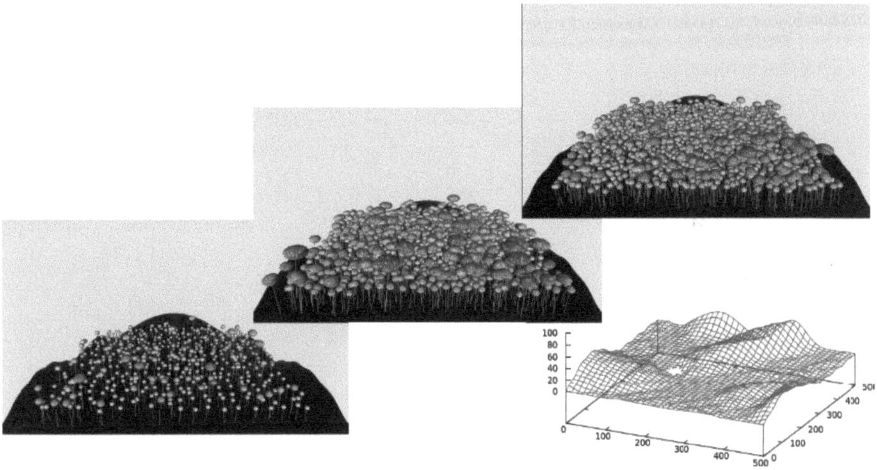

Figure 9.8. *Illustration of forest scenes generated via MIPERS to simulate SAR data from a given digital terrain model (bottom right) associated, for example, at different growth levels of a forest*

In addition to the validation through fundamental theoretical tests (especially concerning polarimetric properties of symmetry and interferometric phase wrapping [VIL 09]), MIPERS results have also been compared with experimental measurements over temperate forests using spaceborne data from ERS SAR (C-band), airborne data over the Ebersberg test site in Germany (L-band from E-SAR instrument), over the Nezer test site in France (P-band from RAMSES), and over tropical forests in French Guiana (P- and L-bands from SETHI) and in the Congo basin (L-band from ALOS-PALSAR), with mostly a specific focus on cross-comparisons along with forest ages or biomass.

9.2.2. *Relevant SAR measurements and techniques for biomass*

The potential of radar remote sensing for forest biomass estimation has been highlighted since the early 1980s. For the first SAR systems, which were neither polarimetric nor interferometric, only the backscattering intensity was used. More recently, studies have demonstrated the relevance of polarimetry and coherence of the backscattered fields, which paved the way for the development of new SAR imaging processing methods [FER 16]: Polarimetry SAR (PolSAR), Polarimetric Interferometric SAR (POLInSAR) and SAR Tomography (TomoSAR) (Figure 9.9).

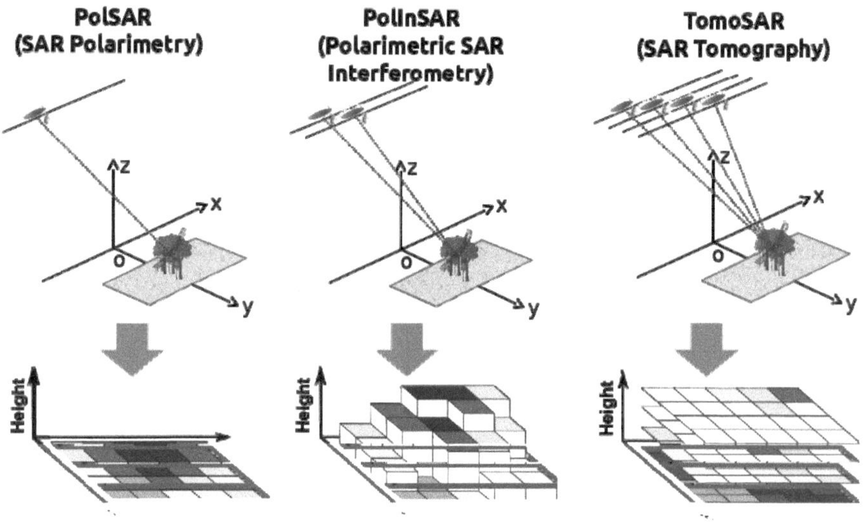

Figure 9.9. *Geometries of PolSAR acquisitions (SAR polarimetry) POLInSAR (polarimetric interferometric SAR) and TomoSAR (SAR tomography) with, below the kind of information that can be retrieved. For a color version of this figure, see www.iste.co.uk/baghdadi/3.zip*

9.2.2.1. *Interest of cross-polarization and the contribution of PolSAR*

9.2.2.1.1. Interest of HV cross-polarization

As detailed previously, SAR images provide 2D maps of backscattering measurements. Considering that these measurements can be correlated to biomass through the relation between the backscatter and the overall volume of scatterers, there is however no relation between backscatter and AGB that can be considered as universal, because of the impact of other factors such dielectric constant (related to the water content) forest structure and coupling terms with the ground.

The value of cross-polarization (HV) for estimating biomass lies precisely in the fact that this polarization minimizes the contribution of coupling terms with the ground. Indeed, the backscattered power in HV mainly from the depolarizing part (oriented branches) represents a small proportion of biomass (typically around 20%), but is highly correlated with the total biomass.

To model the relationship between backscatter and AGB (t/ha), the backscatter coefficient γ_{hv}^{0} is generally used. Expressed in *dB*, it is defined from the RCS coefficient σ_{pq} by:

$$\gamma_{pq}^{0}\left[dB\right]=10.\log_{10}\left(\frac{\sigma_{pq}}{A_s}\tan\left(\theta_i\right)\right) \qquad [9.10]$$

where the division by the term $A_s/\tan(\theta_i)$ is a normalization factor integrating resolution surface A_s following the radar line of sight and the local incidence angle θ_i. Note that this coefficient is used specifically to take into account the angular dependence of the backscattering of the vegetation, in comparison with the more general coefficient $\sigma_{pq}^{0}[dB]$ and directly derived from the concept of RCS adapted to the distributed targets (as opposed to individualized), hence the notation and the following definition:

$$\sigma_{pq}^{0}\left[dB\right]=10.\log_{10}\left(\frac{\sigma_{pq}}{A_s}\sin\left(\theta_i\right)\right) \qquad [9.11]$$

where the term $A_s/sin(\theta_i)$ represents the ground surface intercepted by the radar resolution cell (the reader may refer to [VIL 15b] for a more detailed description of the backscatter coefficients).

Considering power laws (log–log) of the type:

$$\gamma_{qp}^{0}\left[dB\right]=a_1.\log10\left(AGB\left[t/ha\right]\right)+a_0 \qquad [9.12]$$

linking the backscatter coefficient $\gamma_{hv}^{0}[dB]$ to the AGB (t/ha), it is interesting to study the dependence of the parameters a_1 and a_0 on the forest sites being considered. Figure 9.10 shows these differences for multiple sites, from tropical forests (La Selva in Panama, Paracou in French Guiana), boreal forests (Maine and Alaska in the US, Remningstorp in Sweden), temperate plantation forest sites (Landes in France and Queensland in the United States). In tropical and boreal forests, it is interesting to note (Figure 9.10) that if the offset is significant (a_0 parameter [9.12]), the slope coefficients (parameter a_0 [9.12]) representing the sensitivity of the backscattering coefficient to biomass are very similar.

Electromagnetic simulations (from the MIPERS model) show that the sensitivity of the backscattering coefficient to biomass is related to the degree of openness of forests, depending in particular on the number of stems per hectare and its evolution with biomass.

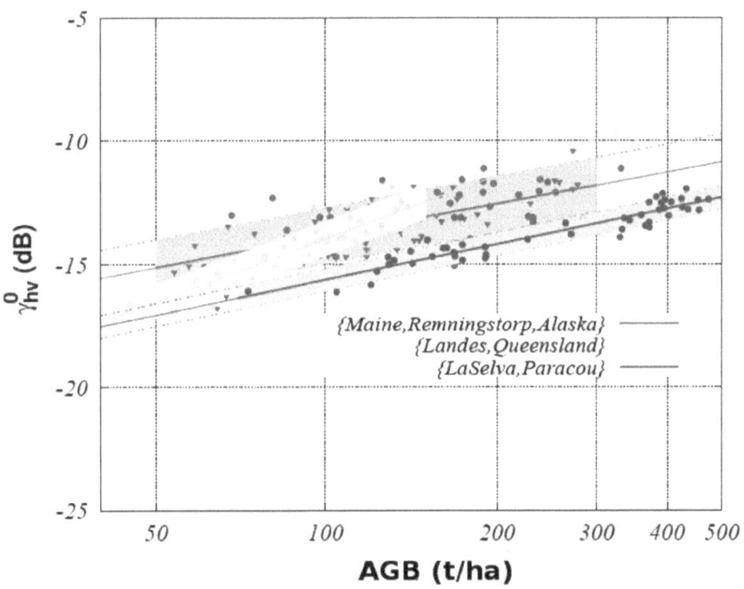

Figure 9.10. *Illustration of the power law between cross-polarized backscattering coefficient (dB) and above-ground biomass (t/ha) in logarithmic scale. For a color version of this figure, see www.iste.co.uk/baghdadi/3.zip*

9.2.2.1.2. Interest of PolSAR

The use of polarimetry for biomass estimation is not only limited to the cross-polarized backscatter. Indeed, polarimetric SAR (PolSAR) data not only allows classifications according to several relevant types of vegetation, for example for mapping and land use [LAR 09, NIA 13] or before processing backscatter coefficients into biomass, but it also allows the minimization of effects related to terrain topography [VAN 93, VIL 10], being able to mask the sensitivity of the backscatter to forest biomass [VIL 15a].

To facilitate the analysis of coherent polarimetric information, the **S** Sinclair matrix is usually decomposed from the group of modified Pauli matrices [CLO 96, CLO 02], defined in the monostatic case by:

$$\{\varphi_P\} = \left\{\sqrt{2}\begin{bmatrix}1 & 0\\0 & 1\end{bmatrix}, \sqrt{2}\begin{bmatrix}1 & 0\\0 & -1\end{bmatrix}, \sqrt{2}\begin{bmatrix}0 & 1\\1 & 0\end{bmatrix}\right\} \qquad [9.13]$$

By noting k_p the target vector resulting from the projection of S following the base of Pauli matrices, we have (still for the monostatic case):

$$k_p = \frac{1}{\sqrt{2}}\begin{bmatrix}S_{hh}+S_{vv}\\S_{hh}-S_{vv}\\2S_{hv}\end{bmatrix} \qquad [9.14]$$

allowing the introduction of the coherence matrix T defined by:

$$T = \langle k_p k_p^{\dagger}\rangle =$$
$$\frac{1}{2}\begin{bmatrix}\langle(S_{hh}+S_{vv})(S_{hh}+S_{vv})^*\rangle & \langle(S_{hh}+S_{vv})(S_{hh}-S_{vv})^*\rangle & 2\langle(S_{hh}+S_{vv})S_{hv}{}^*\rangle\\ \langle(S_{hh}-S_{vv})(S_{hh}+S_{vv})^*\rangle & \langle(S_{hh}-S_{vv})(S_{hh}-S_{vv})^*\rangle & \langle(S_{hh}-S_{vv})S_{hv}{}^*\rangle\\ 2\langle S_{hv}(S_{hh}+S_{vv})^*\rangle & 2\langle S_{hv}(S_{hh}-S_{vv})^*\rangle & 4\langle S_{hv}S_{hv}{}^*\rangle\end{bmatrix} \qquad [9.15]$$

For example, the use of the simple Pauli decomposition based on the three diagonal elements of the coherence matrix T (Figure 9.11(a)) highlights the difference between the backscattering from forests on moderate ground topography (western part of the image, more in red) and that of a forest with rugged terrain (eastern part of the image), where double bounce mechanisms also come from scatterers composed of the crowns of trees, not just the trunks. To quantify this phenomenon, it is interesting to note that even for dense tropical forests, the P-band can penetrate the canopy, making it possible to use the POA angle (Polarization Orientation Angle [LEE 00]) to assess the azimuthal component of sloped angles (Figure 9.11(b)).

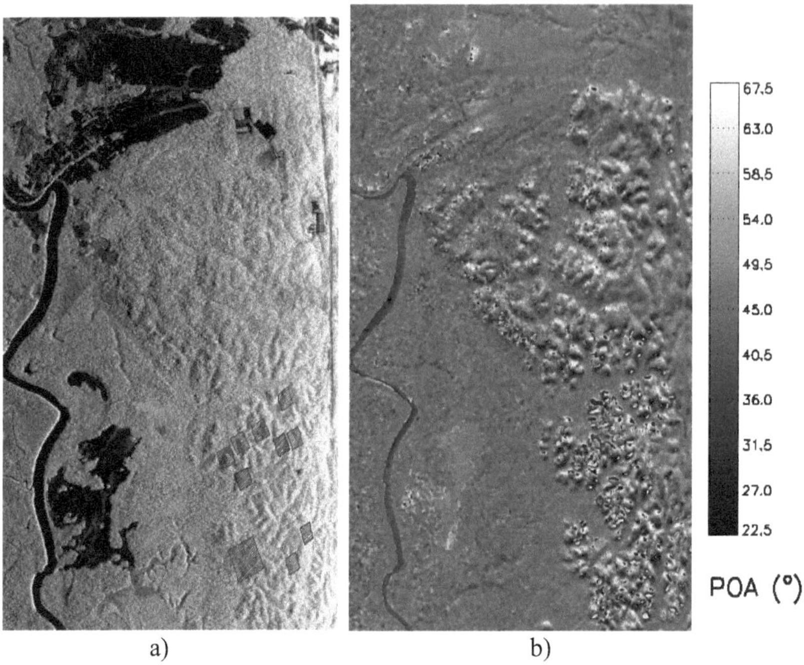

Figure 9.11. *a) Composite RGB image resulting from the Pauli decomposition (|HH–VV|, |HV|, |HH+VV|); b) image of POA in degree, which represents the azimuthal component (parallel to the flight line) of slope. For a color version of this figure, see www.iste.co.uk/baghdadi/3.zip*

9.2.2.2. *POLInSAR: estimating canopy height for biomass*

Polarimetric Interferometric SAR (POLInSAR) data enable us to derive interferometric coherence for each combination of polarizations, from repeat-pass or simultaneous SAR acquisitions, but geometrically separated by a distance called the baseline (Figure 9.12).

As detailed in [FER 16], a multiplicative model can be used to decompose the experimental interferometric coherence in terms of system (γ^{syst}), geometric (γ^{geo}), temporal (γ^{temp}) and voluminal (γ^{vol}) decorrelation. This decomposition can be generalized in polarimetric cases, namely for polarization states p and q transmitted and received:

$$\gamma_{pq} = \gamma_{pq}^{syst} \cdot \gamma_{pq}^{geo} \cdot \gamma_{pq}^{temp} \cdot \gamma_{pq}^{vol} \qquad [9.16]$$

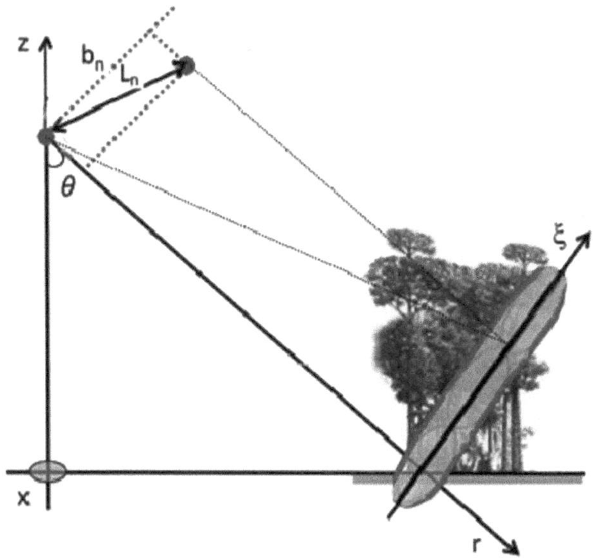

Figure 9.12. *POLInSAR geometry, acquisitions separated by the baseline L_n, under the angle of incidence θ. For each scatterer in a given resolution cell (circled in red), the phase difference because of different distances between antennas is directly related to the scatterer height. For a color version of this figure, see www.iste.co.uk/baghdadi/3.zip*

If the volume decorrelation term is considered as a perturbation for InSAR applications (Interferometric SAR), it is precisely the sensitivity of that term to vegetation volume above the terrain surface that is exploited by POLInSAR to estimate vegetation height. This volume decorrelation term being dependent to the combination of polarizations selected, POLInSAR coherences can then allow us to obtain several independent observations.

For each polarization, the module and phase of the complex POLInSAR coherences measure a decorrelation magnitude and the phase center height, resulting from the contributions of the various mechanisms involved in backscatter, but do not directly measure forest height. This height is actually estimated through the use of theoretical models linking the PolInSAR coherences to descriptive parameters of forest models. These models are based on an analytical calculation of the backscattered electromagnetic fields, and therefore require simplifying assumptions of the forest description. Based on previous theoretical work [ISH 78], it has been shown [TRE 96] in fact that the decorrelation because of a set of identical scatterers independent and uniformly distributed within a layer at height h could be expressed as:

$$\gamma^{vg} = \frac{\int_0^h \rho_d(z) e^{ik_z z} dz}{\int_0^h \rho_d(z) dz} \quad [9.17]$$

where $\rho_d(z)$ represents the attenuation for a scatterer located at height z above a reference surface and seen from the incidence angle θ, through a homogeneous medium of extinction coefficient κ:

$$\rho_d(z) = \exp\left(\frac{2\kappa z}{\cos\theta}\right) \quad [9.18]$$

The parameter k_z represents the sensitivity of the phase with respect to the scatterer height, which can be expressed from the parallax geometry (Figure 9.12) by:

$$k_z = \frac{4\pi}{\lambda r \sin\theta} b_n = \frac{2\pi}{h_a} \quad [9.19]$$

where r represents the distance along the line of sight, b_n the normal component of the baseline, and where we introduce the height of ambiguity denoted h_a. Essential in interferometry, this parameter can be interpreted as the layover height beyond which the interferometric phase must be unwrapped (to correct for the 2π shift).

We then understand how the decorrelation introduced by a vertical distribution of scatterers is related to layer height and extinction, which can be represented on the circle complex united according to these two parameters (Figure 9.13).

This formulation, independent of polarization (in the case of an isotropic medium) only describes decorrelation because of simple scattering from vegetation elements, and does not take into account either the direct contribution from the underlying surface or the coupling effects between ground and volume scatterers. Considering next a uniform underlying surface whose contribution is independent of the vegetation, the volume decorrelation (vegetation + ground) is given by:

$$\gamma_{pq}^{vol} = e^{j\Phi_0}.\left[\gamma^{vg} + L_{pq}\left(1-\gamma^{vg}\right)\right] \qquad [9.20]$$

with:

$$L_{pq} = \frac{\mu_{pq}}{1+\mu_{pq}} \quad ; \quad 0 \le L_{pq} \le 1 \qquad [9.21]$$

where we introduce the reference phase of the soil ϕ_0 as well as the relationships noted μ_{pq} between the intensity of ground backscattering and volume backscattering.

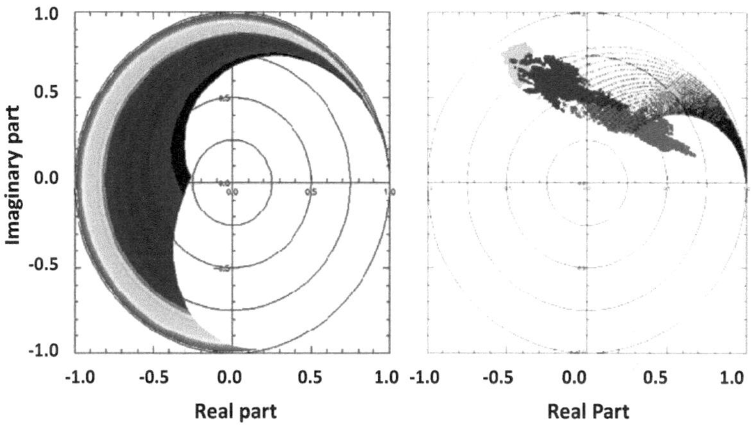

Figure 9.13. *On the left, loci of volume decorrelation because of vegetation, for heights ranging from 0 to 50 m (and a height ambiguity h_a of 60 m) and a variable extinction coefficient from 0 to 2.2 dB/m (from red to black). On the right, unit circle representing places of POLInSAR coherences (HH, HV and VV, respectively in red, green, blue) simulated with the MIPERS model in order to reproduce the dispersion (especially because of speckle) around theoretical decorrelation mean values. For a color version of this figure, see www.iste.co.uk/baghdadi/3.zip*

By considering the possible values L_{pq} (bounded between 0 and 1) based on backscatter mechanisms involved in each polarization, one can then understand the interest of polarimetric interferometric coherences, in the case of a polarization state which maximizes scattering mechanisms because of vegetation scatterers (typically HV for which $\mu_{hv} \sim 0$), and a polarization state maximizing scattering mechanisms involving the ground contribution (typically HH $\mu_{hh} >> 1$).

Assuming that the combinations of polarization are sufficiently discriminating (cloud of points spread enough), the phase reference, the forest extinction coefficient and height can be estimated from the POLInSAR coherences, using a rather simple three-step inversion algorithm [COL 03], at the origin of the infatuation for the POLInSAR technique.

Although highly effective in its simplicity, this formulation linking decorrelation and height leads to varying levels of accuracy, especially when it comes to estimating tropical forest heights with about 20% of accuracy, in order to deduce their biomass from allometric relationships (given the decrease in sensitivity of the height–biomass relation with increasing biomass values).

Various improvements to this formulation called "RVoG" (Random Volume over Ground) have been proposed to better model the vertical structure of a forest and its reflectivity, which are usually not uniform, either using layered models [VIL 09] or continuous functions [GAR 10]. However, these improvements also drive an increased complexity of the descriptive model through the use of new input parameters, which accordingly implies an increase of the unknowns to be addressed by the inverse problem. More robust methods against these disturbances based on advanced statistical analysis [ROU 11] and specific decomposition of the POLInSAR coherency matrix (detailed in [TEB 09]) are therefore preferred.

Among perturbation sources which hinder the estimation of volume decorrelation multi-pass POLInSAR coherences, temporal decorrelation is certainly the most critical, which explains why no spatial mission using multi-pass POLInSAR has yet been launched (single-pass POLInSAR being only possible in bistatic or quasi bistatic configuration, coming, however, other constraints).

To better understand the phenomenon of temporal decorrelation, especially at P and L frequencies which would allow forest height estimation from multi-pass POLInSAR coherences accurate enough to estimate and to derive forest biomass, ground-based experiments from flux towers have been recently carried out in tropical rainforests. Detailed in [HAM 14], the time series of the TropiScat experiment (consisting of a scatterometer operating from a guyaflux tower in French Guiana) made possible quantitative and qualitative analyses of the temporal decorrelation causes. Based on EM simulations from the MIPERS model, the impacts of temporal decorrelation because of diurnal variations of the water content in vegetation, or because of the motion of the scatterers caused by convective winds could be quantified separately (Figure 9.14). Given the very good adequacy of these simulations with experimental results presented in [HAM 14], multifactorial causes of temporal decorrelation could be identified and separated according to changes in water content or convective motions. Furthermore, this analysis has also been extended to the lower part of L-band frequencies, as shown in Figure 9.15 in which one can quantitatively appreciate how much the decorrelation at L-band is stronger than at P-band. These results come not only from the motion of the scatterers because of convection, but also from the variations in time of the dielectric values, considering for both frequencies the same gradient of water content in the vegetation.

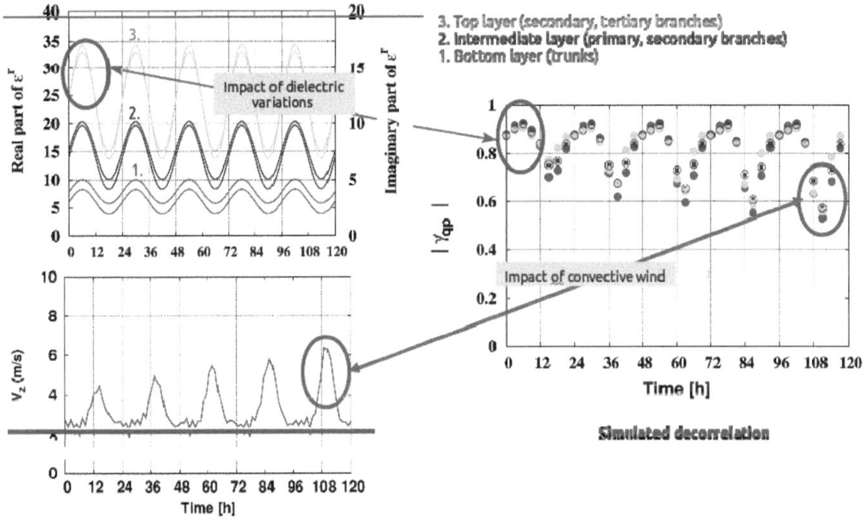

Figure 9.14. *Simulated results of temporal decorrelation at P band (right), considering the coherence magnitudes noted γ_{qp} for HH (red), HV (green) and VV (blue) polarizations, based on simulated diurnal variations of the water vertical gradient within the vegetation (expressed in the top left graph through complex dielectric permittivity values for each of the three layers describing forest structure), and based on simulated diurnal variations in the vertical velocities of winds related to convection in the canopy (bottom left). For a color version of this figure, see www.iste.co.uk/baghdadi/3.zip*

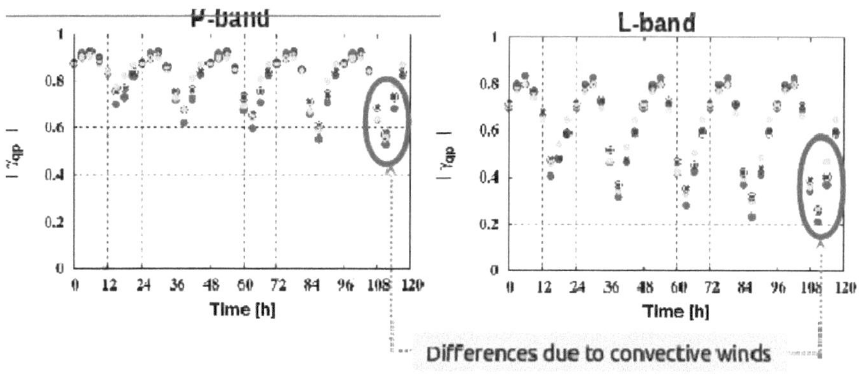

Figure 9.15. *Comparisons between simulated temporal coherences depending on the time of day (reference data at 6 am) at P-band (left) and L-band (right), and for HH (red), HV (green) and VV (blue) polarizations. Temporal variations of water content and convective winds are identical to those shown in Figure 9.14. For a color version of this figure, see www.iste.co.uk/baghdadi/3.zip*

9.2.2.3. *TomoSAR: providing 3D polarimetric backscattering of cover*

The potential of SAR tomography for characterizing vegetation cover of forests has been highlighted by numerous airborne campaigns. Various methodological developments based on this technology have resulted in numerous scientific publications since the first results [REI 00] with more advanced approaches in terms of spatial and radiometric resolutions [TEB 09], and the estimation of biomass, especially in the case of dense tropical forests [MIN 14].

SAR tomography is based on the coherent combination of N SAR 2-D measurements acquired in interferometric configuration and which allows for the estimation of 3D reflectivity, $|f(z,r,x)|^2$, for complex environments. As shown in Figure 9.16, the 3D focusing by tomography allows for the natural separation in a vertical direction of the various constituents of a volumic medium while SAR 2D information results from the sum of contributions which are very difficult to separate even with modes of diversification (polarimetric, frequential) without very strong assumptions about the characteristics of the observed medium.

To explain the basic principle of SAR tomography, consider the geometry of Figure 9.16, representing N parallel lines of flight along the x axis. The N

SAR 2D measurements (n ranging from 1 to N, where N is the number of images) can be expressed in terms of the distance r and azimuth x by:

$$y_n(r,x) = \int S(\xi,r,x)e^{jk_z^n \xi} d\xi \qquad [9.22]$$

where the only dependence of the acquisition *n* comes from the term, the wave number projected along the orthogonal projected baseline, if it is assumed that *N* images were resampled according to the same “master” image, phase corrected to take into account the local topography, as well as positioning errors [BAM 98, TEB 10], and for reflectivity in a stationary scene *S*(*z*,*r*,*x*). The estimation of this reflectivity is then similar to a spectral problem, and the most immediate way to solve it is to use an inverse Fourier transform, thus:

$$\hat{S}(\xi,r,x) = \sum_{1 \le n \le N} y_n(r,x)e^{-jk_z^n \xi} \qquad [9.23]$$

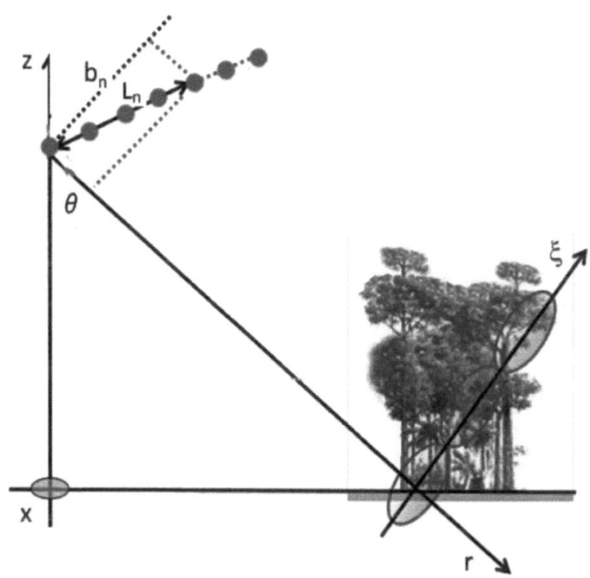

Figure 9.16. *Geometry of multi-baseline TomoSAR acquisition (line of sight according to the incidence θ from the axis r), allowing for the discrimination of different voxels (circled in red) along the axis ξ and to obtain vertical resolution along z. For a color version of this figure, see www.iste.co.uk/baghdadi/3.zip*

The resulting vertical resolution is therefore given by:

$$\Delta z = \frac{\lambda}{2} \frac{r \sin\theta}{b_{max}} \quad [9.24]$$

with b_{max} the normal component of the maximum length of the baselines, also considering a sufficient sample relative to the Nyquist criterion, requiring a minimum distance between passes $\frac{\lambda}{2} \frac{r \sin\theta}{H}$ in order to avoid aliasing phenomena in a layer height H. While this method provides an accurate estimate of the reflectivity, it is, however, costly in terms of the number of acquisitions needed to obtain relevant vertical resolutions to identify the structure of forest cover, resulting in more optimal methods as detailed in [GIN 02], but at the expense of the radiometric resolution.

Preserving the radiometric resolution is indeed important to the extent that the backscattering value layers resolved by the TomoSAR processing may be connected to the biomass.

On a study site in dense rainforest (Paracou, French Guiana), it was shown in [MIN 14] that the reflectivity of the layer centered on 30 m and at ±5 m is strongly related to biomass. This result is interpreted by the lack of soil contribution (causing major disruptions on SAR backscatter coefficients) and the existence of a strong relationship between the biomass of this layer and the total above-ground biomass. The observation led to the development of a method that provides an estimate of biomass at 10% of the biomass *in situ* and at a resolution of 1.25 ha. For the biomass spatial mission (see section 9.4), the feasibility of the tomographic method has been demonstrated [MIN 15], including under the constraint of a frequency band reduced to 6 MHz (thus limiting the spatial resolution along the distance and vertical axes).

In Figure 9.17, a sensitivity analysis was performed by assessing the correlation between HV backscatter at different height layers (using tomography) and the measures of AGB for 112 plots of 1 ha each. At ground level, the backscatter is very weakly correlated with biomass because of the high extinction of signals in the upper canopy. However, the correlation is very high (about 0.75) for an intensity evaluated at the upper layer of vegetation (30 m), for AGB values ranging from 150 to 600 t/ha. This implies that:

– the disruptive effect of soil contribution is minimal, the TomoSAR focus separating the contributions of soil and volume, allowing a better characterization of the diffusion by vegetation;

– there is a strong correlation between the biomass contained in the section of heights ranging from 20 to 40 meters and the big trees, which largely contribute to the total AGB of a forest plot [MIN 14].

Figure 9.17 clearly shows that the correlation between the intensity of the radar signal and the biomass is much lower than that obtained by tomography for the upper layers of the canopy (Pearson coefficient of 0.37 with the use of backscatter intensity and 0.75 with the use of TomoSAR intensity sampled at a canopy height of 30 m).

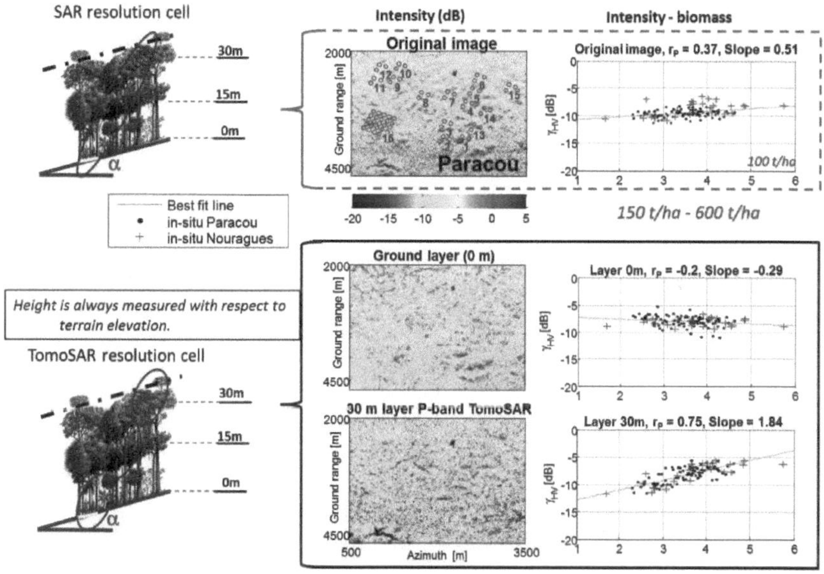

Figure 9.17. *Correlation (Pearson coefficient denoted r_p) between P-band backscatter and AGB over 112 plots (85 in Paracou and 27 in Nouragues, French Guiana). The upper graphs show the HV backscatter, the others correspond to TomoSAR backscatter sampled at different heights. For a color version of this figure, see www.iste.co.uk/baghdadi/3.zip*

In order to study the robustness of this approach [MIN 16] the TomoSAR method developed on the site of Paracou was applied on the neighboring site of Nouragues (~100 km south, also in French Guiana), but with a rather different canopy structure and a more rugged topography. Figures 9.18 and 9.19 show the estimated results of the height and biomass. The results show the strong potential of the SAR tomography to calculate biomass and the height of dense tropical forests, including sites with significant topography and high biomass values (up to 500 t/ha). Estimates of the height and biomass of dense rainforests could be performed with an average relative error of about 10% and 15%, respectively.

9.3. Biomass from SAR L-band

9.3.1. *Introduction*

As stressed before, large wavelengths like the ones at L- or P-bands are more for the monitoring of forests. The Japanese space agency, JAXA, was the first to explore that at L-band and continues to do so successfully. First, the JERS satellite in orbit from 1992 to 1998 was equipped with an SAR L-band (1270 MHz, $\lambda = 23.6$ cm), polarized uniquely in HH. The PALSAR instrument of the ALOS satellite has been a great milestone in the history of environmental applications from SAR remote sensing, including forest monitoring. Launched in January 2006 by JAXA and operational until May 2011, it had the ability to function in a fully polarimetric manner, with a 46-day revisit period. The ALOS mission is considered to have been a pioneer in systematic data acquisition which allowed for the imaging of the planet twice a year in high resolution, in the same way, in the same seasons, each area being imaged within a limited time frame to minimize seasonal heterogeneities. This strategy of systematic acquisition enabled the production and free distribution by JAXA of "mosaic" productions in up to 25 m resolution. These mosaics are radar images preprocessed by the JAXA SIGMA-SAR processor, available in tiles of 1 degree squared in HH and HV polarizations, for the years 2007–2010. They are constructed from PALSAR images acquired using the Fine-Beam Double polarization (FBD) method. For the first time, homogeneous SAR data, preprocessed and thus easily exploitable data worldwide, became accessible to everyone[6].

6 http://www.eorc.jaxa.jp/ALOS/en/palsar_fnf/fnf_index.htm.

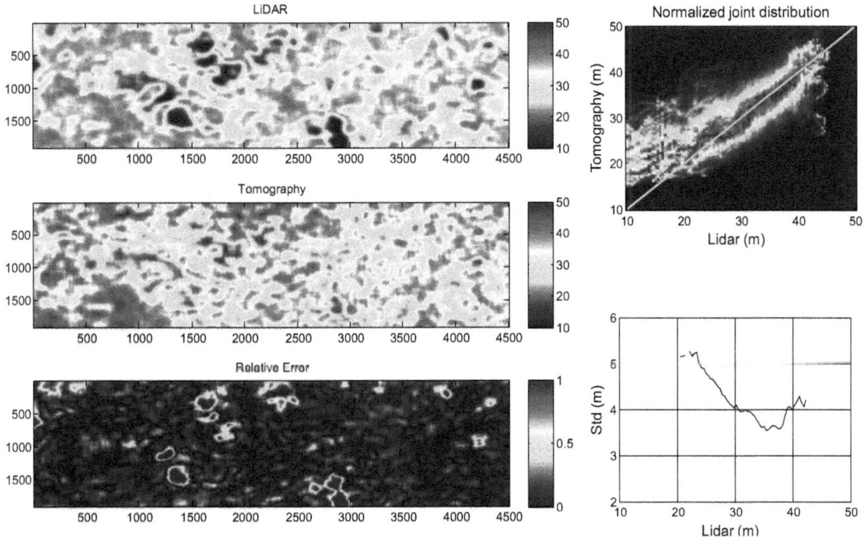

Figure 9.18. *Comparison between heights from tomography and heights computed from airborne LiDAR data (Guianese company ALTOA). The alignment between the two estimates is represented by their joint distribution (top right), the relative error* $|H_{tomography} - H_{LiDAR}|/H_{LiDAR}$ *on the bottom left, and the standard deviation* $H_{tomography}$ *relative to* H_{LiDAR} *on the bottom right. For a color version of this figure, see www.iste.co.uk/baghdadi/3.zip*

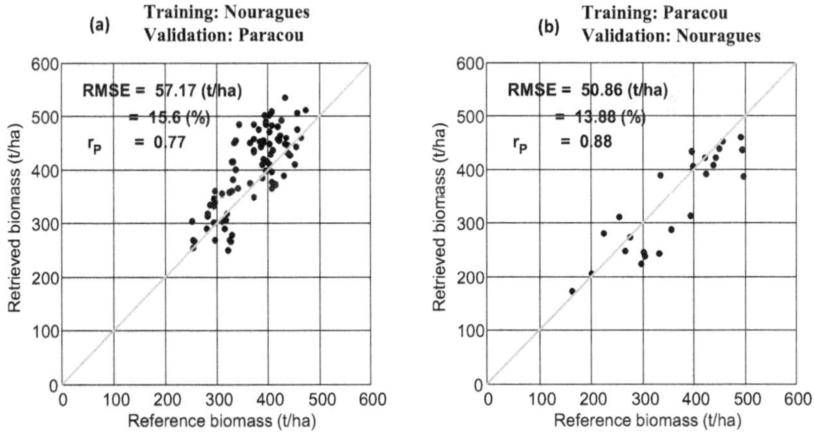

Figure 9.19. *Comparison between biomass derived from tomography and biomass in situ: a) the method is carried out on plots in Nouragues and validated on Paracou, b) the method is carried out on the plots in Paracou and validated on plots in Nouragues*

The ALOS/PALSAR data have been widely used to estimate the above-ground biomass [LUC 10] or wood volume [SAN 09]. Regarding boreal and temperate forests, certain authors used PALSAR mosaic directly, such as Antropov *et al.* [ANT 13], and estimated the volume of wood on two Finnish sites using a semi-empirical model (RMSE = 43%). Others, like Cartus *et al.* [CART 12] have used PALSAR FBD images to estimate the biomass of the northeastern United States, based on the method of Santoro *et al.* [SAN 11]. Sun *et al.* [SUN 11] have also mapped the biomass in a part of Maine in the United States jointly using PALSAR and LiDAR data. In China, Sarker *et al.* [SAR 12] benefitted from the texture of ALOS data to estimate biomass, while Ni *et al.* [NI 13] exploited the minimum distances between the simulated intensities from a forest growth model, and the intensities measured by PALSAR. PALSAR is also widely used to estimate the biomass of tropical forests. Cartus *et al.* [CART 14], for example, produced a Mexican national biomass map at 30 m resolution based on FBD data and the product Landsat tree cover. Morel *et al.* [MOR 11] also exploited the FBD data to estimate the biomass of forests threatened by palm oil plantations in Malaysia, while Thapa *et al.* [THA 15] use the textures from PALSAR data in Sumatra. Tropical and subtropical savannas are also studied with PALSAR since their lower carbon stocks than dense forest per unit area allow the L-band to be used to draw some of the data that saturates between 70 and 150 t/ha. Mermoz *et al.* [MER 14] produced a national map of Cameroon's savannah biomass at 25 m and the associated uncertainties from mosaics. Other national or sub-national maps have been produced: Carreiras *et al.* [CARR 12] produced a biomass map of Guinea-Bissau, Avtar *et al.* [AVT 13] a map of Cambodia, and Michelakis *et al.* [MIC 14] part of Belize.

9.3.2. *The relationship between radar measurements (L-band) and biomass*

Most of the above-cited studies based on L-band data are built on the fact that the intensities are positively correlated with biomass. The signal sensitivity decreases as the biomass increases, up to a point called the saturation point for which the sensitivity is lost. A few studies [RAU 94, WOO 06, LUC 07] have discussed a weak negative correlation of intensity with biomass after the saturation point. This correlation was theoretically and experimentally highlighted by Mermoz *et al.* [MER 15] for the case of dense forests (high number of trees per unit area). In this study, the MIPERS model was used for the theoretical demonstration of the relationship between the radar intensity and biomass. The forest, that is to say the trunks and branches (the intensities in the L-band being very insensitive to leaves), has been described using simple geometric shapes, dielectric cylinders. A forest growth model was developed based on the allometric relations of the article by Asner *et al.* [ASN 11], developed using 130 biomass plots of 0.28 ha in Peru. This growth model allowed the calculation of the size of the dielectric cylinder used in the MIPERS model with the number of trees and the ratio of canopy height/main trunk. The full description of the method is detailed in the article by Mermoz *et al.* [MER 15]. Simulation results are presented in Figure 9.20.

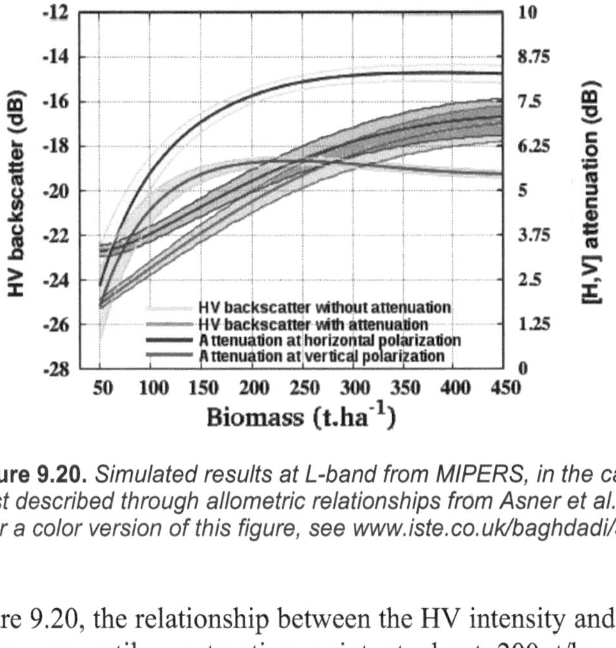

Figure 9.20. *Simulated results at L-band from MIPERS, in the case of rainforest described through allometric relationships from Asner et al. [ASN 11]. For a color version of this figure, see www.iste.co.uk/baghdadi/3.zip*

In Figure 9.20, the relationship between the HV intensity and biomass (in green) increases until a saturation point at about 200 t/ha. Indeed, the number of scatterers in the canopy increases with biomass and causes an increase in the backscatter intensity. The attenuation describing the losses of the signal through the canopy also increases with the biomass (in blue and red for the horizontal and vertical polarizations, respectively). After the saturation point, a slight decrease in intensity in HV is observed. Indeed, the intensity no longer increases with the number of scatterers, while attenuation continues to increase, resulting in lower intensity with biomass. To illustrate this phenomenon, the signal strength was also plotted with attenuation omitted (in cyan). It is observed that the intensity decreases more with biomass and that all sensitivity to biomass is lost.

The relationship between intensity and biomass has been extensively studied experimentally in the L-band. An example, at HV polarization and over the entire range of biomass up to 550 t/ha, is shown in Figure 9.21. The red dots are from 26 biomass plots of 1 ha, measured in the framework of the European FP7 project REDDAF [HAE 12] in the areas of Adamawa and Central Cameroon. The black dots are from 632 plots of biomass all greater than 1 ha, measured through commercial inventories of forests in the Central African Republic and then processed to improve the biomass estimate. These plots have been selected based on field slope criteria and homogeneity of plots. Details about the selection of plots are published in Mermoz *et al.* [MER 15]. In Figure 9.21, significant growth was observed in intensity with low biomass levels, followed by a loss of sensitivity and a slight decrease in the signal beginning at 150 t/ha (black dots). The analysis of experimental data therefore corroborates the simulation results of the MIPERS model.

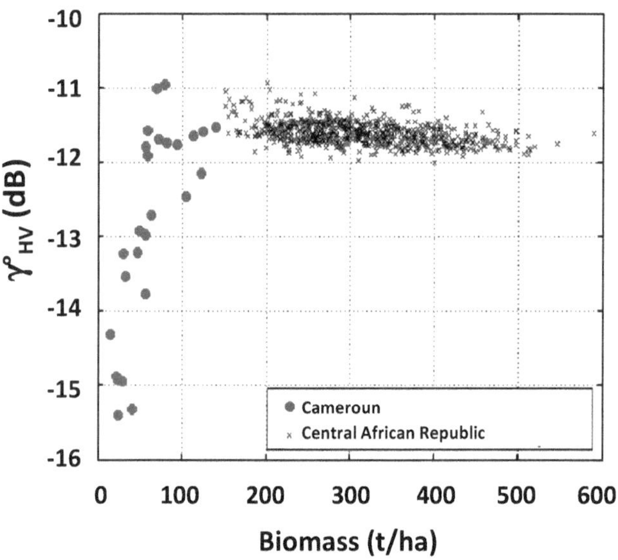

Figure 9.21. *Backscatter coefficient for L-band and HV polarization relative to biomass. The red dots are from 26 biomass plots of 1 ha measured in the framework of the European FP7 project REDDAF [HAE 12], in Cameroon. The black dots are from 632 biomass plots of 1 ha, measured through commercial inventories of forests in the Central African Republic. For a color version of this figure, see www.iste.co.uk/baghdadi/3.zip*

Based on theoretical and experimental results of sensitivity of the SAR L-band intensity in relation to biomass, it is possible to use the SAR data in the L-band to estimate biomass for up to a maximum of 150 t/Ha. For biomass at higher values, the low sensitivity of biomass to this intensity appears much more difficult to exploit. The choice of models linking the L-band intensity with low biomass is essential for proper inference of biomass from these models. For example, a logarithmic model as used by Mitchard *et al.* [MIT 09] or Englhart *et al.* [ENG 11] assuming that the intensities do not saturate and are sensitive to very high biomass values, is not suitable and is even less so when the sensitivity of this intensity is projected at high biomass values. Many studies use increasing asymptotic regressions to describe the relationship between biomass and intensity, and which describe the loss of sensitivity of the intensity when biomass exceeds a critical value (usually 150 t/ha maximum). In all cases, the use of a model having a physical base leads to better results inference that empirical regressions. The Water–Cloud model [ATT 78] allows us, for example, to decompose the intensity weighted in terms of scattering mechanisms. It allows us to express in a simple manner the fact that the signal is mainly composed of diffusion from the ground for very low biomass values, of volume diffusion for high biomass values, and two simultaneous diffusion mechanisms for intermediate biomass values. The Water-Cloud model connects the backscatter coefficient γ^0 to the AGB biomass as follows:

$$\gamma 0 = a\, e^{-c\, AGB} + b\,(1 - e^{-c\, AGB}) \qquad [9.25]$$

where:

– a represents the contribution of the soil and holes in the forest (zero biomass) and varies according to the moisture, roughness and topography to the ground, and the biomass of small vegetation;

– b represents the value of the maximum intensity when the signal is saturated, corresponding to very high biomasses;

– c is the attenuation coefficient of the forest and depends on the structure and moisture of vegetation.

The component $a\, e^{-c\, AGB}$ represents the backscattering because of the ground, attenuated by the forest, whereas $b\,(1 - e^{-c\, AGB})$ represents backscatter because of vegetation.

The impact of the variations of parameter *c* on the relationship with biomass intensity when using the *Water-Cloud model* is shown in Figure 9.22. Part of the data is from the REDDAF project at five different dates as well as data published by Mitchard *et al.* [MIT 11]. Different regression models in Figure 9.22 correspond to the values of *c* equal to 0.017, 0.02, 0.025, 0.029, 0.035 and 0.04 ha/t, the best regression (solid black line) corresponding to a *c* value equal to 0.029 ha/t (r_p = 0.89 and *p*-test = 3 × 10^{-7}). The parameter *c* is often considered fixed when inverting biomass, despite the impact of this parameter on models linking intensity to biomass, as highlighted in Figure 9.22.

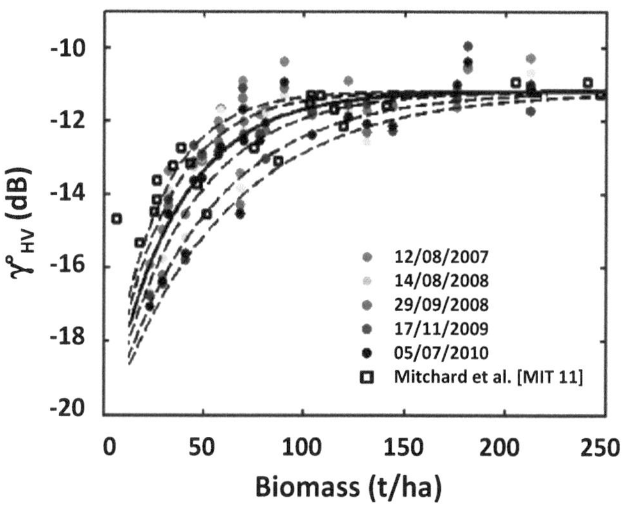

Figure 9.22. *Backscattering coefficient at L-band and HV polarization at five different dates in relation to biomass. The round points are from 21 plots of biomass of 1 ha measured in the framework of the European FP7 project REDDAF [HAE 12], in Cameroon. The black squares are from plots measured in Cameroon and published in Mitchard et al. [MIT 11]. The solid black line results from the regression using all points. For a color version of this figure, see www.iste.co.uk/baghdadi/3.zip*

The parameters *a*, *b* and *c* can be estimated in different ways:

– we can estimate the three parameters using statistical regressions between the SAR and *in situ* data, provided that they are distributed evenly over a wide range of biomass;

– if field data are not distributed evenly over a wide range of biomass, the parameters *a* and *b* can be approximated using the SAR data and, optionally, auxiliary data, as Santoro *et al.* did [SAN 11]. The parameter is estimated from the pixels corresponding to bare ground and parameter *b* from those of dense forests, for example using the Landsat tree cover map [SEX 13], calculated from optical data. Santoro *et al.* [SAN 11] almost arbitrarily then set the value of *c*. The parameter *c* can also be estimated using statistical regression as in Mermoz *et al.* [MER 14];

– the variability of the three parameters *a*, *b* and *c* can be estimated using the simulation results of the MIPERS model, using a growth model of adequate forest to properly define the inputs.

9.3.3. *Inversion in the relationship between radar measurements (L-band) and biomass*

Many inversion methods exist to infer the biomass from SAR data, often associated with optical data and/or others. Some studies use empirical methods [PER 13], based on Bayesian theory, learning machine methods such as regression trees [WIL 14], neural networks, support vector machines or the method of maximum entropy. It is also possible to invert a semi-empirical physical model as presented in equation [9.25]. In this case, biomass was estimated using the equation below:

$$AGB = (1/c) \times \ln\left[\frac{b-a}{b-\gamma^0}\right] \qquad [9.26]$$

This equation, which constitutes the basic relationship between biomass and radar indicator, is integrated later in the detailed Bayesian formulation which aims to minimize the propagation of errors associated with measurements (radar and *in situ*) and in modeling (structure and parameterization) when inverting the biomass.

9.4. The BIOMASS mission: forest biomass with P-band SAR

Initiated by CESBIO (*Centre d'Etudes Spatiales de la Biosphère*, France) in 2005 as the result of extensive research work [LET 92] and also fostered by the release to civilian applications of a short but usable part of the frequency spectrum at P-band, the European Space Agency (ESA) has selected BIOMASS in May 2013 as the 7th mission of its Earth Explorer program [ESA 12].

9.4.1. *Scientific objectives of the mission*

The scientific challenge of the BIOMASS space mission, dedicated to the estimation of forest biomass and changes at global scale, has been undertaken in the context of a better understanding and analysis of the participation of forests in the carbon cycle and therefore climate change. The estimated biomass and its variations will indeed enable us to quantify and predict the role of the Earth's surface as a carbon source or sink. There is also an economic dimension, considering the development of a global carbon market. Indeed, the BIOMASS mission will allow developing countries to rely on reliable mapping to meet the requirements of Reduced Emissions through Deforestation and Degradation (REDD), the United Nations program that prompts these countries to compensate for the carbon emissions they are responsible for.

Above all, the focus will be directed at critical areas regarding the terrestrial carbon balance. This is the case of the tropics, for which deforestation is considered as the main source of carbon flux caused by land use change (the reported figure is 98% [FAO 04]), albeit with considerable uncertainty in its estimation, as detailed in section 9.1. It should be remembered that our misestimation of deforested areas and associated quantities of lost biomass (deforestation efficiency) is causing uncertainty in terms of the flux estimates from land use, reported values ranging from 4% to 26% of flux from anthropogenic emissions, around a mean value of 1.1 GtC/year.

Carbon sinks related to the growth of tropical rainforests are indicated as being half of the Earth's carbon sinks. However, terrestrial sink estimates are derived from global carbon levels: their nature, their distribution in the world and their amplitude are very uncertain. In this context, the carbon sink and terrestrial carbon source estimates will be significantly improved with the reliable measurement of biomass in tropical forests. This need for information on the biomass is crucial for developing countries, especially to initiate and validate actions to reduce climate change.

Estimating the biomass of tropical rainforests is a difficult task, whether on the ground, airborne or by satellite. So far, no spatial instrument has been capable of measuring the biomass range of tropical forests (dense forests of a complex structure up to 500 t/ha and more). The work over recent decades has shown that only a radar operating at low frequency would both penetrate through a canopy (40–50 m high), and have the sensitivity for wooded biomass elements necessary for measuring the above-ground biomass of the tropical forest. Hence, the choice of the SAR P-band for the BIOMASS mission, made possible by the recent release to civilian applications of this spectrum (6 MHz bandwidth for a central frequency of 435 MHz).

9.4.2. *Concept of the biomass mission*

The unique sensitivity of SAR P-band measurements to the above-ground biomass of forests, will be exploited using a set of advanced methods to provide estimates of forest biomass and height, and their changes depending on human disturbances or natural cycles. The entire range of biomass will be estimated every 6 months, at a spatial resolution of 4 ha. The resolution and uncertainties of biomass products (shown in Figure 9.23) have been defined not only according to the requirements of international treaties on carbon stocks, but also in close collaboration with scientific expertise in terrestrial carbon models.

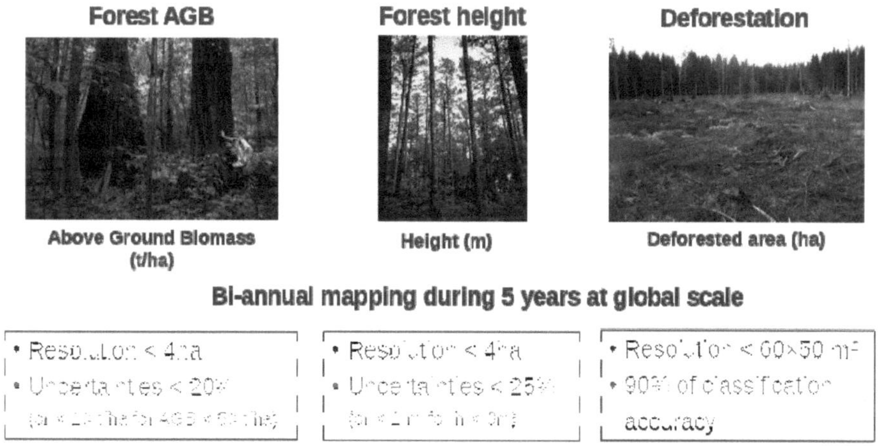

Figure 9.23. *Products from the BIOMASS mission*

BIOMASS will also provide the very first images of the Earth at P-band, giving us the possibility to explore the Earth in an entirely new dimension.

Monitoring of glaciers, geological mapping of underground areas in deserts and mapping the topography under canopies form part of the opportunities identified among these benefits, and many other applications will certainly emerge during the mission timelife.

The mission includes a single satellite operating on a near-polar orbit, sun-synchronous, at an altitude of 637–666 km, depending on the operation phase of the mission. The orbit is defined to allow for multi-pass acquisitions for interferometry and to minimize the impact of ionospheric disturbances.

The recommended use of the Vega launcher led to a reflector antenna (about 12 m in diameter). The mission is scheduled for dawn-dusk acquisitions (06:00/18:00), in particular to limit ionospheric effects.

The retained P-band SAR will have polarimetric and interferometric capabilities, to achieve multi-pass POLInSAR acquisitions. The bandwidth is only 6 MHz, because of restrictions on frequencies allocated to civil applications in this spectrum until recently exclusively for military applications.

The mission will last 5 years and includes a tomographic phase (initially 55 days), followed by the nominal phase with a repeat period of less than 4 days. Recent work in tomography has shown the advances of this new technique. This led to a reconsideration of the operation phases of BIOMASS (planned for 1 year) to allow for greater tomographic acquisition coverage.

Using data from the airborne campaign TropiSAR [DUB 12], the limitations of standard inversion models based on indicators derived from cross-polarization (γ^0_{hv}) and the estimation of canopy height using the POLInSAR technique (h^P) were first highlighted. Indeed, the high biomass range being considered, these indicators were found to be not robust enough regarding the perturbations caused by terrain topography. From statistical and geometric analysis as well as the results of electromagnetic simulations, corrections leading to new indicators, denoted t^0 and h^C, were introduced, significantly improving their relationship with biomass (Figure 9.24). Detailed in [VIL 15b], the t^0 indicator has been developed to minimize perturbations because of topography, using polarimetric corrections resulting from the coherence matrix and more specific geometric corrections such as those used for the coefficient γ^0. With the same objective of reducing topographic effects, the h^C indicator was developed using the standard POLInSAR height (denoted by h^P) to minimize local effects of geometry layover because of the slant range SAR geometry.

As part of the preparatory work for the BIOMASS mission, the relevance of a combined use of PolSAR and POLInSAR indicators, benefitting mostly from the independence of perturbation sources affecting these two types of data, has been demonstrated. In this effort to minimize the propagation of errors in the estimates of indicators for the estimation of biomass, the inversion algorithm is not based on direct use of the relationship $t°$ or h^C-biomass, but on a Bayesian estimation, minimizing mean-squared error using conditional probability law (detailed below). This approach was also chosen because of being particularly well adapted to the joint use of optimized indicators and to optimize the combination of independent error sources.

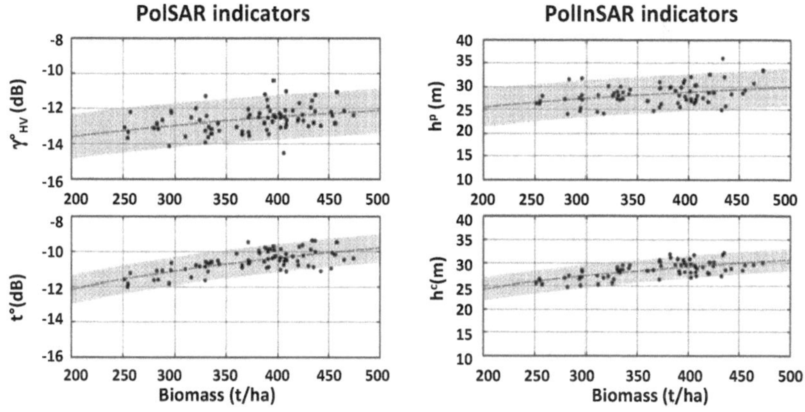

Figure 9.24. *Relationship between biomass values from field measurements and indicators from PolSAR data (γ^0_{HV} and t^0) and POLInSAR (h^P and h^C). The newly developed indicators (t^0 and h^C) show significant improvement in the relationship with biomass. Using equation [9.12] employed as a regression model, the improvement is significant as much in terms of sensitivity (slopes from 3.7 to 6), as for correlation (from 0.36 to 0.68) and dispersion (statistical chi2 from 32 to 15)*

Illustrated by the inversion scheme in Figure 9.25, this approach nevertheless requires the use of reference plots for the parameterization of inversion models connecting PolSAR or POLInSAR indicators to biomass, in order to ensure the required performance on a global scale (taking into account the different biomes and types of forests associated). To do this, the establishment of a global network of reference plots is envisaged, following the base of 50 super-sites made up of AfriTron[7] [LEW 09], RainFor[8] or CTFS[9] networks.

To make the most of these plots of networks, current studies aim to better optimize the tradeoff between plot size and the levels of uncertainty of estimated biomass. These fundamental studies are also dedicated the potential use of land where biomass was estimated via TomoSAR data or LiDAR measurements: the idea is to spatialize larger plots via extrapolation of the reference plots estimated using *in situ* measurements from TomoSAR or LiDAR data.

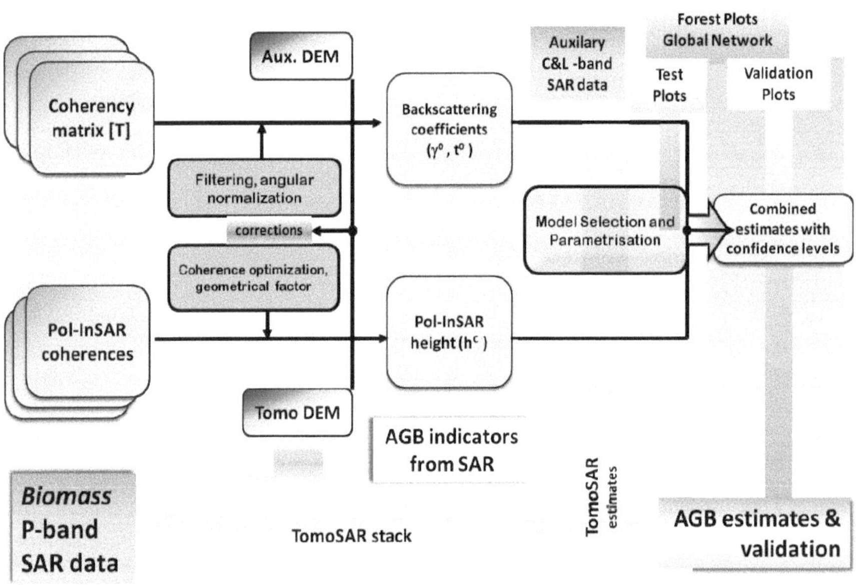

Figure 9.25. *Schematic diagram of the BIOMASS inversion algorithm (developed for the selection phase of the mission)*

7 African Tropical Rainforest Observation Network (*Réseau d'observation des forêts tropicales humides Africaines*), http://www.afritron.org/.

8 Amazon Forest Inventory Network (*Réseau d'inventaire forestier amazonien*), http://www.rainfor.org/.

9 *Centre technique de forêt communale*, Cameroon, http://www.foretcommunale-cameroun.org/.

Figure 9.25 shows the schematic diagram of the algorithm inversion BIOMASS data. It shows the different steps for estimating biomass from PolSAR and POLInSAR measurements. The entry boxes represent the triplet matrices of PolSAR and POLInSAR coherences from three consecutive passages (separated by only 3 to 4 days) constituting an acquisition cycle, allowing a global coverage in 6 months and the generation of so-called Level-2 products. Level-3 products will be the ones obtained from the use of all PolSAR and POLInSAR data archives, benefiting from the sequence of acquisitions, in order to produce consistent time series.

Ahead of the formation of these coherence matrices, it is also worth noting that several levels of raw data processing must be achieved, such as calibration of the backscattering measurements (relative to reference targets) and the correction of ionospheric effects [XU 04]. These are indeed much more important for the P-band than for higher radar frequencies (L or C, for example), hence the choice of sun-synchronous orbit with local passages at 6:00/18:00 because of solar activity, in addition to the desire to afford fully polarimetric acquisitions (for Faraday rotation corrections [CHE 10]).

The choice of triplet acquisition was primarily motivated by the establishment of several pairs of interferometric baselines (for better optimization of POLInSAR coherences), but also by the improved speckle filtering benefitting from the spatial resolution of the final products (see [QUE 01], speckle which is quantified by the number of equivalent views).

Finally, based on a digital elevation model (DEM), the local geometry parameters can be calculated to minimize topographic effects affecting the PolSAR backscatter coefficients like POLInSAR heights. These corrections are used to calculate backscatter indicators (t^0) and vegetation height (h^C). These indicators are then combined by a Bayesian estimator, benefiting from their complementarity (different sources of errors). Based on the conditional expectation of the variable biomass for a given observation of PolSAR or POLInSAR indicators, this estimator minimizes error propagating from SAR measurements or analytical models linking SAR indicators to biomass. This Bayesian formulation also requires training plots, obtained from reference plots estimated by *in situ* measurements or from a research perspective by the TomoSAR measurements during the preliminary phase of BIOMASS (12 months before the PolSAR-POLInSAR-phase).

9.5. Summary and prospects

Through its essential but complex participation in the global terrestrial carbon cycle, forest biomass has been recognized as an essential climate variable, for which the spatial and temporal variability at global scale make *in situ* measurements or the simulations of dynamic vegetation models obsolete, but fully justifies the use of remote sensing. Inverse methods using remote sensing observations to estimate forest biomass estimation are nevertheless only correlative, in the sense that they do not directly measure biomass. Among these methods, the use of radar frequencies below the L-band has been put forward, as the penetration of the microwaves at these frequencies allows optimal interactions with vegetation scatterers large enough to be sufficiently correlated to the total forest biomass for its estimation.

Mainly for technical reasons (increasing antenna size with wavelength but also because of frequency allocations), only space missions using the L-band have been implemented until now. These missions, including ALOS/PALSAR, were able to demonstrate the benefit of L-band SAR for biomass estimates at the global level, although penetration at this frequency is not sufficient to obtain performance levels consistent with user requirements for dense tropical forests.

It is precisely to meet these expectations that the BIOMASS mission project was initiated when a part of the P-band of the spectrum was opened to civilian applications (see Figure 9.26). Supported by many studies (including SAR airborne campaigns), the project has been selected as the seventh ESA Earth Explorer mission. Considering its main scientific objective of an almost biannual global mapping of forest biomass, expectations are highly directed towards the improvement of our understanding of the forest component in the carbon cycle, especially through the assimilation of biomass maps in climate models (continental surfaces, dedicated to climate modeling). Far from being completed, the theoretical and experimental research around the P-band radar data are primarily intended to improve the robustness of estimation methods, particularly to minimize perturbations caused by terrain topography and changes in forest water content (rainy or dry seasons in the tropics, freeze/thaw episodes in the boreal zone). As further prospects, the intrinsic limitations of the side looking geometry (and its associated shadowing and layover effects) will truly benefit from the combined used of multi acquisitions, whether mono or bistatic.

Figure 9.26. *Forest biomass from the BIOMASS eye (drawing by Jacques Faure)*

9.6. Key points

Forest observation constitutes a major application in remote radar sensing, this technique being the most optimal to map forest biomass and changes at a global scale. Recognized as an essential climate variable, forest biomass allows us to quantify the forest component in the carbon cycle and its related increase linked to the uptake of atmospheric carbon through photosynthesis and its related decrease linked to emissions because of deforestation are a sensitive lever in managing climate change.

Although no remote sensing technique allows a direct measurement of biomass, the interest of the radar at low frequencies (below 1 GHz) lies in the sensitivity of the backscatter to the total biomass through the interactions between these wavelengths and a significant part of the canopy, whereas the lower penetration of higher frequencies (radar or optical) does not make it possible to survey a sufficient volume. Equally essential, the coherence of radar measurements does not only allow the generation of images at metric resolutions via the synthetic aperture technique, but also enables the exploitation of various measures in time and lines of sight, making possible the use of interferometric and tomographic techniques in order to extract forest height information 3D backscatter, thereby consolidating the biomass estimates.

Until now, the potential of P-band SAR (frequency of about 400 MHz) has been demonstrated through several airborne campaigns, supporting in particular the future BIOMASS spaceborne mission (scheduled launch towards 2020) which carries very high expectations for a better knowledge of forest biomass at a global scale, and its benefits for the climate and the Earth system.

9.7. Bibliograpy

[ANT 13] ANTROPOV O., RAUSTE Y., AHOLA H. *et al.*, "Stand-level stem volume of Boreal Forests from spaceborne SAR Imagery at L-Band", *IEEE Journal of Selected Topics of Applied Earth Observation and Remote Sensing*, vol. 6, pp. 35–44, 2013.

[ASN 11] ASNER G.P., MASCARO J., MULLER-LANDAU H.C. *et al.*, "A universal airborne LiDAR approach for tropical forest carbon mapping", *Oecologia*, vol. 168, pp. 1147–1160, 2011.

[ATT 78] ATTEMA E.P.W., ULABY F.T., "Vegetation modeled as a water cloud", *Radio Science*, vol 13, pp. 357–364, 1978.

[AVT 13] AVTAR R., SUZUKI R., TAKEUCHI W. *et al.*, "PALSAR 50 m mosaic data based National level biomass estimation in Cambodia for implementation of REDD+ mechanism", *PLoS ONE*, vol. 8, p. e74807, 2013.

[BAM 98] BAMLER R., HARTL P., "Synthetic aperture radar interferometry", *Inverse Problems*, vol. 14, pp. R1–R54, 1998.

[BAR 11] BARBIER N., PROISY C., VÉGA C. *et al.*, "Bidirectional texture function of high resolution optical images of tropical forest: an approach using LiDAR hillshade simulations", *Remote Sensing of Environment*, vol. 115, no. 1, pp. 167–179, 2011.

[BAS 14] BASTIN J.F., BARBIER N., COUTERON P. *et al.*, "Aboveground biomass mapping of African forest mosaics using canopy texture analysis: toward a regional approach", *Ecological Applications*, vol. 24, no. 8, pp. 1984–2001, 2014.

[BEL 11] BELLASSEN V., VIOVY N., LUYSSAERT S. *et al.*, , "Reconstruction and attribution of the carbon sink of European forests between 1950 and 2000", *Global Change Biology*, vol. 17, no. 11, pp. 3274–3292, 2011.

[BER 94] BERENGER J.P., "A perfectly matched layer for the absorption of electromagnetic waves", *Journal of Computational Physics*, vol. 114, no. 2, pp. 185–200, 1994.

[BRO 97] BROWN S., Estimating biomass and biomass change of tropical forests: a primer, FAO (Food & Agriculture Organization) Forestry Paper – 134, 1997.

[BUV 06] BUVANESWARAN C., GEORGE M., PEREZ D. *et al.*, "Biomass of teak plantations in Tamil Nadu, India and Costa Rica compared", *Journal of Tropical Forest Science*, vol. 18, pp. 195–197, 2006.

[CAR 12] CARREIRAS J.M.B., VASCONCELOS M.J., LUCAS R.M., "Understanding the relationship between aboveground biomass and ALOS PALSAR data in the forests of Guinea-Bissau (West Africa)", *Remote Sensing Environment*, vol. 121, pp. 426–442, 2012.

[CAR 12] CARTUS O., KELLNDORFER J., ROMBACH M. *et al.* "Mapping Canopy Height and Growing Stock Volume Using Airborne LiDAR, ALOS PALSAR and LANDSAT ETM+", *Remote Sensing*, vol. 4, pp. 3320–3345, 2012.

[CAR 14] CARTUS O., KELLNDORFER J., WALKER W. *et al.* "A national, detailed map of forest aboveground carbon stocks in Mexico", *Remote Sensing*, vol. 6, pp. 5559–5588, 2014.

[CHA 04] CHAVE J., CONDIT R., AGUILAR S. *et al.*, "Error propagation and scaling for tropical forest biomass estimates", *Philosophical Transactions of the Royal Society of London, Series B: Biological Sciences*, vol. 359, pp. 409–420, 2004.

[CHA 05] CHAVE J., ANDALO C., BROWN S. *et al.*, "Tree allometry and improved estimation of cabon stocks and balance in tropical forests", *Oecologia*, vol. 145, pp. 87–99, 2005.

[CHA 16] CHAZETTE P., TOTEMS L., HESPEL L., "Principle and physics of the LiDAR measurement", in BAGHDADI N., ZRIBI M. (eds), *Optical Remote Sensing of Land Surfaces*, ISTE Press, London and Elsevier, Oxford, 2016.

[CHE 10] CHEN J., QUEGAN S., "Improved estimators of Faraday roation in spaceborne polarimetric SAR Data", *IEEE Geoscience and Remote Sensing Letters*, vol. 7, no. 4, pp. 846–850, 2010.

[CLO 96] CLOUDE S.R., POTTIER E., "A review of target decomposition theorems in radar polarimetry", *IEEE Transactions on Geoscience and Remote Sensing*, vol. 34, no. 2, pp. 498–518, 1996.

[CLO 02] CLOUDE S.R., "Polarimetry in wave scattering applications", in SABATIER R.P.P. (ed.) *Scattering*, Academic Press, London, 2002.

[CLO 03] CLOUDE S.R., PAPATHANASSIOU K.P., "Three-stage inversion process for polarimetric SAR interferometry", *IEEE Proceedings-Radar, Sonar and Navigation*, vol. 150, no. 3, pp. 125–134, 2003.

[COU 02] COUTERON P., "Quantifying change in patterned semi-arid vegetation by Fourier analysis of digitized aerial photographs", *International Journal of Remote Sensing*, vol. 23, no. 17, pp. 3407–3425, 2002.

[DEL 10] DELBART N, CIAIS P, CHAVE J *et al.*, "Mortality as a key driver of the spatial distribution of aboveground biomass in Amazonian forests: results from a Dynamic Vegetation Model", *Biogeosciences*, vol. 7, no. 10, pp. 3027–3039, 2010.

[DEW 02] DE WASSEIGE C., DEFOURNY P., "Retrieval of tropical forest structure characteristics from bi-directional reflectance of SPOT images", *Remote Sensing of Environment*, vol. 83, no. 3, pp. 362–375, 2002.

[DON 03] DONG J., KAUFMANN R.K., MYNENI R.B. *et al.*, "Remote sensing estimates of boreal and temperate forest woody biomass: carbon pools, sources, and sinks", *Remote Sensing of Environment*, vol. 84, no. 3, pp. 393–410, 2003.

[DUB 12] DUBOIS-FERNANDEZ P., LE TOAN T., DANIEL S. *et al.*, "The TropiSAR airborne campaign in French Guiana: Objectives, description and observed temporal behavior of the backscatter signal", *IEEE Transactions on Geoscience and Remote Sensing*, vol. 56, no.8, pp. 3228–3241, 2012.

[ENG 11] ENGLHART S., KEUCK V., SIEGERT F., "Aboveground biomass retrieval in tropical forests – the potential of combined X- and L-band SAR data use", *Remote Sensing of Environment*, vol. 115, pp. 1260–1271, 2011.

[ENG 13] ENGLHART S., JUBANSKI J., SIEGERT F., "Quantifying dynamics in tropical peat swamp forest biomass with multi-temporal LiDAR datasets", *Remote Sensing*, vol. 5, no. 5, pp. 2368–2388, 2013.

[ELF 04] ELFOUHAILY T.M., GUÉRIN C.A., "A critical survey of approximate scattering wave theories from random rough surfaces", *Waves in Random Media*, vol. 14, no. 4, pp. R1–R40, 2004.

[ESA 12] ESA, Report for Mission Selection: BIOMASS, ESA SP-1324/1 (3 volume series), European Space Agency, Noordwijk, The Netherlands, 2012.

[FAO 04] FOOD AND AGRICULTURE ORGANIZATION OF THE UNITED NATIONS (FAO), Global forest resources assessment. Technical Report. FAO, United States, 2004.

[FAO 10] FOOD AND AGRICULTURE ORGANIZATION OF THE UNITED NATIONS (FAO), Global forest resources assessment, Technical Report. Food and Agriculture Association of the United States, 2010.

[FER 16] FERRO-FAMIL L., POTTIER E., "Coherent diversity mode SAR imaging: SAR polarimetry, interferometry and tomography", in BAGHDADI N., ZRIBI M. (eds), *Microwave Remote Sensing of Land Surfaces*, ISTE Press, London and Elsevier, Oxford, 2016.

[FUN 94] FUNG A.K., *Microwave Scattering and Emission Models and their Applications*, Artech House, Inc., Boston, London, 1994.

[HAE 12] HAEUSLER T., Gomez S., Siwe R. *et al.*, "Reducing emissions from deforestation and degradation in Africa (REDDAF) ", *Let's Embrace Space, Volume II – Space Research Achievements under the 7th Framework Programme*, EU Publications, 2012.

[HEN 10] HENRY M., Carbon stocks and dynamics in Sub-Saharian Africa, Doctoral Thesis, Università degli studi della Tuscia, Viterbo, 2010.

[ISH 78] ISHIMARU A., "Wave propagation and scattering in random media", *Vol. II: Multiple Scattering, Turbulence, Rough Surfaces and Remote Sensing*, Academic Press, New York, 1978.

[IPC 13] STOKER T.F., QIN D., PLATTNER G.K.. *et al.*, "Climate change: the physical science basis", in IPPC (eds), *Fifth Assessment Report of the Intergovernmental Panel on Climate Change*, available at http://www.ipcc.ch, 2013.

[GAR 10] GARESTIER F., LE TOAN T., "Estimation of a forest backscatter profile at P-band using Single Baseline Pol-InSAR", *IEEE Transactions on Geoscience and Remote Sensing*, vol. 48, no. 9, pp. 3340–3348, 2010.

[GIN 02] GINI F., LOMBARDINI F., MONTANARI M., "Layover solution in multibaseline SAR interferometry", *IEEE Transactions on Aerospace and Electronic Systems*, vol. 38, no. 4, pp. 1344–1356, 2002.

[GCO 06] GLOBAL CLIMATE OBSERVING SYSTEM (GCOS), Systematic observation requirements for satellite-based products for climate, Supplemental details to the satellite-based component of the Implemention plan for the Global Observing System for Climate in support of the UNFCCC, GCOS 107, WMO/TD No. 1338, 2006.

[HAM 14] HAMADI A., ALBINET C., BORDERIES P. *et al.*, "Temporal survey of polarimetric P-band scattering of tropical forests", *IEEE Transactions on Geoscience and Remote Sensing*, vol. 52, no. 8, pp. 4539–4547, 2014

[HAN 13] HANSEN M.C., POTAPOV P.V., MOORE R. *et al.*, "High-resolution global maps of 21st-century forest cover change", *Science*, vol. 342, no. 6160, pp. 850–853, 2013.

[KAR 86] KARAM M., FUNG A.K., "EM scattering from a randomly oriented circular dielectric, finite-length cylinder", *International Union Radio Science Commission F.: Wave Propagation and Remote Sensing*, University of New Hampshire, Duham, New Hampshire, 1986.

[KRI 05] KRINNER G., VIOVY N., DE NOBLET-DUCOUDRÉ N. *et al.*, "A dynamic global vegetation model for studies of the coupled atmosphere-biosphere system", *Global Biogeochemical Cycles*, vol. 19, no. 1, pp. 1–33, 2005.

[LAR 09] LARDEUX C., FRISON P.L., TISON C. *et al.*, "Support vector machine for multifrequency SAR polarimetric data classification", *IEEE Transactions on Geoscience and Remote Sensing*, vol. 47, no. 12, pp. 4143–4152, 2009.

[LEC 03] LECKIE D.G., GOUGEON F.A., WALSWORTH N. *et al.*, "Stand delineation and composition estimation using semi-automated individual tree crown analysis", *Remote Sensing of Environment*, vol. 85, no. 3, pp. 355–369, 2003.

[LEE 00] LEE J.S., SCHULER D., AINSWORTH T., "Polarimetric SAR data compensation for terrain azimuth slope variation", *IEEE Transactions on Geoscience and Remote Sensing*, vol. 3855, pp. 2153–2163, 2000.

[LEQ 13] QUÉRÉ C.L., ANDRES R.J., BODEN T. *et al.*, "The global carbon budget 1959–2011", *Earth System Science Data*, vol. 5, no. 1, pp. 165–185, 2013.

[LET 11] LE TOAN T., QUEGAN S., DAVIDSON M.W.J. *et al.*, "The BIOMASS mission: mapping global forest biomass to better understand the terrestrial carbon cycle", *Remote Sensing of Environment*, vol. 115, pp. 2850–2860, 2011.

[LET 92] LE TOAN T., BEAUDOIN A., RIOM J. *et al.*, "Relating forest biomass to SAR data", *IEEE Transactions on Geoscience and Remote Sensing*, vol. 30, pp. 403–411, 1992.

[LEW 09] LEWIS S.L., LOPEZ-GONZALEZ G., SONKE B. *et al.*, "Increasing carbon storage in intact African tropical forests", *Nature*, vol. 457, no. 7232, pp. 1003–1006, 2009.

[LUC 07] LUCAS R.M., MITCHELL A.L., ROSENQVIST A. *et al.*, "The potential of L-band SAR for quantifying mangrove characteristics and change: case studies from the tropics", *Aquatic Conservative Marine Freshwater Ecosystem*, vol. 17, pp. 245–264. 2007.

[LUC 10] LUCAS R., ARMSTON J., FAIRFAX R. *et al.*, "An evaluation of the ALOS PALSAR L-Band backscatter—above ground biomass relationship Queensland, Australia: impacts of surface moisture condition and vegetation structure", *IEEE Journal of Selective Topics on Applied Earth Observations and Remote Sensing*, vol. 3, pp. 576–593, 2010.

[MER 14] MERMOZ S., LE TOAN T., VILLARD L. *et al.*, "Biomass assessment in the Cameroon savanna using ALOS PALSAR data", *Remote Sensing Environment*, vol. 155, pp. 109–119, 2014.

[MER 15] MERMOZ S., RÉJOU-MÉCHAIN M., VILLARD L. *et al.*, "Decrease of L-band SAR backscatter with biomass of dense forests", *Remote Sensing Environment*, vol. 159, pp. 307–317, 2015.

[MIC 14] MICHELAKIS D., STUART N., LOPEZ G. *et al.*, "Local-scale mapping of biomass in tropical lowland Pine Savannas using ALOS PALSAR", *Forests,* vol. 5, pp. 2377–2399, 2014.

[MIN 14] MINH D.H.T., LE TOAN T., ROCCA F. *et al.*, "Relating P-band synthetic aperture radar tomography to tropical forest biomass", *IEEE Transactions on Geoscience and Remote Sensing*, vol. 52, no. 2, pp.967–979, 2014.

[MIN 15] MINH D.H.T., TEBALDINI S., ROCCA F. *et al.*, "BIOMASS tomography: temporal decorrelation performance in the tropical forest", *IEEE on Geoscience and Remote Sensing Letter*, vol. 12, no. 6, pp. 1297–1301, 2015.

[MIN 16] MINH D.H.T., TEBALDINI S., ROCCA F. *et al.*, "Capabilities of BIOMASS tomography for investigating tropical forests", *IEEE Transactions on Geoscience and Remote Sensing*, vol. 53, no. 2, pp. 965–975, 2015.

[MIT 09] MITCHARD E.T.A., SAATCHI S.S., WOODHOUSE I.H. *et al.*, "Using satellite radar backscatter to predict above-ground woody biomass: A consistent relationship across four different African landscapes", *Geophysical Research Letter*, vol. 36, p. L23401, 2009.

[MIT 11] MITCHARD E.T.A., SAATCHI S.S., LEWIS S.L. *et al.*, "Measuring biomass changes due to woody encroachment and deforestation/degradation in a forest–savanna boundary region of central Africa using multi-temporal L-band radar backscatter", *Remote Sensing Environment*, vol. 115, pp. 2861–2873, 2011.

[MOR 11] MOREL A.C., SAATCHI S.S., MALHI Y. *et al.*, "Estimating aboveground biomass in forest and oil palm plantation in Sabah, Malaysian Borneo using ALOS PALSAR data", *Forest Ecological Management*, vol. 262, pp. 1786–1798. 2011.

[MUI 01] MUINONEN E., MALTAMO M., HYPPÄNEN H. *et al.*, "Forest stand characteristics estimation using a most similar neighbor approach and image spatial structure information", *Remote Sensing of Environment*, vol. 78, no. 3, pp. 223–228, 2001.

[NI 13] NI W., SUN G., GUO Z. *et al.*, "Retrieval of forest biomass from ALOS PALSAR data using a lookup table method", *IEEE Journal of Selected Topics of Applied Earth Observation and Remote Sensing*, vol. 6, pp. 875–886, 2013.

[NIA 13] NIAMIEN D.N., Evaluation du potentiel des données radar multi-paramètres pour la cartographie en milieu tropical humide : Applications en Guyane française et en Côte d'Ivoire, PhD thesis, University Paris-Est, France, 2013.

[PAN 11] PAN Y., BIRDSEY R.A., FANG J. *et al.*, "A large and persistent carbon sink in the World's forests", *Science*, vol. 333, pp. 988–993, 2011.

[PEK 02] PEKKARINEN A., "Image segment-based spectral features in the estimation of timber volume", *Remote Sensing of Environment*, vol. 82, pp. 349–359, 2002.

[PER 13] PEREGON A., YAMAGATA Y., "The use of ALOS/PALSAR backscatter to estimate above-ground forest biomass: a case study in Western Siberia", *Remote Sensing of Environment*, vol. 137, pp. 139–146, 2013.

[PEP 95] PEPLINSKI N.R., ULABY F.T., DOBSON M.C., "Dielectric properties of soils in the 0.3-1.3-GHz range", *IEEE Transactions on Geoscience and Remote Sensing*, vol. 33, no. 3, pp. 803–807, 1995.

[PRO 07] PROISY C., COUTERON P., FROMARD F., "Predicting and mapping mangrove biomass from canopy grain analysis using Fourier-based textural ordination of IKONOS images", *Remote Sensing of Environment*, vol. 109, no. 3, pp. 379–392, 2007.

[QUE 01] QUEGAN S., Yu J.J., "Filtering of multi-channel SAR images", *IEEE Transactions on Geoscience and Remote Sensing*, vol. 39, no. 11, pp. 2373–2379, 2001.

[RAN 95] RANEY R.K., "SAR systems", in IKEDA M., DOBSON F.W. (eds), *Oceanographic Applications of Remote Sensing*, CRC Press, Boca Raton, 1995.

[RAU 94] RAUSTE Y., HAME T., PULLIAINEN J. *et al.*, "Radar-based forest biomass estimation", *International Journal of Remote Sensing*, vol. 15, pp. 2797–2808, 1994.

[REI 00] REIGBER A., MOREIRA A., "First demonstration of airborne SAR tomography using multibaseline L-band data", *IEEE Transactions on Geoscience and Remote Sensing*, vol. 38, no. 5, pp. 2142–2152, 2000.

[ROU 11] ROUEFF A., ARNAUBEC A., DUBOIS-FERNANDEZ P. *et al.*, "Cramer-Rao lower bound analysis of vegetation height estimation with random volume over ground model and polarimetric SAR interferometry", *IEEE Geoscience Remote Sensing Letters*, vol. 8, no. 6, pp. 1115–1119, 2011.

[SAN 09] SANTORO M., FRANSSON J.E.S., ERIKSSON L.E.B. *et al.*, "Signatures of ALOS PALSAR L-Band Backscatter in Swedish Forest", *IEEE Transactions of Geoscience and Remote Sensing*, vol. 47, 4001–4019. 2009.

[SAN 11] SANTORO M., BEER C., CARTUS O. *et al.*, "Retrieval of growing stock volume in boreal forest using hyper-temporal series of Envisat ASAR ScanSAR backscatter measurements", *Remote Sensing Environment*, vol. 115, pp. 490–507, doi:10.1016/j.rse.2010.09.018, 2011.

[SAR 12] SARKER M.L.R., NICHOL J., AHMAD B. *et al.*, "Potential of texture measurements of two-date dual polarization PALSAR data for the improvement of forest biomass estimation", *ISPRS J. Photogramm. Remote Sens.*, vol. 69, pp. 146–166, 2012.

[SES 08] SESSA R., DOLMAN H., Terrestrial essential climate variables for climate change assessment, mitigation and adaptation, FAO GTOS-52, 2008.

[SEX 13] SEXTON J.O., SONG X.P., FENG M. *et al.*, "Global, 30-m resolution continuous fields of tree cover: LANDSAT-based rescaling of MODIS vegetation continuous fields with LiDAR-based estimates of error", *International Journal of Digital Earth*, vol. 6, pp. 427–448, 2013.

[SUN 11] SUN G., RANSON K.J., GUO Z. *et al.*, "Forest biomass mapping from LiDAR and radar synergies", *Remote Sensing Environment*, vol. 115, pp. 2906–2916, doi:10.1016/j.rse.2011.03.021, 2011.

[TEB 09] TEBALDINI S., "Algebraic synthesis of forest scenarios from multibaseline POLInSAR data", *IEEE Transactions on Geoscience and Remote Sensing*, vol. 47, no. 12, pp. 4132–4142, 2009.

[TEB 10] TEBALDINI S., "Single and multipolarimetric SAR tomography of forested areas: A parametric approach", *IEEE Transactions on Geoscience and Remote Sensing*, vol. 48, no. 5, pp. 2375–2387, 2010.

[THA 15] THAPA R.B., WATANABE M., MOTOHKA T. *et al.*, "Potential of high-resolution ALOS–PALSAR mosaic texture for aboveground forest carbon tracking in tropical region", *Remote Sensing Environment*, vol. 160, pp. 122–133, 2015.

[TRE 96] TREUHAFT R.N., MADSEN S.N., MOGHADDAM M. *et al.*, "Vegetation characteristics and underlying topography from interferometric radar", *Radio Science*, vol. 31, no. 6, pp. 1449–1485, 1996.

[TSA 00] TSANG L., KONG J.A., DING K.H., "Scattering of electromagnetic waves", *Vol. 1: Theory and Applications*, 2000.

[TSA 85] TSANG L., KONG J.A., SHIN R.T., *Theory of Microwave Remote Sensing*, Wiley, New York, 1985.

[TSA 92] TSANG L., CHAN C.H., KONG J.A. *et al.*, "Polarimetric signatures of a canopy of dielectric cylinders based on first and second order vector radiative transfer theory", *Journal of Electromagnetic Waves and Applications*, vol. 6, no. 1, pp. 19–51, 1992.

[ULA 13] ULABY F.T., MICHIELSSEN E., RAVAIOLI U. et al., *Fundamentals of Applied Electromagnetics*, Instructor, 2013.

[VAN 93] VAN ZYL J., "The effect of topography on radar scattering from vegetated areas", *IEEE Transactions on Geoscience and Remote Sensing*, vol. 31, no. 2, pp. 153–160, 1993.

[VIL 07] VILLARD L., BORDERIES P., "Backscattering border effects for forests at C-Band", *Piers Online*, vol. 3, no. 5, pp. 731–735, 2007.

[VIL 09] VILLARD L., Forward & inverse modeling for synthetic aperture radar observables in bistatic configuration: applications in forest remote sensing, PhD thesis Institut Supérieur de l'Aéronautique et de l'Espace, France, 2009.

[VIL 10] VILLARD L., BORDERIES P., LE TOAN T. *et al.*, "Topography effects on forest radar scattering, consequences on biomass retrieval", *Proceedings in IGARSS 2010 (International Geoscience and Remote Sensing Symposium)*, 25–30 July 2010, Honolulu, Hawaii, USA, pp. 60–63, 2010.

[VIL 15a] VILLARD L., BORDERIES P., "On the use of virtual ground scatterers to localize double and triple bounce scattering mechanisms for bistatic SAR", *Journal of Electromagnetic Waves and Applications*, vol. 29, no. 5, pp. 626–635, 2015.

[VIL 15b] VILLARD L., LE TOAN T., "Relating P-band SAR intensity to biomass for tropical dense forests in hilly terrain: t0 or γ0", *IEEE Journal of Selected Topics in Applied Earth Observations and Remote Sensing*, vol. 8, no. 1, pp. 214–223, 2015.

[WIL 14] WILHELM S., HÜTTICH C., KORETS M. *et al.*, "Large area mapping of Boreal growing stock volume on an annual and multi-temporal level using PALSAR L-band backscatter mosaics", *Forests*, vol. 5, pp. 1999–2015, 2014.

[WOO 06] WOODHOUSE I.H., "Predicting backscatter-biomass and height-biomass trends using a macroecology model", *IEEE Transactions of Geoscience and Remote Sensing*, vol. 44, pp. 871–877, 2006.

[XU 04] XU Z.W., WU J., WU Z.S., "A survey of ionospheric effects on space-based radar", *Waves in Random Media*, vol. 14, no. 12, pp. 189–279, 2004.

Glossary

Aerodynamic temperature: Theoretical temperature of the average source of sensitive heat flux for a complex surface, corresponding to the extrapolation within the canopy of the air temperature profile observed above the canopy.

Agro-hydrosystems: Hydros ystems under influence of agricultural activity.

Angle of incidence: Angle between the incident radar beam on a surface and the normal to the surface at the point of incidence.

Apex of a tree: Top of a tree.

Available water capacity: Maximum amount of plant available water a soil can provide. It is an indicator of a soil's ability to retain water and make it sufficiently available for plant use. Available water capacity is the water held in soil between its field capacity and permanent wilting point.

Backscatter radar coefficient (denoted σ^0): Designates the RCS (Radar cross-section) by observed surface unit, and is expressed in (m^2/m^2). It is preferred instead of the RCS in the case of large surfaces. Due to its strong dynamics, it is usually expressed in decibels: $\sigma^0_{dB} = 10 * \log_{10}(\sigma^0)$.

Backscattering: The reflection of waves back to the direction from which they came (sensor), depending on the surface properties and sensor parameters.

Basal area of a tree: Average area occupied by tree stems. It is defined as the total cross-sectional area of all stems in a stand measured at breast height, expressed as per unit of land area.

Basic density of wood: Ratio between the dry weight and the volume of green wood or at a water-saturated state. This parameter is typically used for the establishment of allometric equations (see *Allometry*).

Bidirectional Reflectance Distribution Function (BRDF): The ratio of outgoing radiance and incident irradiance.

Biomass: In ecology, biomass is the amount of living material in a given habitat. Often, it is estimated in a surface unit or volume rather than absolute mass. The biomass of vegetation cover is usually estimated by removing some plants which are weighed fresh and dry (after passing in an oven) in order to obtain dry biomass. In practice, we distinguish above-ground biomass (trunk, branches, leaves) from below-ground biomass (roots).

Bistatic scattering: The electromagnetic scattering phenomenon considered when the radar receiver is distinct of the radar transmitter, contrary to backscattering which is considered when using a monostatic radar system where the transmitter and receiver are collocated.

Brightness temperature: Temperature obtained by converting the radiance measured according to Planck's Law at the level of the sensor, weighted by its transfer function; it can be defined for a measurement in a given direction (called directional brightness temperature).

Canopy Height Model (CHM): For a given portion of territory, the mapping (in raster format) of the maximum height of the forest area compared to the ground within each pixel. It is a surface resulting from the calculation of the difference between a Digital Surface Model (DSM) and a Digital Terrain Model (DTM). Excluding forest environments, it is commonly called Digital Elevation Model (DEM). Subtracting the DTM makes it possible to eliminate the effect of the terrain, and any area not composed of objects above-ground has zero height.

Carbon balance: An environmental indicator which assess the amount of greenhouse gases caused by human activities. On the scale of agricultural plot, the carbon footprint is calculated on a crop year or on the scale of crop rotation, considering: i) the net exchange of CO_2 between the agricultural plot and atmosphere; ii) carbon inputs in the plot (via organic fertilization and the contribution of seeds at the moment of planting); iii) the loss of carbon at the time of harvest (export of grain, straw). The carbon footprint of crops represents the evolution of the organic carbon stock in the soil between the beginning and end of the agricultural period. The carbon footprint is controlled mainly by environmental conditions (temperature, air and soil moistures soil) and agricultural practices (tillage, organic fertilization, straw management). The carbon footprint is negative if the plot acts as a carbon sink and positive if it is a carbon source.

Carbon sinks: Natural or artificial reservoir of carbon, which contributes to carbon sequestration and limit global warming.

Cartographic reference system: Coordinate system allowing to locate a point on the surface of the Earth.

Chemometrics: Set of statistical, graphic or symbolic methods to better understand the information obtained in the field of chemistry.

Circumference (of a tree) at breast height: See *Diameter (of a tree) at breast height (DBH)*.

Classification (method of): Technique to determine the affiliation of an object (for example, a pixel of an image) to a group (class) using the object's attributes (for example, its spectral signature). It can be supervised or unsupervised.

Clustering: Technique separates datasets into homogeneous groups (as defined by previously established criteria).

Confusion matrix: Double entry table used to analyze the quality of a classification; the rows correspond to the class of the evaluated objects and the columns correspond to the assigned class; the boxes in the table indicate the number of objects.

Copernicus: A European Earth spatial monitoring program, previously known as Global Monitoring for Environment and Security (GMES).

Crop coefficient: The ratio of evapotranspiration (ET) observed for the crop studied (potential ET) over that observed for the well calibrated reference crop (reference ET) under the same conditions. This ratio is denoted by Kc. Kc can be decomposed into several components: Kcb for fraction transpired by the plant and Ke for the fraction evaporated from the soil. Hydric stress in the root zone is expressed by multiplying Kc or Kcb by a stress coefficient Ks between 0 and 1 (where 1 is without stress).

Crop modeling: Representation in the form of computer coding of a set of equations representing the main processes that regulate the functioning of a crop (growth, development, water balance and nitrogen) in interaction with the climate, soil and agricultural practices (date and seeding rate, characteristics of varieties planted, water and nitrogen inputs). These are dynamic models where processes are usually described with daily time.

Crop rotation: The division of agricultural land for different crops (different soils) for given seasons; also refers to the geographical organization of the cultivated plots.

Data assimilation: mathematical methods that serve to optimally correct the previsions of a dynamic model (for example a TSVA model) by using available observations (for example field measurements or measurements issued from remote sensing).

Defoliation: Partial or total loss of foliage from trees, leaves in the case of leaved species or needles in the case of coniferous species. It is a component of the normal cycle of development and renewal of the foliage. In the case of deciduous species, defoliation is total at the end of the seasonal cycle (fall for the temperate forests of the northern hemisphere, for example). Defoliation can be of accidental origin (water stress, disease, insect attack).

Dempster–Shafer: Theory that allows modeling and combining uncertain information.

Dendrometry: Concerns the measurement of the various dimensions of trees, such as their diameter, size, shape, age, overall volume, thickness of the bark, etc., as well as the statistical properties of tree stands, including measures of central tendency and dispersion of these quantities, wood density, or yearly growth, for instance.

Density and tree spacing: Nonlinear transformation of the density assuming each tree at the top of an equilateral triangle: $= \sqrt{\frac{20000}{\sqrt{3}\,Nha}}$, *Sp* being the space (in meters) and *Nha* the density (number of trees per hectare).

Diameter (of a tree) at breast height (DBH): Measurement associated with the evaluation of the size of a tree, usually expressed through three interrelated variables: diameter, circumference and basal surface. The diameter and circumference are directly measured from the perimeter of the section of the tree "at breast height" (1.3 m on average), while the basal surface corresponds to the surface of this section.

Dielectric constant: Known physical quantity, also known as complex permittivity. It characterizes the electrical properties of the soil in terms of electrical losses due to transportation and absorption of the energy provided by the radar wave. In the case of soil, depends on moisture content and soil composition.

Differential Global Positioning System (DGPS): Instrument that allows measuring geographic positions in a differential manner using a constellation of satellites and A network of fixed ground-based reference stations.

Differential Radar Interferometry (DinSAR): Particular application of radar interferometry in order to measure the displacement of the ground in using the phase difference of two radar images acquired at different dates, combined with topographical information provided by an Digital Surface Model (DSM).

Diffuse reflection: A reflection of waves in different directions and with a distribution of energy following these directions. The distribution between specular and diffuse reflection processes is determined by the roughness and the dielectric properties of that surface.

Digital Elevation Model (DEM): 3D representation of a terrain's surface. It represents the Earth's surface and all objects on it.

Digital Height Model (DHM): 3D description of our environment. It describes the ground profile while integrating what is on the ground surface (buildings, vegetation, etc.).

Digital soil mapping: Is the production of a geographically referenced soil databases (soil types and soil properties) generated at a given resolution by using field and laboratory observation methods coupled with environmental data through quantitative relationships.

Digital Surface Model (DSM): Model depicting elevations of the top of reflective surfaces, such as buildings and vegetation.

Digital Terrain Model (DTM): See *Digital Elevation Model.*

Direct modeling: Direct modeling aims to predict the return of an electromagnetic wave to the sensor using a set of physical characteristics (shape, size, dielectric properties).

Dominant height: It is generally defined as the average height of the hundred largest trees per hectares.

Eco-physiological model: Describes the growth and development of crops interacting with their agro-environmental conditions (soil, climate and environment).

Electromagnetic radiation: An energy wave that is propagated without any hardware support (in vacuum) and is characterized by a period (cycle length or wavelength – in seconds) and a frequency (number of cycles per second – in hertz). The electromagnetic spectrum is the decomposition of electromagnetic radiation according to its wavelength or frequency (Hz). The long wavelengths (low frequencies) are the radio and microwaves while smaller wavelengths (high frequencies) are nuclear or gamma rays.

Emissivity: Ratio between brightness temperature and surface temperature.

Enhanced Vegetation Index (EVI): Defined by $\mathrm{EVI} = G \frac{\mathrm{R_{NIR}} - \mathrm{R_{Red}}}{L + \mathrm{R_{NIR}} - C_1 \mathrm{R_{Red}} - C_2 \mathrm{R_{Blue}}}$ where $\mathrm{R_{NIR}}$, $\mathrm{R_{Red}}$ and $\mathrm{R_{Blue}}$ denote the reflectance in wavelengths near infrared, red and blue. It measures the intensity of photosynthetic activity of the surface. For very dense covers such as tropical forests it is preferred to the Normalized Difference Vegetation Index (NDVI). L is in relation to the effects of the soil, C_1 and C_2 to those of aerosols and G is a gain factor. In MODIS products, L=1, C_1=6, C_2=7.5 and G=2.5.

Entropy (Cloude decomposition): Represents the random nature of the mechanisms of scattering present in the pixels whose average was calculated.

Evapotranspiration (ET): Amount of water transferred to the atmosphere by plant covers resulting from plant transpiration and soil evaporation. ET represents the actual water consumption of a crop; it depends on the needs of the crop (themselves dependent on its state of development and climatic demand) and water available in the soil. Its knowledge is essential for monitoring crop water balance.

Exogenous data: Data not originating from images to be processed which are used to assist in extracting information, for exemple, a DEM, soil maps, weather data.

Faraday rotation: Rotation of a beam of polarized microwaves traversing an isotropic medium along the lines of force of a magnetic field.

Feature: Digital or symbolic information characterizing an object (for example, a pixel of an image); there are features that are spectral, texture, temporal, among others.

Finite-difference time-domain (FDTD): A numerical analysis technique in the time domain. This method is commonly used in electromagnetism to solve Maxwell's equations.

Fluorescence: The emission of light by a substance that has absorbed light or other electromagnetic radiation. It is a form of luminescence. In most cases, the emitted light has a longer wavelength, and therefore lower energy, than the absorbed radiation.

Fraction of Absorbed Photo synthetically Active Radiation (fAPAR): Fraction of photo synthetically active radiation absorbed by vegetation.

Fraction of Vegetation Cover (fCover): Corresponds to the fraction of ground covered by green vegetation. Practically, it quantifies the spatial extent of the vegetation.

Geographic Information System (GIS): A geographic information system or geographical information system (GIS) is a system designed to capture, store, manipulate, analyze, manage, and present all types of spatial or geographical data (source Wikipedia).

Geo-referencing: Operation which consists of properly positioning geographical data (like that from a satellite image) in a geographic coordinate system.

Global Positioning System (GPS): Geolocation system operating at a global scale and based on a measurement of receipt electromagnetic signals transmitted by satellites in orbits around the Earth.

Greenhouse Gases (GHG): A gas in an atmosphere that absorbs and emits radiation within the thermal infrared range. This process is the fundamental cause of the greenhouse effect. The primary greenhouse gases in Earth's atmosphere are water vapor, carbon dioxide (CO_2), methane (CH_4), nitrous oxide (N_2O) and ozone (O_3).

Ground truth: In image processing, information that is available on a set of pixels.

Hot Spot: Corresponds to the maximum reflection observed on a rough surface (soil, vegetation, etc.) in the direction of backscatter (illumination and viewing directions are collinear). When we deviate from this direction, the shadow fraction viewed increases and the reflected radiation in this direction decreases.

Hydrosystem: A system composed of water and all associated aquatic environments deplace it in a defined geographic sector, especially a watershed.

Hyperspectral remote sensing: Remote sensing from a sensor allowing the observation in many narrow spectral bands (a few nanometers) in the wavelength range from the visible, near infrared, mid-infrared to the infrared thermal.

Ice, Cloud and land Elevation satellite (ICEsat): NASA's spatial mission launched in 2003. The mission embarked a backscatter LiDAR system for the study of surfaces, primarily polar caps (GLAS).

Infrared: Electromagnetic radiation of a wavelength between 0.7 μm and 100 μm.

Inputs (for agricultural plots): The products brought to cultivated plots in order to improve productivity: fertilizers (organic or mineral), amendments to improve the physicochemical properties of soil, phyto sanitary products (herbicides, fungicides, insecticides, etc.) and irrigation water.

Interferogram: Image of the phase difference between 2 or more radar images.

Inverse modeling: Provides the physical characteristics of the observed medium from data extracted from extracted from radar, optical or other remote sensing signals.

Inversion: Estimation method of a parameter based on the minimal error between measured and simulated signals.

Kappa coefficient: A quality estimator. It measures the level of agreement between the classification and the actual data. It varies between 0 and 1. The closer it is to 1, the more the agreement is important.

Kriging: Geostatistical spatial interpolation method which provides the value of a variable in unsampled locations by a linear combination without bias and with a minimum variance of the observations of the variable in neighboring sites.

Land cover: Description of the presence of physical and biological objects identified on a mapped landscape.

Land use: Description of the uses and functions of physical or biological objects identified in a mapped landscape

Leaf area index (LAI): A dimensionless quantity that is defined as the one-sided green leaf area per unit ground surface area (LAI = leaf area/ground area, m^2/m^2); the "green" LAI is the part of the green vegetation LAI.

Legend (of a map): List of thematic classes in a map, such as types of vegetation, infrastructure, etc. Each object belongs to one classe.

Light Amplification by Stimulated Emission of Radiation (Laser): Emits coherent light thanks to an amplified stimulated emission. The time between transmission and reception is converted into a distance by multiplying by the speed of light.

Local incidence angle: Angle between the incident radar beam on a surface and the line perpendicular to the illuminated surface in case of an inclined surface (local slope).

Microwaves: A form of electromagnetic radiation with wavelengths ranging from one meter to one millimeter; with frequencies between 300 MHz (100 cm) and 300 GHz (0.1 cm).

Multiple scattering (optical): A photon released outside of the medium after its interaction with several particles.

Multispectral remote sensing: Remote sensing from a sensor allowing to observe several wide enough spectral bands (of the order of several tens of nanometers) in wavelengths ranging from the visible, near infrared, mid-infrared to the thermal infrared.

Multistatic Interferometric Polarimetric Electromagnetic model for Remote Sensing (MIPERS): An electromagnetic model to simulate radar data from plant-covered scenes.

Multi-view radar rata (N views): Radar data where each measurement corresponds to an average (in amplitude or intensity) of "N" statistically independent samples. Radiometric accuracy is better than when the N number of views is bigger.

Natura 2000 Network: Set of natural or semi-natural sites of the European Union having a great heritage value due to their exceptional wildlife and flora therein.

Net Ecosystem production (NEP): Defined as the difference between gross primary production and total ecosystem respiration, it represents the total amount of organic carbon in an ecosystem available for storage, export as organic carbon, or non-biological oxidation to carbon dioxide thro.

Nomenclature (for landcover map): Hierarchical organization of thematic categories used for the description of a landscape; each object can belong to multiple categories.

Normalized Difference Vegetation Index (NDVI): The vegetation index is calculated from the reflectance in the Near Infrared (NIR) and in the red (R): NDVI= (NIR – R)/(NIR+R). It measures the intensity of photosynthetic activity of the surface. It is efficient to distinguish the types of covers provided that they are not too dense as tropical rainforests.

Normalized Difference Water Index (NDWI): The water index is denoted by $\mathrm{NDWI} = \frac{R_{NIR} - R_{MIR}}{R_{NIR} - R_{MIR}}$, where R_{NIR} and R_{MIR} are the reflectances in the wavelengths of the near infrared and middle infrared, respectively. Other names given to this index, depending on the applications and sometimes according to the MIR spectral band used and the source sensor, are: NDII, LSWI, NBR, for example, where W of LSWI refers to the information provided on the contents in water of the soil and the vegetation and the B of NBR (normalized burn ratio) refers to the biomass burned by fires.

Object-oriented classification: Image classification method aiming to segment an image into distinct polygonal objects. The objects are classified using the radiometric properties of the image as well as the geometric properties of the objects such as their size or form.

Optical reflectance: Proportion of incident light (solar illumination) reflected by the Earth's surface. It is expressed as a percentage.

Optical thickness: The optical thickness of a layer (atmosphere, vegetation) measuring the degree of transparency of the medium. It characterizes the fraction of electromagnetic radiation (or light radiation) scattered or absorbed by the components of the layer traversed.

Panchromatic: A panchromatic image is a comprehensive record of the radiation reflected by the Earth's surface in the range of wavelengths of the visible spectrum without color separation. For example, a SPOT 5 panchromatic image is acquired between 0.48 and 0.71 µm. This data is more spatially resolved that the multispectral image acquired by the same satellite.

Penetration depth: Depth at which a wave, radar or optical, penetrates in the medium. It corresponds to the depth of a medium which issues 63% of energy. It depends of the medium's properties.

Perpendicular Vegetation Index (PVI): A vegetation index established to minimize the influence of the soil under vegetation. It is a linear combination of the reflectances in the near infrared (R_{NIR}) and the red (R_{Red}). Mathematically, it is equal to the distance perpendicular to the line of the soil, defined by the linear relationship between the reflectances of the bare soil within two spectral bands: $R_{NIR,bare\ soil} = a\, R_{Red,bare\ soil} + b$. The PVI index: $PVI = \frac{1}{\sqrt{1+a^2}}(R_{PIR} - a\, R_{Red} - b)$, where a and b are the slope and the ordinate at the origin of the straight soil, respectively.

Persistent Scatterer Interferometry (PSI): Radar interferometry using individual pixels whose response is dominated by the presence of a largely preponderant scatterer compared to others and whose characteristics change very little during the period followed.

Phenological stages: Development phases of a vegetation cover corresponding to the succession of periodic events during its growth cycle (for example, flowering, heading, budding, etc.); in remote sensing, it can be correlated to the evolution of NDVI.

Phenology: The study of the rhythm of life of plants.

Photo interpretation: Image analysis by a human operator, eventually assisted by image processing tools.

Photo synthetically Active Radiation (PAR): The spectral range (wave band) of solar radiation from 400 to 700 nm that photosynthetic organisms are able to use in the process of photosynthesis.

Pixel: Unitary element constituting an image (deriving from the English phrase "picture element"). Depending on the image type and the encoding used, the pixel "contains" one or more values corresponding to the digital pixel counts in different image channels.

Plant Area Index (PAI): The surface developed by all plant materials (includes leaves, stems, branches, trunks, reproductive organs, regardless of their photo synthetic potential). PAI is equal to half the surface of the convex envelope? of all plant materials by horizontal ground surface unit (units: m^2/m^2).

Polarimetric decomposition: Behavior of scattered signals by a target, measured by polarimetric radar by decomposing it into simpler components.

Polarimetric radar: A polarimetric radar transmits with two orthogonal polarizations, often linear horizontal (H) and linear vertical (V), and receives the backscattered wave on the same two polarizations. This results in four received channels, i.e. HH, HV, VV and VH, where both the amplitude and relative phase are measured.

Polarimetric SAR interferometry (PolInSAR): An advanced technique of radar remote sensing that combines the advantages of polarimetry and interferometry. This combination leads to the seperation of scattering mechanisms within a resolution cell.

Polarization: The orientation of the electric field vector in the plane orthogonal to the wave propagation direction.

Potential evapo transpiration: Theoretical evapo transpiration of a well water-fed surface submitted to real climatic conditions.

Precision agriculture: Designates a set of agricultural land management practices oriented towards optimizing the returns on inputs while preserving resources. These practices are based on technological innovations (precision GPS positioning, proxy sensors – or remote sensing of soil and crops properties, tools of soil work or the spreading of input variable rate, etc.) which allow to take into account the intra-parcel heterogeneity of yield potential (due to the heterogeneity of soils, topography, etc.).

Primary soil properties: Set of properties used in pedological studies to describe and map the soils which are usually indicated in the soil databases derived from these studies. These properties are determined by field observations (for example, porosity, soil depth) or by laboratory analysis (for example, organic matter, grain size, calcium carbonate, iron, pH, etc.).

Principal Components Analysis (PCA): A statistical procedure that uses an orthogonal transformation to convert a set of observations of possibly correlated variables into a set of values of linearly uncorrelated variables called principal components. The number of principal components is less than or equal to the number of original variables (source Wikipedia).

Radar bands X, C, L, P: Frequency bands used in microwaves, from 8 to 12.5 GHz for the X-band, 4 to 8 GHz for the C-band, from 1 to 2 GHz for the L-band and from 0.3 to 1 GHz for the P-band.

Radar interferometry: Consists of rebuilding the relief from two radar images of the same area acquired simultaneously by two synthetic aperture radars, or by a single radar during different times. Afterwards, phase diffrences point to point of the generated images are studied to find the vertical dimension of the ground, known as InSAR.

Radiative transfer: Theory based on the conservation of energy flux through an elementary volume. It can describe the behavior of an electromagnetic wave passing through a scattering medium composed of dielectric particles.

Radiative surface temperature: Emission temperature of the surface considered as a gray body; it can be defined for a measurement in a given direction (referred to as directional radiative temperature).

Radiometer: A radiometer is a device for measuring the radiant flux (power) of electromagnetic radiation. Generally, a radiometer is an infrared radiation detector or ultraviolet detector.

Radiometric resolution: It describes the ability of a sensor to distinguish small differences in the intensity of the received energy. The maximum number of levels of intensity depends on the number of bits used. For example, a RapidEye image is encoded in 16 bits or 65,536 numeric values available to transcribe the recorded intensity; a SPOT image is encoded in 8 bits or 256 digital values.

Real evapotranspiration: Vertical density of flux of water vapor corresponding to the sum of evaporation from the soil, rain intercepted by the foliage and the transpiration.

Reference data (in classification): Set of objects or parts of the landscape to map where the class (land cover) is known; reference data is used for validation of produced maps and also for machine learning in supervised methods.

Reference ellipsoid: In geodesy, a reference ellipsoid is a mathematically defined surface that approximates the geoid, the truer figure of the Earth, or other planetary body. Because of their relative simplicity, reference ellipsoids are used as a preferred surface on which geodetic network computations are performed and point coordinates such as latitude, longitude, and elevation are defined (source Wikipedia).

Remote sensing: The acquisition of information about an object or a phenomenon without making physical contact with the object.

Riparian forest: Forested or wooded area of land adjacent to a body of water such as a river, stream, pond, lake, marshland, estuary, canal, sink or reservoir.

SAR tomography: SAR tomography is the extension of conventional two-dimensional SAR imaging principle to three dimensions. A real three-dimensional imaging of a scene is achieved by the formation of an additional synthetic. aperture in elevation by a coherent combination of images acquired from several parallel flight tracks.

Scale: The ratio of the distance on a map to the corresponding actual distance. Often expressed as a fraction, a large scale is a ratio which is approximated to the integer 1.

Scatterometer: Sideways-looking radar instrument used to perform an accurate measurement of the radar reflectivity of observed surfaces. The improvement on the radiometric precision is thus obtained at the expense of spatial resolution (on the order of tens of kilometers).

Segmentation: Delimitation of a polygon composed of a set of similar pixels from a spectral point of view.

Selection of features: Set of data analysis techniques that allow for sorting out the attributes of a set of objects in order of relevance for a given problem, for example, a classification.

Sensor (satellite): Device embedded on the satellite that collects and measures incident radiation from the observed surface.

Simple scattering (optical): A photon is disseminated outside of the medium traversed as a result of its interaction with a single particle.

Soil texture: Distribution of minerals in the soil by category of size indepentently of the nature and composition of these minerals. The most used size categories categories: clay (diameters < 0.002 mm), silt (diameters between 0.002 mm and 0.05 mm) and sand (diameters > 0.05 mm). These categories are combined in a classification of soil textures in a triangle, called texture triangle whose three sides correspond to the percentages of sand, silt and clay.

Soil: Soil is defined as the top layer of the earth's crust. It is formed by mineral particles, organic matter, water, air and living organisms.

Spatial resolution: The spatial resolution of a geographical document is the actual size of the smallest element shown in that document. On a satellite image, the spatial resolution corresponds to the size of a side of a pixel.

Specific leaf area: A measure of leaf thickness, calculated by dividing the area of a portion of a leaf by the dry weight of that same portion of leaf. It is expressed in grams of dry matter per m^2 of leaves.

Speckle (radar): Speckle is produced by random interference, constructive and destructive, coming from multiple scattering that occurs in every resolution cell of a radar image. This effect manifests itself as a salt and pepper texture on images.

Spectral index (or radiometric index): Combination of spectral bands in optical imaging to bring out a certain type of material (vegetation, water, etc.).

Spectral resolution: Describes the finesse of a spectral band used by a sensor. If the spectral resolution is finer, the windows of the different channels of the sensor are narrower.

Specular reflection: The mirror-like reflection of light (or of other kinds of wave) from a surface, in which light from a single incoming direction (a ray) is reflected into a single outgoing direction (source Wikipedia).

Supervised classification: Automatic learning technique that produces a classification map from a ground truth.

Supervised learning: Learning method where we look to automatically generate rules from a learning database containing examples (cases previously treated and validated).

Support Vector Machines (SVM): Name given to the techniques used to find a separator (linear or not) between several groups of variables. The separator depends only of a few points called "support vectors".

Synthetic Aperture Radar (SAR): A coherent mostly airborne or space borne side-looking radar system which utilizes the flight path of the platform to simulate an extremely large antenna or aperture electronically, and that generates high-resolution remote sensing imagery.

Temporal resolution: Time frequency with which a remote sensing image acquisition system can acquire an image of the same point on the Earth's surface.

Terrestrial laser scanner: Equipment operating on the basis of a laser allowing measurement and digitization of objects or large complex spaces in three dimensions.

Texture (of an image): Spatial arrangement of the image signal levels.

Texture index (image): Local statistic calculated in the neighborhood of a pixel of an image which takes into account spatial organization.

Thermal conductivity: Physical quantity characterizing the behavior of materials during heat transfer by conduction. It represents the energy (quantity of heat) transferred per surface and time unit under a temperature gradient of 1 Kelvin per meter. It is expressed in watts per meter-Kelvin ($W \cdot m^{-1} \cdot K^{-1}$).

Tree allometry: Tree allometry establishes quantitative relations between some key characteristic dimensions of trees (usually fairly easy to measure) and other properties (often more difficult to assess). In a forest medium, allometric equations are used to estimate the volume and biomass from forest inventory data, such as the diameter and height of trees.

Unsupervised learning: Automatic learning method, sometimes referred to as "clustering", which divides a heterogeneous group of data in subsets of similar data.

Visible: Part of the electromagnetic spectrum visible to the human eye, between 380 and 780 nm.

Volumetric soil moisture: Volume of water contained in a volume of soil (measured in m^3/m^3). It is often discussed used in soil moisture index ranges between 0 and 1,which represents the saturation rate in water of the soil.

Voxel: A unit of graphic information that defines a point in three-dimensional space.

Water stress (index of): Water availability indicator; occurs when the demand for *water* exceeds the available amount during a certain period or when poor quality restricts its use.

Wavelength: Each electromagnetic wave is defined by its wavelength, which represents the spatial periodicity of oscillations. The wavelength is inversely proportional to frequency and is expressed in meters.

List of Authors

Nicolas BAGHDADI
TETIS – IRSTEA
Montpellier
France

Frédéric BARET
EMMAH
INRA
Avignon
France

Sébastien BAUWENS
BIOSE
University of Liège
Gembloux
Belgium

Julie BETBEDER
CESBIO
Toulouse
France

Stéphanie BONNET
BIOSE
University of Liège
Gembloux
Belgium

Alexandre BOUVET
CESBIO
Toulouse
France

Nathalie BRÉDA
EEF
INRA
Nancy
France

Samuel CORGNE
COSTEL
University of Rennes 2
France

Dominique COURAULT
EMMAH
INRA
Avignon
France

Valérie DEMAREZ
CESBIO
University of Toulouse III
France

Sylvain FERRANT
CESBIO
Toulouse
France

Cécile GOMEZ
LISAH
IRD
Montpellier
France

Martine GUÉRIF
EMMAH
INRA
Avignon
France

Dominique GUYON
ISPA
INRA
Bordeaux
France

Dinh Ho TONG MINH
TETIS
IRSTEA
Montpellier
France

Laurence HUBERT-MOY
COSTEL
University of Rennes 2
France

Jordi INGLADA
CESBIO
CNES
Toulouse
France

Jaco KEMP
Stellenbosch University
South Africa

Philippe LAGACHERIE
LISAH
INRA
Montpellier
France

Philippe LEJEUNE
BIOSE
University of Liège
Gembloux
Belgium

Michel LE PAGE
CESBIO
IRD
Toulouse
France

Thuy LE TOAN
CESBIO
CNRS
Toulouse
France

Stéphane MERMOZ
CESBIO
Toulouse
France

Adrien MICHEZ
BIOSE
University of Liège
Gembloux
Belgium

Vincent SIMONNEAUX
CESBIO
IRD
Toulouse
France

Pierre TODOROFF
AIDA
CIRAD
La Réunion
France

Amanda VELOSO
CESBIO
Toulouse
France

Ludovic VILLARD
CESBIO
Toulouse
France

Mehrez ZRIBI
CNRS – CESBIO
Toulouse
France

Index

E, F

G, H, I

K, L, M

N, O, P, R

S

T, V, W

Scientific Committee

Marie WEISS, INRA, France

Adrien MICHEZ, University of Liège, Belgium

Sébastien BAUWENS, University of Liège, Belgium

Stéphanie BONNET, University of Liège, Belgium

Philippe LEJEUNE, University of Liège, Belgium

Thierry BELOUARD, INRA, France

Nicolas BARBIER, IRD, France

Maxime REJOU, French Institute of Pondicherry, India

Grégoire VINCENT, IRD, France

Nicolas BAGHDADI, Irstea, France

Mehrez ZRIBI, CNRS, France

Kolibri 360 GmbH
Ferdinand-Jühlke-Straße 7
99095 Erfurt, Deutschland
produktsicherheit@kolibri360.de